Praise for *The Attentive Public*

"A major contribution to the literature on empirical modifications of democratic theory."

—Samuel Kirkpatrick, Texas A&M University,
Midwest Journal of Political Science

"An excellent example of a fresh look at one of the most basic questions in the empirical study of democracy, the relationship between popular demands and outputs."

—Arend Lijphart, University of Southern California,
San Diego, *Comparative Politics*

Praise for *The Political Culture of the United States*

"Few books serve more admirably the dual tasks of introductory survey and monograph."

—Glenn R. Parker, Miami University,
American Political Science Review

"One of the most distinguished pieces of secondary analysis in the field."

—Gabriel Almond, Stanford University,
Analytic Studies in Comparative Politics

"A thorough examination of American political culture based upon survey data."

—Alan Monroe, Illinois State University,
Public Opinion in America

"A valuable piece of synthesizing."

—James Danielson, Morehead State University,
The Journal of Politics

Praise for *Does Freedom Work?*

"Refreshing and thought provoking."

—George Carey, Georgetown University,
The Political Science Reviewer

"Wonderfully useful."

—William F. Buckley Jr., *National Review*

Praise for *Reagan's Terrible Swift Sword*

"An extraordinarily creative combination of inside information and sophisticated social science analysis."

—Aaron Wildavsky, University of California, Berkeley

"This is an important first-hand book about democracy and bureaucracy."

—Richard Nathan, State University of New York, Albany

Praise for *America's Way Back*

"At last someone has produced a worthy successor to Frank Meyer's classic *In Defense of Freedom*."

—Edwin J. Feulner, The Heritage Foundation

"Too many conservatives have forgotten that freedom and tradition, limited government and conservative morality, stand or fall together. Devine's fine new book, a vigorous defense of fusionism, provides a much-needed reality check."

—Edward Feser, Pasadena City College

"A philosophical, historical, and political tour de force. Must read—an outstanding book."

—Lee Edwards, author of *William F. Buckley Jr.: The Maker of a Movement*

"*America's Way Back* follows on the work of my father, Frank Meyer, showing that libertarianism and traditionally oriented conservatism, while sometimes conflicting in specific policy areas, are complementary and vital as bulwarks of the West."

—Eugene B. Meyer, The Federalist Society

"Both enjoyable and compelling. Informed by the insights of James Madison, Adam Smith, and Alexis de Tocqueville—and William F. Buckley Jr., Frank Meyer, and Ronald Reagan—Devine shows how we can recover what we have lost. An intellectual journey to be savored."

—Mark Levin, nationally syndicated radio host

"Reagan had a heckuva lieutenant in Donald Devine. It is good to see him now mentoring the next generation of conservative leaders."

—L. Brent Bozell III, Media Research Center

"The solution for the modern GOP... This book provides plenty of intellectual ammunition."

—Senator Rand Paul

"A trenchant critique of the flaws and failures of statist progressivism—and a powerful case for a revived conservative alternative. Conservatives of all stripes—and progressives too—will benefit from a close reading of this discerning and very timely book."

—George H. Nash, author of *The Conservative Intellectual Movement in America Since 1945*

THE ENDURING TENSION

DONALD J. DEVINE

THE ENDURING TENSION

CAPITALISM AND THE MORAL ORDER

New York • London

First American edition published in 2021 by Encounter Books, an activity of Encounter for Culture and Education, Inc., a nonprofit, tax-exempt corporation. Encounter Books website address: www.encounterbooks.com

Manufactured in the United States and printed on acid-free paper. The paper used in this publication meets the minimum requirements of ANSI/NISO Z39.48-1992 (R 1997) (*Permanence of Paper*).

FIRST AMERICAN EDITION

Library of Congress Cataloging-in-Publication Data

Names: Devine, Donald John, 1937– author.
Title: The enduring tension: capitalism and the moral order by Donald J. Devine.
Description: New York: Encounter Books, [2020]
Includes bibliographical references.
Identifiers: LCCN 2020012444 (print) | LCCN 2020012445 (ebook)
ISBN 9781641771511 (cloth) | ISBN 9781641771528 (epub)
Subjects: LCSH: Capitalism. | Income distribution.
Civilization, Western. | Economic history.
Classification: LCC HB501.D439 2020 (print)
LCC HB501 (ebook) | DDC 330.12/2—dc23
LC record available at https://lccn.loc.gov/2020012444
LC ebook record available at https://lccn.loc.gov/2020012445

Interior page design and composition by Bruce Leckie

Dedicated to my family, to my intellectual home
at The Fund for American Studies, and to the numerous friends,
associates and groups who have assisted me in this endeavor,
especially Bob Luddy, Bill and Berniece Grewcock,
and my spouse and top supporter Ann Devine.

CONTENTS

INTRODUCTION

Western civilization under its capitalist economic order has produced a worldwide cornucopia of consumer goods and a remarkable improvement in human health. But few among the beneficiaries are grateful. Many in the West are instead reconsidering socialism as an alternative economic order and way of life.

Capitalist civilization, and the freedom and wealth associated with it, have long been objects of criticism—from thinkers such as Jean-Jacques Rousseau, Charles Dickens, and Karl Marx, down to politicians including Bernie Sanders and sometimes even Donald Trump. The economic historian Joseph Schumpeter, while sympathetic to capitalism, argued in the 1940s that its civilizational walls were "crumbling," and he predicted a probable collapse.[1] A person as sensitive as Pope Francis has called the Western civilization of capitalism and "limitless" freedom a "fundamental terrorism against all humanity."[2] On the right in the United States today, "national conservatism" rejects aspects of free-market capitalism by supporting industrial policy, tariffs, and federal child and family welfare programs. There have even been promises to lay to rest the conservative consensus that prevailed since the presidency of Ronald Reagan.[3]

These critiques demand a response, for there is much to say on the other side of the question. This book examines the historical, scientific, moral, and philosophical assumptions that underlie the criticisms of capitalism and its civilization from various quarters. Then it considers what would be required for a new generation to build the scaffolding that might hold the walls and preserve the culture's positive characteristics.

The award-winning financial columnist Robert Samuelson called Schumpeter "the prophet" for his description of capitalism as a form of "creative destruction" and for his understanding of its true strengths and limitations.[4] Schumpeter's classic work, *Capitalism, Socialism and Democracy*, was among the most influential and broadly read treatises of the later twentieth century, not only in history and economics but perhaps even more so in my own field of political science, where it was required reading for many years. Schumpeter had many valuable insights—on where Marx was wrong and right, on the creativity of capitalism, on the importance of feudalism in forming capitalism, on Western democracy, on elections and elites, and on the possibilities of a freer form of socialism than what Marx envisaged. But the essence of his thought was that the capitalism and freedom that generated productivity and widespread prosperity in the West required supporting institutions to legitimize the moral beliefs and legal principles that undergirded this success. Indeed, Schumpeter predicted that capitalism was likely to be superseded by some form of administrative socialism if it did not develop and maintain a countervailing legitimacy.

Schumpeter found Marx to be wrong in asserting that capitalism would be brought down by its weaknesses—by overproduction, by a concentration of capital in a few hands, by the rise of labor unions and democratic political parties, or even by revolution from the proletarian masses. Instead, Schumpeter predicted that capitalism would be destroyed by its own success. As widening prosperity and freedom corrupted the social discipline supporting thrift and work and encouraging procreation, the dynamic bourgeois entrepreneurship that was the engine of capitalism would stagnate. The resulting social decline would be exploited by intellectuals promoting the ideal of scientific socialism to political elites, who at best would transform it into a populist democratic socialism, or what Americans would call the welfare state.

Schumpeter was a prophet because what he predicted seems to be pretty much what has happened even in the redoubt of freedom and capitalism, the United States. Capitalism survives today only in what Schumpeter called a "fettered" form, shackled by bureaucratic regulations that impede productivity, compound the problems they were designed to fix, and dissolve the moral structure that underlay Western civilization's creativity and gave it legitimacy. A reversal back to a more unfettered capitalism would seem to require such fundamental change in culture as to render it unlikely.

Your author comes to the discussion from the academic field of political science and two decades of teaching at the University of Maryland and Bellevue University. One competency was in normal politics, government, and democratic theory, but I also taught philosophy of science, specifically scope and methods. Another specialty was public administration, put to practical use as director of the U.S. Office of Personnel Management in 1981–1985. These interests and experiences provide the basis for the development of the argument.

Part I reevaluates the historical roots of Western civilization, looking closely into Marx and John Locke—at what they actually wrote, rather than what many wish they had written. We then go back further, into prehistory, myth, and the primitive beginnings of human society, and trace key institutions into the modern era of Enlightenment, nationalism, and progressivism. Part II examines what the Nobel laureate F. A. Hayek called the modern "superstition" that empirical science has the answers to all social problems. We start with an analysis of modern physics, artificial intelligence, and scientific proof. Then we move into modern scientific administration, constitutions, social divisions, and some possible ameliorations. Finally, we go to the center of the problem and evaluate the various ways that capitalism has been legitimized in the past, and consider the practical and philosophical issues that need to be addressed in order to legitimize capitalism and indeed all Western civilization.

To provide a more detailed roadmap: Chapter 1 looks at multiple conceptions of the enigma that is capitalism, beginning with that of Karl Marx, who coined the term. It empirically tests his and Schumpeter's resting of capitalism's moral justification in feudalism, and ends up with a conclusion closer to Marx than to most supporters of capitalism. The chapter examines the central role of Locke in the history of capitalism, and looks at the highly divergent modern interpretations of his philosophy—as immoral, anarchical, possessive, and dark, or alternatively as pluralist, medieval, tolerant, and Christian. One cannot overemphasize the importance of feudalism, particularly concerning property rights and the growth of individualism—though later theorists (Renaissance, Reformation, Counter-Reformation, Enlightenment) all had reasons to downplay the medieval contributions to launching the modern age.

Chapter 2 delves into early beginnings of social organization, following Rousseau. Political analysis normally starts with recorded history, but it is clear that the origins of social order long predate written records. Anthropology and archeology provide some evidence for how order first developed and what was passed on to agricultural society, through feudalism and down to modern institutions. Successors to the feudal order have included divine-right monarchy, nationalism, free-trade capitalism, and the expert-led bureaucratic nation-state. Recently the nation-state has come under serious challenge due to wars, strains on state treasuries, frustration with limited success in controlling events, doubts about national legitimacy, and a waning sense of common citizenship.

Chapter 3 reviews attempts to tame unbridled capitalism, beginning with the efforts of the British philosophers J. S. Mill and T. H. Green to reduce the economic and social inequalities resulting from capitalist productivity and creative destruction. In the United States, the project of minimizing economic inequality, beginning with the War on Poverty, has entailed redistributing $22 trillion over a half century. It has also affected workforce participation, marriage, reproduction, and social satisfaction. Efforts toward equalization by race, ethnicity, gender, and sexual orientation have had marginal success and have even increased social discord in some ways. The one undoubted success in ending capitalist inequality—the Soviet Communist regime—turned out to be not very pleasant. So we use Pope Francis's moral critique of capitalism, freedom, and Western morality to consider how we might ameliorate what he calls a fragile world.

Following the great sociologist Max Weber, Chapter 4 investigates the tools of policy analysis and public administration as means of rationalizing market capitalism. The author's own experiences in government, along with multiple other kinds of evidence, throw into question their adequacy to the task. The instrument that the pope would rely upon to rescue the world from capitalist markets is found not up to the responsibility.

Opening Part II of the book, Chapter 5 examines the hope that scientific calculation can simplify human interaction sufficiently to make the market's social calculation superfluous, and finally deliver the good life to all. Two views of science are compared: a "constructivist" one that aims to reduce everything to rational abstractions and predictability, and a "critical rationalist" conception that is more open to complexity and different kinds of reasoning. An analysis of the quintessential science of physics reveals uncertainties at the heart of the scientific enterprise, while philosophers of science—Michael Polanyi, Karl Popper, Friedrich Hayek, Thomas Kuhn, Albert Einstein—help shed light on the nature of scientific knowledge. After looking into the science of the human mind, we conclude with Hayek's observation that much modern social analysis has degenerated from a critical-rationalist scientific tool into little more than superstition.

Chapter 6 surveys various attempts to legitimize Western capitalism and freedom morally. Pope Francis's moral critique is distinguished from his political one, and answered from the viewpoint of his own moral tradition. The challenge for capitalist legitimacy is met with the epistemological method of synthesis, drawing from the writings of Adam Smith, Edmund Burke, F. A. Hayek, and Kenneth Minogue, as well as popularizers: William F. Buckley Jr., Frank S. Meyer, and even Ronald Reagan. The influence of their pro-market teachings waned during the presidencies of George W. Bush and Donald Trump, however. What is needed, as Minogue emphasized, is a deeper understanding of human freedom and a positive vision of government that is compatible with both free-market capitalism and a moral society.

Chapter 7 begins with the necessity of the nation-state for capitalism, but acknowledges a general American belief that the U.S. government is dysfunctional. The normal political science responses tend to be rather superficial and to show misunderstanding of the U.S. Constitution. The expert-led central bureaucracy offered as a solution to the flaws of capitalism is in fact the main problem, even from a purely administrative point of view. Moralizing capitalism requires understanding the pluralism that is inherent in the Constitution, with market-like principles in governance as well as in economics. Expert administration is contrasted with the institutional diversity of federalism, localism, and voluntary governance, which are offered as the more compatible solutions.

Chapter 8 confronts the challenge of providing the moral restraints needed for social order, but still consistent with pluralist capitalism. It opens with the obvious decline of a cultural consensus in the United

States today and the loss of belief in what Walter Lippmann called the natural law of civility, which is necessary to make a nation-state work, especially a pluralist one. Lippmann located the source of the problem in the "Christian heresy" of Rousseau, which transferred the search for perfection from the next world to this, leading to the excesses of Jacobinism, Nazism, Leninism, nationalism, and pseudoscientific bureaucratic expertise. All these have raised impossible expectations for capitalism, and in so doing have set the stage for its decline. Lippmann proposed a new rationalistic stoicism as the basis for social order.

John Nicholas Gray likewise favors a kind of stoicism as a modern moral code, but he finds the materialist worldview to be fundamentally illogical, and science to be incapable of providing restraints on human behavior. If all other social solutions are illusory, that may ultimately leave only the amoral use of power to control society. But power alone has limits as a means of establishing social order. Professor David Martin concluded that there can be no power logos without a moral mythos to support it, and vice versa, even in modern times. Indeed, there is voluminous evidence that religion is among the most persistent and deeply held social values worldwide.

Alternative mythos stories are evaluated as ways of defining a Western and particularly an American future. As Hayek and Locke argued, a moral legitimizing ideal must have at least a symbolic truth. If we are to rebuild the scaffolding that can hold the walls and preserve a capitalist civilization, some such solution must be explained and implemented.

We end where we began: with Schumpeter's observation that not only capitalism but even a humane socialism requires a moral justification rather similar to that which initially cemented the forces of Western civilization.

PART I

How We Got Here

CHAPTER 1

WHICH CAPITALISM?

W hy would someone like the late Stephen Hawking, the renowned theoretical physicist at the University of Cambridge, have regarded Western capitalism and its untamed freedom as far more dangerous to the future of humanity than quantum particles, robots, and aliens combined?[1] Coming from such a penetrating mind, this judgment on what appears to be the engine of unparalleled prosperity suggests there must be a serious intellectual enigma at the heart of capitalism.

The Capitalist Enigma

"Capitalism" is one of those big concepts concocted by their ene-mies—in this case, named as such by its chief antagonist, Karl Marx, in his definitive book on the subject, *Das Kapital*. Even the supporters of capitalism cannot agree on what its essence is, or when it began. Marx targeted the fourteenth century as the end of feudalism and the beginning of the capitalist stage of history. Others point to the Renaissance (fifteenth century), the Reformation (sixteenth century), the Enlightenment (seventeenth century), or even the start of the Industrial Revolution (eighteenth century).

Marx's timeline has capitalism emerging from a long feudal conflict between lord and serf that ended with owners of capital exploiting workers in the modern era. Though intrigued by capitalism's longevity,

Marx was still certain it would fall. But he described the rise of capitalism in a surprisingly positive way, noting that "serfdom had practically disappeared" in England by the late fourteenth century. "The immense majority of the population consisted then and, to a still larger extent in the fifteenth century, of free peasant proprietors, whatever was the feudal title under which their right of property was hidden."[2] The fact that serfs had wrested more and more control over their own labor from lords allowed new capitalists to attract the freed labor and create "more massive and colossal productive forces than have all previous generations together."[3]

Little systematic data has been available to document when capitalist productivity first emerged or why. Angus Maddison's study for the Organization for Economic Cooperation and Development (OECD), *Historical Statistics for the World Economy*, collected the best data available for examining the question empirically.[4] Tom G. Palmer used this data to locate "the sudden and sustained increase in income" that marked Europe's capitalist takeoff above the rest of the world. His analysis of Maddison's data points to "around the middle of the eighteenth century" as the beginning of the takeoff that produced capitalism. The change is illustrated dramatically in a chart beginning with year 1 of the Common Era.[5]

The big spike in wealth per capita appears first in western Europe in the years around 1800, as Palmer claimed, but even more dramatic is the rise of the upstart United States a few years later, shooting past its mother country and all of its European rivals to new heights of wealth. By World War I, eastern Europe began its takeoff, followed by Japan, although both of these remained well behind the United States and western Europe. India, China, and Africa begin rising in the 1950s, but today remain well below what the U.S. had attained in 1900.

Because of this timeline, Palmer credited Adam Smith's *Wealth of Nations* (1776) and his classical liberalism generally as the source of capitalism, market freedom, and modern prosperity. Yet the man who did perhaps the most to revive capitalism and classical liberalism in modern times, F. A. Hayek, had stated in *The Constitution of Liberty*, in 1960, that the origins of liberal capitalism, and especially the central idea of the rule of law, lay in the Middle Ages.[6] In fact, Palmer acknowledged that it took time to build the necessary "legal

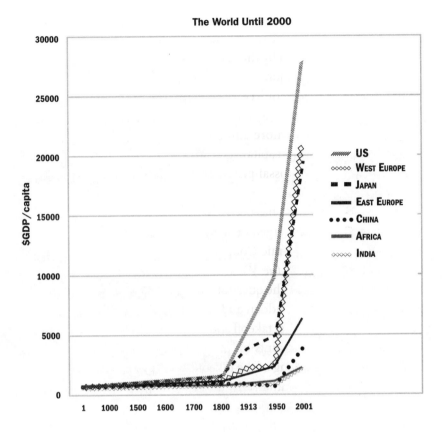

orders characterized by well-defined, legally secure, and transferable property rights, with strong limitations on predatory behavior," and these originated long before the eighteenth century.[7]

In *Suicide of the West*, Jonah Goldberg likewise identified the eighteenth century as the turning point. "For 100,000 years, the great mass of humanity languished in poverty," with innovation stifled by authorities, and then something "miraculous" happened and changed everything. What Goldberg called "the Miracle" occurred spontaneously in the 1700s with the emergence of a middle-class ideology of "merit, industriousness, innovation, contracts, and rights," which produced "the most cooperative system ever created for the peaceful improvement of peoples' lives." He even asserted that "nearly all of human progress has taken place in the last three hundred years." Yet he went on to clarify that his review of history showed the eighteenth-century Miracle to be "the climax of a very long story."[8]

Niall Ferguson, in *Civilization: The West and the Rest*, linked the takeoff with the invention of the printing press, which facilitated communication and trade, and with the sixteenth-century Reformation and the resulting competition between religious institutions. He found the "path to the Scientific Revolution and the Enlightenment" to have been "long and tortuous," however, and saw it beginning in "the fundamental Christian tenet that church and state should be separate." He specifically mentioned the importance of St. Augustine's theology and of the investiture crisis during the papacy of Gregory VII in the eleventh century.[9]

In *How the West Grew Rich*, Nathan Rosenberg and L. E. Birdzell traced the beginnings of the legal order underlying capitalism to the thirteenth century or before.[10] Christopher Dyer, in *An Age of Transition: Economy and Society in the Later Middle Ages*, identified the critical period for the emergence of capitalism as the twelfth century, although he also pointed to earlier foundations.[11] In *The Medieval Machine: The Industrial Revolution of the Middle Ages*, Jean Gimpel attributed the Western growth in population and wealth partly to the use of water-powered mills, recorded in England as early as the Domesday Book of 1086, and later used heavily by Cistercian monasteries, which also developed scientific agricultural practices in the twelfth century. Banking institutions originated in Italy during the same period.[12]

Harold Berman, in *Law and Revolution: The Formation of the Western Legal Tradition*, meticulously traced the crucial legal and moral developments back to the Abbey of Cluny (founded in 909) and to the reforms of Pope Gregory VII, who had been a monk in a Cluniac monastery. The many Cluniac foundations across Europe were governed through an innovative corporate structure under the abbot of Cluny, which enabled Gregory VII to reassert the independence of the church from secular powers, particularly in his conflict with Henry IV, the Holy Roman Emperor. As Berman's comprehensive study suggests, the history of capitalism is simultaneously the history of Western civilization.[13]

Data from the OECD can help settle the matter of when the capitalist takeoff began. Michael Cembalest of JP Morgan and Derek Thompson of the *Atlantic* used this data to plot GDP for various parts

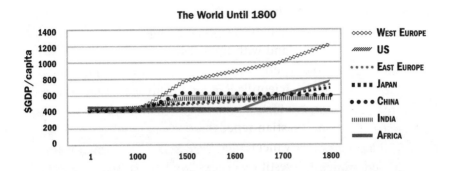

of the world over nearly two millennia up to the year 1800, as shown in the chart above.[14]

This data, the best available, suggests that the economic takeoff of western Europe actually originated as far back as the tenth or eleventh century. In the Palmer chart, this early beginning is obscured by the compressed scale, but still visible. Another distinct uptick appears in the sixteenth century, followed by a more obvious increase in the rate of growth in the eighteenth century, leading up to the dramatic takeoff in Europe and the United States emphasized by Palmer and others, including Deirdre McCloskey.[15]

Many scholars reject the idea of a medieval takeoff as a myth. Palmer attributes the idea to a "common yearning for a past 'golden age,' a yearning that is still with us." He aligns himself with the classical liberals who "have persistently worked to debunk the false image of the past—common to socialists and conservatives alike, in which happy peasants gamboled on the village green, life was tranquil and unstressed, and each peasant family enjoyed a snug little cottage."[16]

Yet the medieval era is more often labeled a dark age rather than a golden one. It was Renaissance classicists who first described the era preceding their own time as benighted—discounting the real accomplishments of medieval thinkers and inventors. Reformation theologians could not have made a case for reforming what was not dark and decadent, and Catholic intellectuals wished to leave the past behind once they had embarked on their modernizing Counter-Reformation. Later, Voltaire could not have proclaimed a new age of "enlightenment" if what came before had not been an age of ignorance. From the fifteenth century to the twentieth, there was little interest in seeing an inventive and prosperous Middle Ages.[17] Yet the best scholarship points to the Middle Ages in Europe as the launching pad for the Western economic takeoff.

The empirical Index of Economic Freedom compiled by the *Wall Street Journal* and the Heritage Foundation shows that the legacy of those medieval roots is still visible in the global economic map. Of the thirty-seven nations rated "free" or "mostly free" in the 2020 ranking, nineteen are European or, like the United States, are former colonies with a predominantly European culture and population. Even among the eighteen non-European nations, most were European colonies or protectorates for a substantial period of time—Hong Kong for a century. Mauritius was unpopulated until occupied by the Portuguese and the Dutch. Japan did not achieve widespread prosperity and freedom until after U.S. occupation. Taiwan was not under direct European control before achieving economic freedom and prosperity, but its free-market reformers were Western-inspired. The others are all very recent additions to the top two categories.[18]

The columnist Fareed Zakaria disputed the connection between prosperity and culture, arguing that although "the key driver for economic growth has been the adoption of capitalism and its related institutions and policies," it has taken place "across diverse cultures," including India and China.[19] But as the charts demonstrate, they lag behind the West economically, and both are rated "mostly unfree" on the worldwide economic freedom list—India in 120th place and China at number 103. Yes, they have seen considerable growth as they have adopted some market reforms, but they are still mainly poor and not very free. Hong Kong's capitalism is straining under mainland Chinese control. Singapore has only a quasi-free political system, threatened by conflict between Chinese, Malay, and Muslim inhabitants. Bahrain was ranked eighteenth on the economic freedom list in 2016, and its economy had many market elements that allowed it to prosper, but it is a centralized Sunni monarchy that was forced to call in Saudi troops in 2011 to quell a restive Shia-majority popular uprising, and was ranked 63rd in 2020.

Top-ranked Singapore and Hong Kong might have appeared to support the idea that economic prosperity doesn't depend on broader political and cultural forms. But how comfortable can one be that non-Western nations now rated relatively free economically can remain so? After all, Mali was once rated mostly free but now it is "mostly unfree," at number 126. If current prosperity can be ascribed to the adoption of scientific materialism, skeptical rationalism, or even classical liberalism, then it might be possible to create a formula that will build and sustain a capitalist system. But if deeper beliefs and attitudes, customs and social institutions are essential, capitalism beyond the West is extremely problematic—a point made convincingly by Svetozar Pejovich in *Law, Informal Rules and Economic Performance.*[20]

Note that Zakaria's statement quoted above has a major qualification: it was not only accepting capitalism that was essential, but also adopting Western civilization's "related institutions," the ones that took centuries to develop before capitalism could take root, as Hayek and the others found. Western nations themselves may not be able to sustain capitalism without these traditional institutions. Italians once led Europe in economic development, but modern Italy has recently ranked only 74th on the Index of Economic Freedom and showed a growth rate of only 1.6 percent. In the past few years, the Index has excluded the United States from its highest "free" category because of restrictions on contract rights in the bailouts of banks, Chrysler, and GM in 2008.

So in pointing to the feudal age, Marx identified the roots of the capitalist civilization more accurately than many of its defenders. Might he likewise be a better prophet? While his economic predictions seem far off the mark, his cultural findings resonate today. Marx found that capitalism had "put an end to all feudal, patriarchal, idyllic relations" by the mid-nineteenth century.

> It has pitilessly torn asunder the motley feudal ties that bound man to his "natural superiors," and has left remaining no other nexus between man and man than naked self-interest, than callus "cash payment." It has drowned the most heavenly ecstasies of religious fervor, of chivalrous enthusiasm, of philistine sentimentalism in the icy water of egotistical calculation.... It has stripped of its halo every occupation...torn away from the family its sentimental veil....All that is solid melts into air, all that is holy is profaned.[21]

As we have seen, the great economic historian Joseph Schumpeter predicted that capitalism, even with all its freedoms and with the market's efficient calculation and allocation, could not survive much longer without the moral values and cultural institutions that had nurtured and sustained it. Indeed, he raised the possibility that capitalism may be better classified as the final stage of feudalism than as a separate stage of history. As the medieval morality has faded, capitalism has tended to be replaced by the next of Marx's historical stages, a form of socialism—today called the welfare state—and perhaps it will eventually disappear altogether.[22]

Immoral Capitalism

Pope Francis devoted his first pastoral letter, *Evangelii Gaudium* (2013), to condemning capitalism and its unrestricted freedom as an affront to a religious or moral conscience.[23] The pope criticized those who "continue to defend trickle-down theories which assume that economic growth, encouraged by a free market, will inevitably succeed in bringing about greater justice and inclusiveness in the world." In the pope's view,

> This opinion, which has never been confirmed by the facts, expresses a crude and naïve trust in the goodness of those wielding economic power and in the sacralized workings of the prevailing economic system. Meanwhile, the excluded are still waiting as a "culture of prosperity deadens" the fortunate to the sufferings of the many. (§ 54)

The experience of his native country had a searing effect on Jorge Mario Bergoglio, who would become Pope Francis. Argentina had developed a successful capitalist economy by the dawn of the twentieth century and was ranked among the ten wealthiest nations per capita. A hundred years later it had dropped to around seventieth place, and little of its wealth was trickling down into the wider population. Argentina seemed to confirm Marx's observations and Schumpeter's predictions.

Argentina had been the "Wild West" even within the chaotic Spanish empire. A colonial reorganization in 1776 made Buenos Aires the capital of a new viceroyalty and a free port, with a lively transatlantic trade, and a rich agriculture developing around it. After Napoleon deposed the king of Spain in 1808, civil war ensued in Argentina, leading to a revolutionary war and finally a declaration of independence in 1816. But civil war continued for decades. A classical liberal constitution established a republic in 1853, followed by a constitutional revision in 1860 and substantial changes in power and fortunes. In an agricultural oligarchy, with a major port and developing industry, free trade was the norm and central government controls were light, resulting in a credible form of capitalism leavened by traditional Spanish moral norms. By 1910, even with much political instability, Argentina's per capita gross domestic product was greater than that of France, twice that of Italy, and five times that of Japan.[24]

This prosperity attracted a large European immigrant population, so that 30 percent of Argentineans were foreign-born by 1930, mostly from Italy and Spain. While providing the skilled labor for its growing industrial economy, these newcomers also brought the fascist and communist ideologies then current in their homelands to the inevitable clashes between the new capitalists (many from landed families) and the muscular labor unions. These conflicts also spread to the feudal agricultural economy of the Pampas and added fuel to existing grudges between aristocratic factions.

Military coups in 1930 and 1943 promised order and led to the long influence of the nationalist Juan Perón, who cashiered the supreme court, politicized the education system, nationalized industry, controlled agricultural prices and distribution, and established an export monopoly favoring both unions and businessmen in a generally syndicalist rather than capitalist regulatory environment. He vastly increased government spending to attract and support client groups, which resulted in enormous debt. Other factions had intervals of power until another military coup in 1976. The Falklands War of 1982 disgraced the military regime, which was replaced by a democratic government the next year. Market reforms were adopted under Carlos Menem in the 1990s, but high social spending continued, and so did mounting debt. Corruption, currency devaluation, and unpopular austerity programs were answered by the Kirchner presidencies with their traditional Perónist policies, resulting in even slower growth.

This is the background for the pope's thinking in *Evangelii Gaudium*. Francis charged that a "deified market" was held to be "the only rule" in the modern world. The result was inequality, "the root of social ills," the pope declared.

> As long as the problems of the poor are not radically resolved by rejecting the absolute autonomy of markets and financial specula-tion, and by attacking the structural causes of inequality, no solu-tion will be found for the world's problems or, for that matter, to any problems. (§ 202)

The pope maintained that economic growth and market forces cannot bring about justice for the poor:

> We can no longer trust in the unseen forces and the invisible hand of the market. Growth in justice requires more than economic growth, while presupposing such growth: it requires decisions, programmes, mechanisms and processes specifically geared to a better distribution of income, the creation of sources of employment and an integral promotion of the poor which goes beyond a simple welfare mental-ity. (§ 204)

Pope Francis modified his comments somewhat in a subsequent interview with the Italian daily *La Stampa*, where he explained that in criticizing "trickle-down" capitalism he "did not talk as a specialist but according to the social doctrine of the church." His reproach was personal, aimed at those "wielding economic power," for their lack of charity and justice. When he was a bishop in Argentina, if he saw a businessman he thought was unjust to his workers coming up to receive communion, he would maneuver another priest into administering the sacrament.[25]

But is "wielding economic power" what constitutes capitalism? As for charity and justice, one should note that Perón created what was perhaps the Western Hemisphere's first comprehensive welfare state, offering aid to the masses in exchange for their political support. Since there were never enough funds for everyone, a rationalized state capitalism under strong political control steered benefits to powerful clients, such as labor unions and big businesses, but without fettering enterprise so much as to deprive the regime of revenue. Under Argentina's many forms of repressive government, capitalists could survive only by being political partners of the state—sometimes as the power behind the throne, but more often deferring to the state. In the United States this is called "crony capitalism," where capitalists collude with the government to get favors for themselves, and it is precisely the opposite of free-market capitalism.

What happened to Argentina as it moved away from markets? By the twenty-first century, it had sunk to number 73 in the world in per capita GDP, estimated by the World Bank at $12,043, only one-third that of France. Its inequality ratio, comparing the richest 10 percent against the poorest 10 percent, was 31.6 to 1, while that of the United States was only 15.9 to 1.[26] With all its government welfare, Argentina is now poorer and more unequal than the more free-market United States. The pope might argue that the market reforms instituted in the 1990s made things worse for the poor in Argentina. Yet the numbers show they did bring improvement for a while, and in fact all three parties in the next election ran as pro-market. In the end, however, the reformers could not defeat the crony-capitalist and union corruption, nor meet public welfare expectations, nor control the swelling debt that still hobbles Argentina today.

The capitalism that Pope Francis knew best can indeed be considered an affront to a moral conscience. But was it market capitalism, or rather crony capitalism? Actually, the pope's criticism went much deeper than the faults particular to Argentina's brand of capitalism. He saw affluence and freedom themselves as breeding mindless commercialism, prurient escapism, and selfish indifference to the needs of others. In the pope's view, even if capitalism was successful on its own terms, generating material prosperity, its possessive individualism and unrestricted freedom made it impossible to defend as a moral system.

Two Capitalist Freedoms

To examine the nature of capitalist freedom, and consider the possibility of a moral defense, a good place to begin is Jacob T. Levy's very important book *Rationalism, Pluralism, and Freedom*. Levy argued that two contrary "strains" of liberal freedom are possible under Western capitalism, and he called them *rationalist* and *pluralist* freedom. Rationalist freedom is designed from the top, with experts developing uniform laws to promote a freedom that is positive and just. Pluralist freedom emerges from the bottom, through independent associations that provide choices for their members, advance their various group interests, and act as buffers between the individual and the state.[27]

Levy found pluralist freedom in the Middle Ages resulting in the development of capitalist property rights. But local associations and the regional powers were eventually overwhelmed by centralizing, rationalizing divine-right kings. This royal claim to total power over subjects was contested with appeals to ancient freedoms, such as those codified in Magna Carta, in an effort to restore pluralist legal and commercial rights. This effort culminated in the powerful writings of Montesquieu, whom Levy regards as the first true pluralist philosopher. On the opposite side was Voltaire, the founding rationalist, himself an entrepreneur and early supporter of capitalism. But Voltaire regarded pluralist associations as the source of the ignorance and superstition that had frustrated the social and economic progress necessary for the development of markets—which were finally instituted rationally under the "enlightened absolutism" of Louis XIV, Frederick the Great, and Henry VII.[28]

Further rationalist progress was stymied by the French Revolution and the rise of mass politics, in Levy's telling. But then arose the modern centralized nation-states, whose form was best described by "two unarguably canonical and liberal theorists": John Stuart Mill, a proponent of rational authority, and Alexis de Tocqueville, an advocate of pluralist associational interests. Today, Mill is generally represented as the archetypical rationalist liberal, and Tocqueville as the prototypical pluralist conservative. By the later nineteenth century, Mill's view had won the day, as centralized, rationalist welfare states dominated the Tocquevillian parochial associations.

Levy writes that his "personal preference" is for pluralist freedom, but acknowledges that parochial groups can limit the freedom of their members through rigid internal rules, and can lobby to have those rules instituted as state policy. Therefore, rationalists like Voltaire had good reason to fear powerful plural associations like the church even being allowed to provide private education, for example. Indeed, both modern secular France and religious India have justified state control of education as necessary to preserve a liberal order. The state likewise had reason to fear that the family would inculcate illiberal ideas (about women, for example). On the other hand, it is clear that centralized state rationalism can go too far in restricting freedom.[29]

Pluralists maintained that possible abuses of power over group members by their leaders should be tenable as long as an "exit" is available. However, Levy believes that a large size and wide geographical reach would inherently restrict exit from a group, and that decisions originally made freely can become entrenched over time, so that individuals have "no place to go to exit the groups into which they are born." One generation can choose, but their children may be trapped if the group or cult becomes very popular over an extensive territory. Levy's main example is the medieval Catholic Church, although in another place he portrays the church as a complex institution that generally includes means of exit.[30]

The family presents an even greater restriction on exit, since young children are unable to leave at all, or to escape the benefits or disadvantages passed down from generation to generation. Likewise, an inferior education can have inescapable consequences. A child born poor and marginalized or biologically limited has little "exit" from these disadvantages. The same applies to many other areas of life in a pluralist society.

Levy reluctantly suggests putting associational freedom under rational state oversight, with the power to limit group autonomy. This might be done if a group were too dominant in a particular place, such as a one-company town, the Mormons in Utah, or Catholic and Orthodox Christianity with their large national majorities in some countries. Limits on autonomy might also be applied if the actions of an association undermined the "basic structure of society," such as when families perpetuate inequality, or public accommodations deny access to outsiders, or private schools and housing covenants thwart justice (for example, impeding desegregation in the South).

But what is left of pluralist associational freedom when the exceptions are religion, family, education, property, housing, and commercial enterprise?

Reviewers of Levy's book have treated his term "rationalism" as a synonym for rational centralization, by contrast with plural decentralization, on a one-dimensional scale.[31] Yet Levy emphasizes at the very beginning that he uses the term "rational" as the great sociologist Max Weber did, in the sense of "bureaucratic rationalization" as it takes place in a modern scientific bureaucracy. He even advises the reader to keep this meaning in mind throughout. But then why not use the term *rationalization* instead of *rationalism*, which to some extent misleads even sound reviewers?[32]

The word *rationalization* is critical to understanding Levy's reliance on Weber as the key to comprehending his "rationalized" (but not necessarily "rationalist") state. Weber is the father of American scientific administration, one of the "eminent German writers" whom Woodrow Wilson credited for his new rationalized liberalism.[33] To Levy, "the political program of liberalism is one about how to direct and limit the power of the modern state in ways that are only comprehensible after the state has taken form, the wars of religion have ended and the attractions of commerce came into focus." Levy therefore does not "think it makes sense to talk about liberalism before about 1700," when he sees the emergence of the "early modern Weberian state," whose features all require "a world of strong executive and security capacity" and limited pluralist independence.[34]

But all this comes historically after what Levy calls a "prehistory of liberalism," with its "tremendous proliferation" of pluralist organizations—universities, cities, the papacy, bishoprics, regional churches, independent military orders, commercial enterprises, guilds, and monastic orders.[35]

Interestingly, Levy acknowledges that J. S. Mill, in *On Liberty*, regularly placed "the permanent interests of man as a progressive being" ahead of "individual freedom" and "self-determination."[36] Mill, like Voltaire, found small-group loyalties and customs stultifying to man's development as a moral and cognitive being. Commenting on Tocqueville's pluralism, Mill offered vaguely positive statements about decentralization in *Representative Government*, but then said, "The principal business of the central authority should be to give instruction, of the local authority to apply it." Of course, Mill ends at least in moderate socialism.[37]

Levy spends his final chapters arguing the impossibility of resolving the "genuine tension" between the two pairs of thinkers and their ideals—Montesquieu vs. Voltaire, Tocqueville vs. Mill. Rather than try to synthesize pluralism and rationalism, we need to live "with a degree of disharmony in our social lives."[38] Yet one is left wondering whether Montesquieu was chosen as the founder of pluralist liberalism to force his rationalizing opposite to be the less attractive Voltaire. To highlight Voltaire extolling "enlightened absolutism" under Louis XIV, Frederick the Great, and Henry VII is really not the best way to build a case for freedom.[39] In any case, Levy's rationalist/pluralist distinction is most useful, and especially his insistence that the search for the roots of capitalism must begin with the idea and history of pluralist freedom.

Possessive Individualism

If one is looking for a consensus figure to characterize pluralist capitalism, it is almost impossible to avoid John Locke, who preceded Montesquieu with his systematic justification for a civilization of freedom, individualism, property rights, and capitalism, primarily in his *Second Treatise of Government*. Locke is certainly viewed as America's founding intellectual source. Louis Hartz, in his widely influential book *The Liberal Tradition in America* (1950), wrote that history seemed to be "on a lark" to identify Locke's philosophy with American reality.[40]

Jonah Goldberg, in *Suicide of the West*, described America's Declaration of Independence and Constitution as "echoes of" John Locke, even saying that U.S. history has been "more Locke than anything Locke imagined."[41] Yoram Hazony, in *The Virtue of Nationalism*, called Locke the author of "modernity's most famous liberal manifesto," one describing individuals "pursuing life, liberty and property in a world of transactions based upon consent." But he faulted Locke for neglecting the importance of "mutual loyalty" within the community.[42] Patrick J. Deneen, in *Why Liberalism Failed*, called Locke "the first philosopher" of pluralist liberal capitalism, the one who initiated the economic individualism and freedom of choice that have now undermined the traditional institutions of family, community, and natural law, which are essential to a good social order.[43]

As these three and others have suggested, Locke's writing invites multiple interpretations.[44] Even Hartz concluded that it is not easy to understand precisely what Locke's philosophy was.[45] Peter Laslett of Cambridge called Locke "perhaps the least consistent of the great philosophers."[46] In his once-dominant political theory text, George Sabine claimed that Locke "stands before all other writers" in explaining America and the West, but admitted that it was "exceedingly difficult to understand exactly what Locke believed to be the philosophical justification for his theory of natural rights."[47] Kenneth Minogue, from the London School of Economics, labeled him "a tricky customer."[48] The philosopher Edward Feser dismissed Locke as ultimately incoherent.[49]

C. B. Macpherson, a Marxist academic, came to the conclusion that Locke's written philosophy was meant to hide his true belief in unrestricted capitalist property accumulation—a belief he could not have made explicit in a world that still valued morality and social cohesion. Macpherson titled his book *The Political Theory of Possessive Individualism.*[50]

The classical philosopher Leo Strauss observed that Locke's claimed beliefs in a personal creator and in life after death were based on revelation, and thus inconsistent with being considered rational, much less being a philosopher. Yet Locke must be rational at some level—how else could he be so widely respected?—so there must be a "hidden" philosophy that is rational. In *Natural Right and History*, Strauss boldly excluded all revelation from Locke's writings to manufacture a Lockean "partial law of nature." The secret, in Strauss's view, is that Locke did not take revelation seriously and was really a pure rationalist and a hedonist whose philosophy was essentially utilitarian, but disguised in a way to appeal to his readers, in a society where virtue rather than pleasure was the highest goal.[51]

Perhaps the most systematically critical interpreter of Locke's thinking was the philosopher Eric Voegelin, who was exceedingly influential on the right. Voegelin most vigorously engaged Locke over his ideas of capitalist freedom and tolerance near the conclusion of Volume 7 of his posthumously published *History of Political Ideas*, where he almost cavalierly dismissed Locke as an unserious person.

The time when Locke was considered by historians a great political philosopher seems to be passing. His thought is recognized
today as the expression of the social and constitutional settlement
of the Restoration and the Glorious Revolution, though it was fixed
in its essential lines before 1688, and the great influence that he
wielded throughout the eighteenth century in England and the
American colonies, and on French political thought, was due to his
very limitations.[52]

Voegelin presented Locke's idea of limited monarchy and separation of powers, his theory of consent, and his beliefs about property as
the reasons for his appeal, but argued that these were merely "ancillary
evocations" not containing "a single idea" that had not been introduced
well before him. "Such importance as Locke's political philosophy has
is not to be sought in this blueprint of government but in those parts
of the work in which he develops his principles of human nature on
which the governmental superstructure is based."

Only when the "incidental subject matter" of government is put
aside can one find "the nucleus of Locke's theory," which is actually a
"new postmedieval anthropology."

> Students of Locke have noticed the inconsistency between his criti-
> cism of innate ideas in the *Essay Concerning Human Understanding*
> and his belief in innate ideas of reason in the *Treatise*. The logical
> incompatibility of the two positions need not, however, be taken
> necessarily as a flaw in the system, but rather as a symptom of Locke's
> easygoing philosophical habits that do not impair seriously the con-
> sistency of his basic attitude. Locke was not a fanatical thinker like his
> two great contemporaries [Hobbes and Spinoza]; he did not attempt
> to penetrate to the elements of human nature but was satisfied with
> a description of man as he appeared to him and the average people
> of his social group.

Locke, suggested Voegelin, "grasped the essence of the type that determined the following centuries of English politics. What may appear
to the philosopher as the unbearable flatness of Locke is the secret of

his effectiveness: he drew the picture of the new man as the new man wanted to see himself." This makes his *Treatise* "the most important" source for understanding English commercial society and related pluralist capitalist societies.[53]

Voegelin argued that the capitalist new man needed especially to be seen as free and tolerant. Locke exemplified this ideal throughout his work, but particularly in the constitution he drafted for the colony of Carolina. It proclaimed (in Voegelin's words) that "no one should be a free man, or even an inhabitant of the colony, who did not acknowledge and publicly worship a God; but that anyone who did so should receive protection for the exercise of his creed irrespective of denomination." While the constitution did not actually go into effect, it did spell out a fundamental difference from other colonies such as Massachusetts.

In his *First Letter Concerning Toleration*, Locke described the tolerant society as one organized by men for the sake of "procuring, preserving, and advancing their own civil interests." Locke enumerated man's civil interests as "life, liberty, health, and indolency of body; and the possession of outward things, such as money, lands, houses, furniture, and the like." The state's function was to secure these secular goods and freedoms for society.[54]

As for the other great traditional institution, the church, Locke regarded it not as the essential partner of government that Voegelin considered it to be, but merely as another "voluntary society of men." Voegelin continued:

> The idea of the church as a private organization within the framework of civil society is the last stage of an evolution the beginnings of which we could discern in Luther and Calvin. The break of the great [medieval] compromise by the Reformation expressed itself in the sectarian insistence on a purified church sphere and in a corresponding neglect of the secular arm. The result was not the desired subordination of the secular sphere to the ecclesiastical organization, but on the contrary the liberation of the secular sphere from the restrictions that the [medieval] religious compromise had imposed. The Reform began with the program of submitting the secular sphere to the control of the saints and ended with the relegation of the saints to the corner of "a free and voluntary society."[55]

The religious sectarians, said Voegelin, "won their freedom of conscience at the price of keeping quiet and not bothering the political community with their affairs." But the wider public has misunderstood this aspect of toleration as it came down from Locke.[56]

Voegelin explained how this form of toleration made religion "impotent" in society, and consequently invited new secular creeds to fill the void in public spiritual life:

> The privatization of the church means, in terms of social effects, that the political sphere has lost its spiritual authority and that the religious sphere, as far as it is coextensive with the tolerated churches, is condemned to public impotence. The toleration society has not only lost its public organs of resistance against inimical creeds but also has deprived itself of organs of public spiritual life in general. Since man does not cease to be man, and spirit does not give up its desire for public status simply because Locke or somebody else tells it to do so, persons who are of a spiritual and at the same time of a political temper have found new avenues by which to reach the public. We see the rise of the intellectual outside the church, ranging, according to temper and circumstance, from the scholar through the publicist to the professional revolutionary who tries to gain political public status for his creed.[57]

These materialist intellectuals, emerging from the disestablishment of traditional religion, "are to a large extent the same that otherwise would have found their way into a spiritual hierarchy." Voegelin pointed out that "a considerable number of eighteenth- and nineteenth-century German scholars and philosophers had Protestant ministers for their fathers, or that three leading Russian statesmen—Stalin, Zdanov, and Mikoyan—were once students of theology." The creeds of the new secular intellectuals give "the masses something that they seriously need: a public form of their spiritual personality."

Returning to Locke, who seemed to believe that man is "the product of divine workmanship," Voegelin examined what the philosopher considered to be the source of the rule that men must not do damage to others. Locke referred to "God, Nature, Reason, and Common Equity," but offered no definition or explanation of those terms. "We can discard, therefore, this enumeration of sources as a collection of mere hieroglyphs. Locke was simply too optimistic to see that here was a problem, and he satisfied himself by throwing at the reader's head any authority that had a good name," God being the obvious choice.[58] Voegelin saw this as the "key" to understanding Locke.

But that "system of legal hieroglyphs comes crashing down" with the invention of money and the ability to store value and exchange it for larger possessions. Here is the turn away from the "egalitarian state of nature" and into "the solid Hobbesian passions: the acquisitive property society is the product not of right, but of passion." In Locke's capitalism, "the grim madness of Puritan acquisitiveness runs amuck," and the "elements of a moral public order that derive from biblical tradition have disappeared. A public morality based on belief in the spiritual substance of the nation is practically absent. What is left, as an unlovely residue, is the passion of property."

Here is the capitalist acquisitiveness that Voegelin's own editor equated with Macpherson's "possessive individualism." Voegelin concluded, "In this respect Locke is the outstanding symbol of the revolution-breeding element within the capitalistic order, foreshadowing the events of the nineteenth and twentieth centuries."[59]

In Voegelin's analysis, "the first cycle of modern political thought comes to a close" with Locke, whose theory of freedom and toleration takes the church out of "the domestic public scene." The medieval order had been overturned by "sectarian religious enthusiasm," which the church was "unable to digest or suppress." It could not avert the "anti-civilizational danger" posed by the belief that "civilization was gained, not destroyed, in the great crisis beginning with the Reformation." According to Voegelin, the "desperateness of the struggle" between religious denominations and the "consequent enormous loss of prestige for the religious organizations" must be understood as the basis for the indictment of the church as the "mortal enemy" of enlightenment by eighteenth-century rationalistic philosophers. The "dark age of religion" story told by Voltaire and others "fixed the sentiment profoundly to this day."

But that was not the end of religion, even if the heirs of the Enlightenment have insisted it was:

The rise of new religions, the appearance of a new Koran in the *Kapital*, of a patristic literature with Lenin as the great church father, of the heresiarch Trotsky, of a new inquisition, are either simply denied by the stalwarts of the [Voltaire] creed as religious phenomena or admitted as such to the extent of being called barbaric relapses from the standards of Civilization into medieval forms that progress has left behind for good. They are in the nature of bad dreams and will pass away; such things do not really happen in the twentieth century.[60]

John Locke's Capitalism

Locke's own formally stated argument for property rights and morality is straightforward.[61] Human individuals were created in a state of nature with an inherent freedom to act for self-preservation—to take from the commons what was necessary to support themselves and their families as long as nothing was taken immorally from others, who also were created free, or was wantonly wasted. All were equally free to act in their own individual interest, to own, accumulate, and control property "in their persons as well as goods," to satisfy their utilitarian needs.

The Creator commanded all not to harm their neighbors, at least if the neighbors did not first injure them or violate their property, and all were morally bound to obey him, yet they were created free. Many therefore disobeyed, and property became "very unsafe, very insecure" in that free state of nature. Since the rights of all were granted by a higher power, they could not morally be taken away by anyone else; instead, each individual had to consent freely to allowing government "all the power necessary" to gain more security. "The great and chief end therefore of man's uniting into commonwealths, and putting themselves into government is the preservation of their property, to which there is much wanting in the state of nature."[62]

Government power should be limited because it should follow "established, settled, known law, received and allowed by common consent to be the standard of right and wrong and the common measure to decide all controversies." This consent would require "a known and indifferent judge with the authority to determine all differences according to the established law." Power should be further limited by dividing it between a legislative assembly that set the rules and an executive that carried them out. Once the law was derived from traditional beliefs, most people would follow the law naturally as coming from themselves. If the government adhered to these rules, there would be significant freedom to live in community much as in the state of nature, but with the security to build a prosperous social order, with individual rights and private property.

But what about Locke's "hieroglyphs," the "inconsistencies" and "easygoing philosophical habits" described by Voegelin? Strauss likewise found inconsistencies, saying that Locke in his easygoing way tried to have both of the two possible, mutually exclusive sources of authority: reason *and* revelation, Athens *and* Jerusalem. Sabine argued in his *History of Political Theory* that Locke was personally religious but that this did not influence his philosophy, which was "as egoistic as that of Hobbes," although it was later thinkers who turned Locke's philosophy into true utilitarianism. A political progressive himself, Sabine characterized Locke as "an empiricist but with a large residue of philosophical rationalism and a firm belief in self-evident principles of right and wrong." Yet he found it impossible logically to "unify" the different elements into an integrated rationalism.[63]

The more conservative Feser, countering Strauss, argued that while "an appeal to God is absolutely necessary" to understand Locke and justify his conclusions, Locke's empiricism "undercuts his argument for God's existence," which in turn is essential to justify his individualist rights. A sound argument, from a rationalist point of view, would require an Aristotelian "right order of inquiry." Because Locke did not base his understanding on this right order of rational inquiry, establishing essentialist concepts and then developing a theory to explain them, he lacked a firm basis for his beliefs. Feser, like others, found Locke to be individualistic, hedonistic, and utilitarian at bottom, trying to justify a possessive capitalism.[64]

Against this broad agreement about Locke, Brad S. Gregory argued in *The Unintended Reformation* that the problem was not with Locke himself but with twentieth-century philosophical thinking generally, which could conceive only of a monist rationalism, a monist

empiricism, or a monist revelation as mutually exclusive logical possibilities.[65] It is certainly difficult to place Locke only with Plato or
Descartes, as a rationalist, since he has also been called the founder of
British empiricism—of "empirical science," as Feser put it. This dualism is exactly why Locke is said to lack coherence. Besides supporting
both rationalism and empiricism, Locke also held that there are "things
above reason," that these are matters of faith and revelation, and that
"an evident revelation ought to determine our assent, even against
probability."[66] Rationalism, empiricism, and traditionalist revelation
were all aspects of Locke's thinking.

In this "easygoing" philosophy, reason is a nondeterministic synthesis of rationalist, empirical, and traditional elements, including
common sense, instinct, and revelation. This way of thinking is what
predominated in those "Dark Ages" that post-Enlightenment philosophers turned so strongly against. The most important high medieval
philosopher, Thomas Aquinas, worked within an explicit synthesis of
Christian revelation and Aristotelian philosophy, which made truth
in medieval scholasticism "a many-sided edifice," as John Courtney
Murray explained in *The Problem of God*.

> I should say rather that there was one universe of truth, within which
> different kinds of truth, and correspondingly different methodolo-
> gies for their pursuit, existed in distinction and in unity. Moreover
> … there prevailed the robust belief that between the valid conclusions
> of rational thought and the doctrines of faith no unresolved clash
> could or should occur.[67]

As Gregory documented, this way of thinking seriously eroded in the
fierce battles between Reformation and Counter-Reformation theorists, each side seeking a unitary solution that must be rational in the
modern monistic sense. Gregory argued that a monist master axiom
is not necessary, or even possible.[68] Indeed, back in 1945, F. A. Hayek
had distinguished between monistic rationalists of the "French and
Continental" type, such as René Descartes and Voltaire, and pluralists
like Locke, Adam Smith, Edmund Burke, and Tocqueville.[69]

In his 1964 essay "Kinds of Rationalism," Hayek described the difference between a "constructivist rationalism" that starts unambiguously

from single monist essences and deduces all conclusions from them, and a "critical rationalism" that employs multiple reasoning methods—rationalism, empiricism, intuition, and traditional common sense.[70] The philosopher of science Karl Popper, in "Towards a Rational Theory of Tradition," even gave tradition the preeminent place in the process of understanding, identifying it as the first reality we can comprehend, from which all else is reasoned.[71]

Strauss and the others knew the Thomist synthesis, of course, but dismissed it as a "dualism."[72] That is why they rejected any attempt to use plural methodologies for what they viewed as undifferentiated reality requiring a single, monist rationality. Feser recognized that Locke was seeking a "middle way" but said it was impossible to avoid the skepticism, subjectivism, and irrationalism that Locke sought to avoid unless it was fully accepted that a natural order "is knowable to reason" in a monist Aristotelian sense. Feser maintained that Locke did not accept enough of Thomism to make his "delicate balance" successful.[73] Voegelin's criticism, on the other hand, was that medievalism itself was a compromise of true philosophy.

There is no question that Locke moved away from Aristotle and the ancients, but so did St. Thomas. The break was located in medieval Christian thinking. While Feser wrote that virtue and morality for the Thomist "need not be determined by an appeal to God's commands" but could be solely based on reason, he was careful to qualify this point by saying that "the Thomistic conception [is] largely (though not wholly) secular." That "not wholly" is crucial. In fact, the very structure of the *Summa Theologica* (St. Thomas did not call it a *Summa Philosophica*) routinely added Christian revelation to Aristotle's philosophy to complete his purely rationalist explanations.

Monistic Aristotelianism has been a difficult sell ever since its physics fell to Newtonian mechanics, but now the absoluteness of mechanics has fallen to probability. Even an explicit Aristotelian like Alasdair MacIntyre utilizes a pluralist interpretation of the great philosopher.[74] For most scientists today, absolutist perfection can no longer be a rational goal. Richard Rudner, a philosopher, has argued that "an adequate reconstruction of the procedures of science would show that every scientific inference is properly construable as a statistical inference."[75]

As Feser recognized, probability is in fact central to Locke, who undermined essentialism by excluding "substantial forms, substance, essence, identity and so forth" from "knowledge in the strict sense." Yet Locke did accept the idea of forms in a less strict sense. More important, Aristotle's idea of forms is more empirical than Plato's, and most important of all, knowledge for Thomists "is always going to be limited in various significant ways," which was Locke's point, and which Feser also understood.

Four decades ago I argued, against the dominant scholarship, that one could not understand Locke unless one viewed him as medieval, as feudal, as Christian.[76] Part of the problem was that many had read only his *Essay Concerning Human Understanding* or the *Second Treatise of Government*.[77] In *Some Thoughts Concerning Reading and Study for a Gentleman*, Locke says that "to give a man a full knowledge of true morality I shall send him to no other book, but the New Testament." Likewise, in *The Reasonableness of Christianity* and *Some Thoughts Concerning Education*, he wrote that Jesus was Messiah and Savior, that this was the center of reasonable religion, and that men were expected to believe this in some way if they were to attain happiness in the hereafter.[78]

Locke was orthodox enough to write a discourse defending miracles, and in his last years he translated and extensively commented upon the Epistles of St. Paul. Even in the *Essay Concerning Human Understanding*, Locke held that the "mere probability" of an afterlife should move reasonable men to follow God's law so that they could enjoy the "infinite eternal joys of heaven" rather than try "to satisfy the successive uneasiness of our desires pursuing trifles" on earth. Locke's beliefs may have been naïve, as Voegelin claimed, but they were sincere—and essential to his philosophical synthesis.

Today, as Paul Sigmund of Princeton noted, it has become almost impossible to deny that revelation is a serious part of Locke's philosophy, in synthesis with both reason and empiricism.[79] Jeremy Waldron of Oxford broke the barrier in academic circles in 2002 with *God, Locke and Equality*, expressing doubt "whether one can even make sense of a position like Locke's apart from the specifically biblical and Christian teaching that he associated with it." Indeed, he argued that without Locke's theological justification there is no support for liberalism's basic premise of moral equality between human beings.[80]

Steven Forde argued in *Locke, Science, and Politics* that with probability so dominant in science today, a purely abstract natural law could not alone bear the weight of morality and that only Locke's introduction of a law imposed from outside nature by a creator could.[81] When even a positivist like Waldron can say, "I actually don't think it is clear we—now—can shape and defend an adequate conception of basic human equality apart from some religious foundation," some reappraisal seems required.[82]

In making a distinction between rationalism and revelation and holding these to be different but complementary methods to approach the truth, Locke was simply following pluralist medieval thought. To Locke (as in Thomism generally), both reason and tradition are needed. It is not a choice between opposites but a harmony. Where Locke and Thomas differed most—where Locke expanded upon and challenged medievalism—was, once again, on the matter of probability, mostly learned from his friend Blaise Pascal, the mathematician who basically conceived it.[83]

In other words, Locke fused probability with a medieval synthesis of reason, tradition, and revelation, producing a more open synthesis that seemed too dualist and plural for classical philosophy, too spiritual and value-oriented for empiricism, and too materialistic and problematic for traditional medievalism. That synthesis may be completely invalid, or faulty in some respect, or at odds with the rationalistic principle of noncontradiction, or simply unfashionable. But it did have a considerable effect on the development of Lockean-influenced capitalism, as critics have complained ever since.

The Property Paradox

Marx identified the essence of pluralist capitalism with a single institution, that of private property. He even bragged that he could put the solution to its evils in "a single sentence" requiring its total "abolition." *The Communist Manifesto* was, as usual, vividly blunt: "You are horrified at our intending to do away with private property. But in your existing society, private property is already done away with for nine-tenths of the population; its existence for the few is solely due to its non-existence in the hands of those nine-tenths." The only solution to capitalist property accumulation with no moral limits was to have property shared by all.

Yet it was Marx himself, in his more empirical *Kapital*, who made the case that the takeoff for the stage of development he called capitalism was when the serfs wrested their property step by step from the landholding lords. As noted earlier, he showed that by the late fourteenth century in England, serfs had largely become property owners and were in fact mostly free from the old feudal bonds, in spite of legal status. The later enclosure laws that coercively turned small, individually owned plots into large holdings—thereby forcing smallholders off their land and into factory labor—merely accelerated a process that was already well advanced. Indeed, the right of property was necessary to enable the new capitalists to become employers, and former serfs and their descendants to become wage laborers.

Harold Berman stayed closer to actual history in tracing Western property law first to the extensive Roman legal inheritance passed down to Europe through the Frankish kings and through the church's canon law (which the eighth-century Lex Ribuaria required to follow Roman law), mingled with Germanic folk law and the traditions of Israel from the Bible. After Louis the Pious divided the empire he had inherited from Charlemagne, these roots atrophied, but the basic elements lingered and were brought together again in canon law, beginning with family and inheritance matters but also covering the Catholic Church's material holdings, which amounted to perhaps one-fourth of medieval Europe's wealth.[84]

As canon law spread through feudal Europe, it came into competition with often erratic baronial courts, which meant that those seeking impartial justice in the handling of property disputes had an alternative. Since many disputes were with the barons themselves, church courts proved more objective and more rational, being based on the best scholarship of the day rather than custom and trial by combat or by ordeal. The more consistent church procedures even won popular support and were later adopted both by the old pluralist monarchies and by the new rationalistic ones, which advanced their principles of property rights into modern times.[85]

Even modern socialists generally express support for personal property, attacking only the productive property held by those characterized as the greedy rich.[86] But concerns about inequality have been growing. Since 1940, the Gallup Poll has asked Americans whether the government should "redistribute wealth by heavy taxes on the rich," and it is only in recent years that a majority have said "yes."[87] The Pew Research Center in 2014 found that 60 percent of Americans believed that "the economic system in this country unfairly favors the wealthy."[88] In 2019, 45 percent of millennials said they supported socialism over capitalism.[89]

Redistributionists like Pope Francis favor an equalization of global resources, but the pope and most other moderate critics of capitalism have explicitly rejected the Marxist call for abolishing private property, which they consider a basic human right. Yet they agree with Marx that property rights under capitalism do not function for all equally, as Locke said they should, but mainly for the benefit of the rich and powerful, and therefore are immoral and unjust.[90]

Those with very little property themselves do not necessarily see it that way. For example, slum dwellers in India have actually agitated for property rights, as *The Hindu* reported a few years ago:

> Members of the district unit of the Karnataka Kolageri Nivasigala Samyukta Sanghatane staged a dharna outside the Deputy Commissioner's office here on Wednesday, demanding issuance of property rights to slum dwellers in the city. The agitators took out a procession from Dr. B.R. Ambedkar circle, which passed through the main streets of the city and ended at the Deputy Commissioner's office. They also raised slogans against the State government and district administration for not fulfilling their demands even after repeated pleas.[91]

In Uganda, a journalist wrote of how the inadequate protection of property rights was detrimental to women's equality:

> [E]ffective statutory laws protecting land, inheritance and property rights of women including the widowed, divorced, separated or those in co-habitation are critically missing....
>
> Generally, there is lack of clear laws to address equality in land ownership, divorce and marriage which affects women's capacity to

enjoy equal rights with men, and this affects their women's health, economic and social rights.[92]

Why might the very poorest be so interested in property rights? Armen A. Alchian, an emeritus economics professor at the University of California, Los Angeles, put it this way: "The fundamental purpose of property rights, and their fundamental accomplishment, is that they eliminate destructive competition." Instead, "well-defined and well-protected property rights replace competition by violence with competition by peaceful means." Alchian explained how property rights actually protect the weak from the depredations of the strong or well-connected, while a restriction on property rights "shifts the balance of power from impersonal attributes toward personal attributes and toward behavior that political authorities approve."[93]

The rich can defend their own property with high walls, secure safes, burly guards, and whatever else money can buy, as well as access to government favoritism. The poor do not have the resources to pay for the counterviolence needed to protect their own property without established legal rights. Moreover, with secure property rights, discrimination on the basis of personal characteristics can be counteracted through pricing: a seller can offer lower prices, or a buyer can offer more. For example, a landlord might prefer to rent to a member of his own race, but a person of another race can offer a higher bid, making it in the owner's interest to prefer him.

The power of property rights to keep the peace can be demonstrated simply at the household level. In the *Sesame Street Parents' Guide*, Katherine Hussmann Klemp recounted how she minimized quarreling among her eight children by assigning clear property rights to toys. Previously she had not intended for toys to belong to a particular child. "Upon reflection," she wrote, "I could see how the fuzziness of ownership easily led to arguments. If everything belonged to everyone, then each child felt he had a right to use anything." So she introduced rules of ownership, and found that they actually promoted sharing rather than selfishness. The children were secure in their ownership and knew they could always get their toys back after sharing them.[94]

When property rights are set up ineffectively, the rich and powerful will resort to private power to guard their own possessions, as

Konstantin Sonin observed in studying the privatization of property following the collapse of the Soviet Union in the 1990s. Once they set up their own private protective mechanisms, the rich gain an interest in maintaining a legal property system that works poorly for everyone else. The wealthy oligarchs become "natural opponents of public property rights." Being powerful, they can bias the law to stifle competition, stunting growth and depressing income for everyone else. This is in fact what happened in Russia after the Communist regime fell.[95]

Hernando de Soto, in *The Mystery of Capital*, demonstrated how a perverted system of property rights harms the poor. His thesis was that the major inhibitor of growth in third-world nations was the lack of fair property rights, and particularly the challenges in documenting legal ownership. When property ownership is unrecorded, long-established de facto owners living on contested land will have difficulty obtaining credit, or expanding or selling small businesses. Without legal title, they cannot protect their property from predation in court. Restricted access to legality in ownership generates two parallel economies, legal and extralegal. An elite minority enjoy the economic benefits of legal commerce, while the majority of entrepreneurs are stuck in poverty as their assets—adding up to more than US $10 trillion worldwide, according to de Soto—languish as "dead capital" rather than being available for use and benefit.

De Soto demonstrated his point creatively in Peru, where he and his research team set up a small garment business with one worker and proceeded to turn it into a legal enterprise. It took his sophisticated team 289 days, at the cost of $1,231—a prohibitive sum for desperately poor people trying to make a living. They then sought legal authority to build a house on long-occupied government land, which took six years and eleven months and required 207 interventions before 52 agencies. To obtain legal title to the occupied homesteads required 728 steps. In the Philippines, they found that if a person wanted to build a house on occupied state or private land and obtain legal title, he would have to form an association with his neighbors and proceed through 168 steps before 53 agencies over a period of 13 to 25 years. In Egypt, registering ownership of a long-occupied lot on state-owned desert land took 77 interventions before 31 agencies over a period of five to 14 years.[96] These requirements mostly burden the poor and powerless.

Property does lead to inequality, and we will evaluate that later, but fair property rights are an economic necessity for those without power. The rich and powerful do not need legal property rights, since they can protect their own. Indeed, market capitalism required a long tradition of property that slowly grew into legal regimes of fairer property rights.[97] But those property traditions had to survive what has often been labeled the Dark Ages of feudalism.

The Final Stage of Feudalism

The term "Dark Ages" to describe the feudal era was not used until the nineteenth century, with its secular rationalization.[98] In large measure, the label reflected hostility to religion, or what Marx called the "opiate of the people." But it is on this subject that concepts of capitalism are most divergent. To Locke and most pluralists, religion was essential for building the fundamental rules of property, individual morality, and the division of powers. To the rationalizers such as Voltaire and Mill, religion was the main obstacle to efficient and rational capitalism, which depended on escaping from the prejudices of a dark religious past into the enlightenment of reason and science. In this view, the creation of the modern nation-state in the seventeenth century, with the Treaty of Westphalia granting the choice of religious denomination to the state, had marked the decisive turn away from the Dark Age of religious wars to an era of rationalism and peace.[99]

The concept of the Dark Ages is contradicted by actual history, however. Europe had superior arms as early as Charles Martel's defeat of the Moors in the eighth century. Stirrups, waterwheels, the heavy plow, horse collars, and three-field crop rotation began in the tenth century. Mechanical clocks, eyeglasses, and the compass appeared in the eleventh century. The university was a great achievement of the twelfth century, eventually leading to Albertus Magnus systematizing botany in the thirteenth century, and then in the fourteenth century, Jean Buridan correcting Aristotle's physics, Nicole d'Oresme anticipating Copernicus in astronomy, and Albert of Saxony providing the basis for Newton's First Law.[100]

But the main charge leveled against the Dark Ages by the rationalizers was moral intolerance and the resulting "religious wars." Karen

Armstrong, fellow of the Royal Society of Literature, offered a convincing correction to this view of religion as a cause of violence, in *Fields of Blood: Religion and the History of Violence.* "The Wars of Religion and the Thirty Years' War may have been pervaded by the sectarian quarrels of the Reformation but they were also the birth pangs of the modern nation state," she argued. Moreover, "casting off the mantle of religion did not bring an end to prejudice," but instead it brought the replacement of sectarian intolerance with the "scientific racism" of nationalism and eugenics.[101]

Armstrong found religion to be universal, and violence to be simply part of a general human intolerance of differences, which she documented worldwide. Throughout history, societies had incorporated religion into governments as a way to give them legitimacy and to "endow everything they did with significance." But the state was always the predominant partner and usually fully dominant, except for periods in early medieval Europe. Until the French and American revolutions, there were no secular states at all, but the end of religious establishment did not necessarily lead to secular enlightenment.[102]

While she found out-group prejudice to be universal, Armstrong identified the particular intolerance of anti-Semitism as most conspicuous in the Christian West, although she ascribed it mostly to reasons of state and the economic importance of Jews in Europe. The Christian Church itself taught tolerance, even if the lesson was often ignored, but the New Testament account of the role that Jews played in Jesus' death did lead to scapegoating. Armstrong blamed the Crusades for the great turn against European Jews: war fever against "infidels" in general led to massacres of Jews in several cities by armed forces en route to the Holy Land to confront Islam, and to the sacking of Jerusalem and the murder of its residents, including Jews.

There certainly is plentiful evidence of irrational Christian prejudice throughout the history of the religion, but very little anti-Semitism during its first millennium. At first, traditional Jews and new Christians alike considered themselves Jewish, since Judaism had been recognized by Rome as a legal religion but Christianity had not. There was no practical reason for Christians to separate themselves publicly until Roman authorities pronounced them legally not Jews and began persecuting them as a hostile element. In 132 CE, when the Mishnah synthesizer

Rabbi Akiva and many other rabbis declared Simon Bar Kokhba the Jewish Messiah, a choice had to be made. Apparently, most Jews rallied to Bar Kokhba, and many supported his revolt against the Roman Empire in 135. Those who held Jesus as the Messiah did not, and they began to separate themselves politically as well as theologically, if only to avoid offending the Roman authorities.

Things began to change after Constantine became emperor in the early fourth century and granted official toleration of Christianity, and later all but establishment as the official religion. First, the Roman legal privileges for Jews were repealed by imperial authorities. Rabbinical legal jurisdiction over Jews was curtailed. Jewish missionary work was prohibited as a threat to Roman law and morality.

Some earlier bishops, most notably Origen of Alexandria and Cyprian of Carthage, had forbade Christians from marrying Jews and even dining together—which were also against Jewish law—and these prohibitions were adopted by the Council of Nicaea in 325. By the fifth century, church leaders such as John Chrysostom in Constantinople and Cyril of Alexandria were derogating and expelling Jews from Byzantium. In the West, Pope Gregory I banned the forced baptism of Jews in 591, and in 600 he ruled that Jews should not "suffer a violation of their rights," although not have special privileges either, mostly undefined.[103]

By then, Christian norms had increasingly been incorporated into Roman civil law, in both East and West. Differences of social practice were viewed as undermining public order and community. Jewish dietary laws forbidding meals in non-Jewish homes were seen as a rejection of community. Jewish law called for no work on Saturday, which was a Roman workday. The sixth-century Byzantine Code of Justinian negated many normal civil rights for Jews, treating them more as sojourners than as full citizens. Yet many Jews prospered and their numbers increased.

Jewish settlement in the West was minimal until the tenth century, when significant legal and economic advances occurred. As Spanish territory began to be recovered from Muslims in the eleventh century, Jews living there were given the choice to settle in northern Europe, often with inducements from existing Jewish settlements. Jews were generally reluctant to do so because of Christian efforts at proselytization. If they did resettle, they drew upon their biblical "badge of honor" as "a people that dwells apart," writes Robert Chazan, a professor of Hebrew and Judaic studies. They maintained their identity in separate communities, often more prosperous than their Christian neighbors.[104]

More dramatic changes for Jews in Europe came in the late eleventh century, when the Seljuk Turks had won big victories over Byzantine forces and captured the great cities of Edessa and Antioch, and then Jerusalem. The Byzantine emperor appealed to Pope Urban II for assistance in his struggle with the Turks. In response, the pope called for a crusade to rescue Christian holy places from Muslim control. Tens of thousands of Christians, both knights and commoners, set out for the Holy Land in 1096. Some of the crusaders massacred thousands of "infidel" Jews in towns along the way, sometimes joined by local mobs, although Jews were often sheltered by local bishops.

There were also mob attacks on Jews in response to the incredibly destructive Black Death in the fourteenth century, for which many people scapegoated the Jews—in spite of a bull from Pope Clement VI declaring that there was no basis for blaming the Jews, some of whom were being struck down too.[105]

In the thirteenth century, Pope Gregory IX organized a religious inquisition in France and Italy, to be conducted by Dominicans. One of the purposes was to convert Jews, which sometimes was done under threat of fines or imprisonment, and occasionally torture. In 1215, the Fourth Lateran Council pronounced the holding of public office by Jews to be an "improper subordination of Christians to Jews." Government policies differed from place to place; there were exclusions from certain occupations and prohibitions against carrying weapons. Most severe was the expulsion of all Jews from England in 1290 (until their readmission in 1656), and from France in 1306, but more temporarily in that case.

Discrimination and repression became more systematic with the Spanish Inquisition, beginning in 1478. While only Granada remained to be recaptured from Muslim rule, Ferdinand and Isabella regarded the Marranos, nominal converts from Judaism, as a threat to their rule. Pope Sixtus IV authorized the monarchs to appoint inquisitors, and attempted to moderate the methods employed. But Ferdinand wielded this instrument of state power mercilessly, later turning it against the

Moriscos, the recent converts from Islam. Machiavelli praised him for using the "pretext of religion" to advance governmental interests.[106]

In the sixteenth century, Martin Luther was frustrated by Jewish refusal to accept his new doctrines, and he denounced Jews in strong terms. He favored the expulsion of Jews from Europe, although little actually resulted.[107]

We can add more depth to this picture of social relations between Christians and Jews during the early phase of pluralist capitalist development. Sara Lipton provides another dimension of evidence in *Dark Mirror: The Medieval Origins of Anti-Jewish Iconography*, her thorough study of pictorial representations of Jews in medieval Europe.

Lipton begins by saying, "For the first thousand years of the Christian era, there were no visible Jews in Western art." While Jewish prophets and armies and kings can be identified as such by context in monuments or manuscripts, they were not "singled out" as conspicuously different from Christians. The general tone of the attitude toward Jews was set by Augustine and those who followed him, who were critical of Jews for not acknowledging Jesus as God, but for centuries this theological criticism "remained metaphorical, a largely literary abstraction" with little connection to the physical presence of Jews in Christendom. As Lipton notes, "Jews in the early Middle Ages were legally free and their lives were considerably more prosperous, secure and comfortable than those of most Christian peasants."[108]

Soon after the year 1000, "the Jew emerged from obscurity," represented with "a host of visual cues" to make the images recognizably Jewish. The first of these cues was a distinctive pointed or peaked hat. Yet Lipton's research led her to conclude that this type of hat was not meant to identify Jews per se, but people of the East generally.[109] It was not until the twelfth century that clearly distinct images of Jews "began to spread across Christian art." It was then that the peaked hat became identified with Jews in particular, primarily in their function as "Jewish witness," referring to the Augustinian idea that their presence testified to Christian truth. "The Jews are indeed for us living letters of scripture, constantly representing the Lord's passion," wrote Bernard of Clairvaux. "They have been dispersed all over the world for this reason: that in enduring just punishments for such a crime wherever they are, they may be witnesses of our redemption."[110]

Being a time of relative economic prosperity and of ecclesiastical reform, the twelfth century brought a movement away from an ascetic Christianity "to make room within spirituality for opulent artworks." Lipton regards the recognizable images of Jews in this art as arising not primarily from "antagonism" or even a wish to make a statement about contemporary Jews, but rather from a desire to "rehabilitate the realm long rhetorically associated with Jews (the external, glorious, temporal world so inimical to early Cistercians) as a valid part of Christianity."[111] The presence of clearly Jewish images in Christian art reflects a new acceptance of the material world.

Ironically, it was liberalizing empiricists such as Roger Bacon in the thirteenth century, encouraging the scientific study of the human form, who inadvertently contributed to the caricaturing of Jews with hooked or prominent noses and jet-black hair, since many came from southern regions where such features were common. Lipton finds caricaturing of Jews beginning to appear in the mid-thirteen century and taking an easily recognizable form by 1340.[112] The most distinctly Jewish representations in art appeared when newly centralized governments were eager to tax and control Jews. Depictions of Jews became more clearly negative in crucifixion artwork.

Lipton notes that "Jewish villains" were often "balanced by positive protagonists also identifiable as Jews," such as scroll-wielding prophets, the Pharisee who accepts Jesus, and the identifiable Jews who lovingly bury Jesus.[113] One reason for the increased number of Jewish portrayals is the greater complexity of design in art and the larger numbers of people included, representing a varied community.

As early as the thirteenth century, Jews were obliged to wear distinctive clothing, but this was an era when badges and identifiable symbols were commonplace for "mainstream" occupations and stations in life, and most laws regulating clothing were meant for Christians. By the late fourteenth century, art showed numerous people in crowds who were easily identified with particular professions or social status. In the later Middle Ages, Lipton observes, group consciousness was heightened for many types of people, including Jews. This was also a time when monarchies feared competition from powerful noble families. The greater attention to and regulation of outward appearance came at a time of intensified surveillance: the first European secret police was organized in 1378, in Florence.[114]

Lipton considers it "far too simple" to see in the images of Jews "a steady progression from positive to negative" attitudes, noting that anyone in search of "a portrait of Hebrew wisdom and dignity would have no trouble finding one" even in late medieval art. Moreover, "at no point" did Christian clerics or artists "consciously set out to create an anti-Jewish visual repertoire, much less inspire anti-Jewish violence or retribution." Still, we bear some responsibility for the effects of stereotypes we create.[115]

In the late Middle Ages, even after the expulsions, there were almost as many Jews in Europe as in the Middle East, and most were more prosperous than the mass of the Christian population.[116] Medieval religion has been overemphasized as a cause of anti-Semitism and of intolerance generally. It was not intrinsic to Christianity. When Peter addressed a crowd of Jews assembled at the Portico of Solomon, he said that while they must repent for demanding Jesus' crucifixion, he understood that they had no idea what they were really doing (Acts 3:17–18). Although Christianity surely can be faulted for not sufficiently opposing anti-Semitism, it was neither planned nor very virulent as long as the early, more religious medieval culture prevailed, as Lipton, Chazan, and Armstrong have demonstrated.

Armstrong concluded that neither religion in general nor medieval Christianity in particular caused hatred of Jews, for the root of such enmity "lies not in the multifaceted activity that we call 'religion' but in the violence embedded in our human nature and the nature of the state."[117] Indeed, medieval religion sometimes inspired serious efforts to suppress violence, as in the Peace of God movement beginning in the later tenth century, followed by the Truce of God initiatives in the eleventh.

The Middle Ages were not all about "gamboling on the greens," but neither were they as dark as often portrayed, or entirely a regression from classical antiquity. There were significant advances in technology and learning, and in the concept of individual rights. Slavery had been a part of all ancient civilizations, but in medieval Europe it was replaced by serfdom, granting limited rights. As Marx himself explained, serfdom was hardly ideal but it was an advance over slavery, and feudalism ended with broadly distributed de facto private property, which prepared the way for wage labor and mature capitalism.

The Root of Unruly Freedom

If the Christian moral order did not make the Middle Ages such a dark era of ignorance and intolerance that it could not have prepared the way for capitalism, there is still an opposing moral argument to address: Did the "limitless" freedom of capitalism sow the seeds of its own destruction by eroding the morality that undergirded it?

Today, a socially conservative commentator like Rod Dreher can sound much like Professor Deneen,[118] Pope Francis, or even Karl Marx, in charging that the "myth of individual freedom"—or possessive individualism—has torn away "the last vestiges of the old order," leaving only the belief that "true happiness and harmony will be ours once all limits have been nullified."[119]

Dreher identifies the "decisive blow" against the last remnants of medieval morality and culture in the legalizing of gay marriage. This new freedom represented "the final triumph of the Sexual Revolution and the dethroning of Christianity, because it denied the core concept of Christian anthropology," the last communal restraint, as embodied in the institutions of traditional marriage and family. Every culture "imposes a series of moral demands on its members, for the sake of serving communal purposes," and the crumbling of these restraints in the West amounts to a "deconversion" away from a Christian ethic, or any moral order, into social and ethical chaos.[120]

Dreher cites the sociologist Philip Rieff, who taught that "the essence of any and every culture" lies in "what it forbids," and argued that "the rejection of sexual individualism" was "very near the center" of Christian culture. As Dreher explains it, Christianity replaced the "sexual autonomy and sensuality of pagan culture" with its own morality as the operational value structure of the West, and then much of the world. But modern society has "inverted the role of culture" by "releasing" individuals from prohibitions and locating freedom in "sexual expression and assertion." We may note that ancient cultures often featured polygamy or concubinage and homosexuality, but usually defended as traditional, not libertine excess. Today, Dreher remarks, unlimited sexual release has become "how the modern American claims his freedom."

Yet the idea of freedom from traditional cultural restraints certainly had roots predating capitalism or even feudalism. When some early Christian converts from Judaism insisted that it was still necessary to follow all the laws of Moses, leading apostles declared that Gentile converts should not be burdened by all those mandates and prohibitions (Acts 15:1–21). Indeed, Paul went to the extent of saying that Christians, being justified by faith, are no longer under the Law (Gal. 3:23–29). "When Christ freed us," said Paul, "he meant us to remain free" (Gal. 5:1). Christianity introduced its own laws and prohibitions, of course, but the new rules could be liberating.

It was Christianity's teaching on sexuality and marriage, Dreher writes, that freed women from the "sexually exploitive Greco-Roman culture." Sexuality in antiquity was controlled by a strict patriarchal social order, which was undermined morally in the late sixth century when Pope Gregory I forbade marriage to close kin, and had marriages made not by extended families or even by clergy, but between individual men and women. The idea of marriage as an individual choice took time to develop, but its seed of free choice was planted here.[121]

In the view of Jean-Jacques Rousseau, Christianity with its individual freedom and the competing moral claims of church and state was "clearly bad," being destructive of "social unity" and civic order.[122] Critics of Christianity such as Machiavelli, Rousseau, and Nietzsche traced the subversive "myth of individual freedom" to Jesus himself or to St. Paul.[123] The founders of Christianity separated the individual from Caesar and the state and from the social order, as more sympathetic scholars have also observed.[124] Jesus, in fact, made a remarkably uncommunal statement about the very central institution of the family: "Do you think I came to bring peace on earth? No I tell you but division. From now on there will be five in one family divided against one another, three against two and two against three" (Luke 12:49–53).

Dreher thought it "pathetic" to regard Christianity as "a therapeutic adjunct to bourgeois individualism." But Christianity was the source of the individualism that characterized both the feudal nobility and Lockean capitalism. Almost everything before had been defined by clan, cult, tribe, or state. It was not modern society that "inverted" culture and freed the individual to question ancient tradition. It was the design within Western civilization itself that alienated the individual conscience from the ancient communal order. But at the same time, Christianity set up an impossibly high moral standard: to be perfect as the Father is perfect.[125]

The tension between individual freedom and a moral standard of perfection was understood even by critics of the Lockean synthesis, like Strauss, to be the source of the energy that turned the West into the world's dominant civilization.[126] But it also prompted the counter-impulse to impose a "human design of perfection upon a world by its nature imperfect."[127] Once Christianity opened Pandora's Box of individual freedom, no culture was safe from this utopian reaction unless it was restrained by the commandment of love—which has generally been considered unrealistic. Individualistic freedom seeking an unachievable perfection fueled Western creativity, but also provoked the modern world's secular utopian plans to bring order to the perceived anarchy.[128]

So the unruly freedom and individualism decried by Rod Dreher and Pope Francis are traceable to their own religion, which dealt the decisive blow to the ancient cosmological order. By the same token, the modern rationalizing project can be seen as an attempt to control the consequences of individualism by re-creating a communal conformity.[129] The tension between freedom and order is fundamental to capitalist civilization, and also the central challenge it faces today. To better understand it, we must delve into the earliest beginnings of human community and social order.

CHAPTER 2

BEGINNING AT THE BEGINNING

To understand the root of society's ambivalence about capitalism, it is necessary to start at the very beginning. Specifically,
we must investigate the question of how human beings first
organized themselves into structured communities. We will begin with
the earliest history, and then reach back even before that.

The Historian's Beginning

For obvious reasons, the serious search for credible evidence about
how humans fashioned social and economic order mostly starts with
"recorded history." The philosopher and historian Eric Voegelin, for
example, basically begins his magisterial multivolume *Order and
History* with the Egyptians and the Homeric Greeks, relying on written sources to describe what he calls the "cosmological civilizations"
of the early agricultural age.[1] Voegelin's popularizer, Frank S. Meyer,
similarly begins with "the first twenty-five hundred years of recorded
history," looking at the cosmological cultures of agricultural civilization.[2] These cultures all "conceived of existence so tightly unified
and compactly fashioned that there was no room for distinction and
contrast between the individual person and social order, between the
cosmos and human order, between heaven and earth, between what is
and what ought to be."[3]

This uniform, close-knit social order existed everywhere that agricultural civilization did—in Egypt, Mesopotamia, Persia, India, China, and even faraway Mesoamerica. There were only two partial exceptions: the classical Greek civilization and the Syriac-Jewish civilization.[4] While these two were quite different in many respects, the philosophers of Athens and the prophets of Jerusalem both made a fundamental distinction between what is and what ought to be, sundering the unity of cosmological civilization. Both of these societies faded under the power of cosmological cultures—Macedon and Rome. Yet the distinction between what is and what ought to be, and between the individual and the social order, would soon be reasserted, decisively placing "the person at the center of being" and freeing the individual from a suffocating cultural uniformity.[5]

Beginning with recorded history is actually quite late in the story, though. The Bible points back to human origins thousands of years before written records appeared. Estimates by anthropologists and ethnologists have humans existing anatomically for perhaps 200,000 years, but behaviorally for somewhere between 50,000 and 80,000 years.[6] The Lascaux cave paintings, dated to between 10,000 and 15,000 BC, show unquestionably human behavior. There are traces of agriculture (and its competitor, pastoralism) dated as early as 11,000 BC, but only as supplemental to wild barley, and without plows. It was not until 4,000 BC or so that cities of a thousand or more inhabitants arose, with a sufficient number of noncultivators to create a system for recording events and traditions, and documenting the development into states.[7]

The story of human society certainly began well before Athens and Jerusalem, and before agricultural civilization and the hierarchical city-state organization under kingship and temple that the standard histories typically start with. Social science usually describes the earlier form of human society as hunter-gatherer (but maybe hunter came later) and starts with families (or maybe bands of males impregnating unprotected females), then clans and tribes, which after a long period evolved into city-states, initially retaining the tribal cosmological form.[8]

A Philosophers' Story of Prehistory

The great philosophers do start with beginnings. Aristotle began with intellect, master and slave, and self-preservation. Hobbes began with a brutish contest, Rousseau with a noble one, and Locke with an "insecure" one. All were mainly interested in the consequences for civilizational order and government, but had little empirical information about the long period of previous development.

At the beginning of his *Discourse on the Origin and Basis of Inequality Among Men*, Rousseau announces that his researches "are not to be taken for historical truths, but merely as hypothetical and conditional reasonings, fitter to illustrate the nature of things, than to show their true origin." He does, however, present examples of primitive or "savage nations" as described by travelers, but notes that these were already corrupted by socialization. So he sketches out what he considers to be the most likely course of social development given his assumptions. His two axioms are that we humans are naturally first interested in "our own preservation and welfare," and that we have a "natural aversion to seeing any other being but especially any being like ourselves suffer or perish."[9]

Far from being aggressive, Rousseau's early man was satisfied with his condition and simply persevered through obstacles with equanimity. He was alone, encountering few others on his solitary excursions; males and females met by chance or in the desire to mate, then parted with ease. This primitive man was unconcerned about children. If challenged by another man, he was likely to move stoically on his way, never seeking revenge. This free but not easy state continued until population growth forced the noble savage into more contact with others.[10]

In Rousseau's telling, the initial socialization went well, given man's natural aversion to violence. When some social "conformities" developed, a man would observe that "all behaved as he himself would have done in similar circumstances," and this gave him the confidence to agree to common rules of a very general form, as interactions with others became more frequent. Over time, commerce and a more "enlightened mind" allowed the stronger or wiser to create separate families and claim property, to which the weak consented, feeling it "safer to imitate them than to dislodge them." Specialized functions and property then generated artificial wants, weakening man's naturally independent spirit. Metallurgy, agriculture, and commerce were his ruin, for they allowed the "sly and artful" to manipulate the rest into an ordered community where they would consent to their own submission.[11]

More empirical investigations into prehistory were not conducted until the late nineteenth century. The numerous ethnological studies of fast-disappearing pre-agricultural social forms provide most of what we know, or think we know, about primitive human life. Professor René Girard of Stanford University organized and expanded these and related studies into a series of analyses starting in literature and proceeding into anthropology, ethnology, psychology, and religion. Whereas Rousseau began from assumptions about man's essential nature, Girard started by looking for empirical evidence of what is different about human beings.

In comparison with other primates, Girard found three distinctive human characteristics: the largest brain, the longest period of immaturity, and a unique opposable thumb with firm-grasping fingers. The newborn human's brain is so large that the female requires a unique enlarging of the pelvis to deliver, after nine months of restricted activity. A long childhood of extreme helplessness requires an extended phase of carrying, protection, and nourishment. Unlike other primates, human males have long periods of restricted mating, and much opportunity to harm infants. The idea of father does not exist among other primates, only dominant and subordinate males, the latter of whom will generally not challenge the former even for food.[12]

The human hand is significant in that it enables holding and throwing rocks and other projectiles—where other primates at best may throw clumsy branches. Along with the big brain, the human hand makes man the most fearsome predator in nature.[13] Human male violence is fueled by vast reserves of adrenaline, which is useful in hunting and war, while most primates are more peaceful omnivores. Unfulfilled desires for goods or females (often kept in common by other primates) can produce rage for dominance, which mostly turns inward even to the family rather than spinning outward, where it meets greater resistance.[14]

Humans seem to have a more aggressive desire for many more things than do apes, not just sex and food but clothing, tools, and even frivolities. Girard traced this "acquisitive desire" to the primal instinct for imitation. Humans learn by social mimicking right from the beginning, as they watch their mothers and others around them. (Blind infants are thus severely impeded in learning.) "Acquisitive desire" arises from mimicking other humans and wanting what they have, or seem to desire. Animals also imitate their elders and have cravings, but the greater human brain function expands these cravings and desires with an exclusively human passion.[15]

While desire comes from observing others and is thus social, it also frustrates sociability, as it overwhelms the normal "animal equilibrium." One human passionately covets what another possesses (whether goods or status), and the other naturally puts up resistance. A fierce struggle may ensue, and even the possibility of dying in the fight doesn't seem to be much of a deterrent for humans, whereas other mammals rarely fight to the death. Acquisitive envy can lead to deep divisions in and between families, until the tribe might be split into warring factions, and its very survival is threatened.[16]

Against this picture of a violent primitive society, there are traditional accounts that describe more peaceful early humans. The Bible has a beginning in the Garden of Eden, but the first dramatic act after the expulsion is the killing of Abel by Cain, who then worried about a violent reprisal and asked for a sign from God to protect him.[17] Whether one begins with the Bible or René Girard or Darwin, early humans appear to have been extremely violent.

Rousseau would seem to be the dissenter from this picture of primal man.[18] But even his solitary man in a state of "nature" puts his own welfare and self-preservation first, and the noble savage is easily corrupted into violence when trade and population growth lead to the formation of a tribe. Indeed, Rousseau seems to consider it natural that the independent noble savage would disappear very quickly. Rousseau is most unconvincing when it comes to children, largely ignoring their need for protection if they are to survive. How then does population increase? The survival of children requires some very early social order.

Creating Primitive Order

Social order must have developed in some way, or we would not be here. So how did early *Homo sapiens* begin to control the violence among themselves? Humans do not seem to have natural patterns of dominance, as apes do, so human fights for supremacy lead easily to death. Simple coercion by the strongest produces even more disorder, instead of the peace enforced by dominant males among other primates.

Girard was intrigued by the fact that today "no one speaks about" the obvious human pattern of passionate envy and desire that so naturally results in violence. The author of the Decalogue did see the problem of covetousness, devoting the last and longest commandment to it—forbidding a *desire*, rather than a directly harmful act. Rationalists like Hobbes saw the problem too, fearing a war of all against all that would require a powerful state to control. Even Locke described an "insecure" period in early social development. Today, one need only watch children quarrel over some trifle to understand the need for a structure to restrain communal violence. If no one has any power greater than another's access to rocks, what could stop a rivalry from escalating to the point of endangering the very survival of the group?[19]

Some other kind of social restraint must substitute for sheer physical force. Sigmund Freud referred to such restraints as *taboos*.[20] In the long period of human society preceding the development of the state, only myth or superstition or religion could keep the peace. Girard observed that the two pillars of religion are prohibition and ritual: "Primitive societies repress mimetic conflict not only by prohibiting everything that might provoke it but also by dissimulating it beneath major symbols of the sacred." Actions that could disturb the public order are not only fenced off by prohibitions, but also sublimated in ritual that reproduces the conflict in a less threatening form. Ritual might even allow taboos to be violated, but only in strictly fixed and circumscribed terms.[21]

What taboo is powerful enough to instill fear sufficient to overcome the passionate desire for goods and suppress destructive rivalries? The most dramatic is taking life, human or animal, in a rite of sacrifice.[22] Freud wrote that all ritual preserves the memory of a murder that founded a culture.[23] Just as animal predators choose the weakest victim, archaic human leaders appear to have chosen the weak and crippled as scapegoats for ritual sacrifice. An outcast can be sacrificed as a way of uniting factions, which will all turn equally against the victim. Girard noted that Caiaphas, the high priest at the time of Jesus' crucifixion, had told his fellow Jewish leaders, "It is better that one man should die for the people, than for the whole nation to be destroyed." Those who did not understand this basic social reality, Caiaphas said, "know nothing at all."[24]

Girard cited myths from nations across the world that involve hidden sacrifice, from which scapegoats emerge as gods: the Ojibwa and Tikopia tales that Claude Levi-Strauss relates in *Totemism*, Euripides' *The Bacchae*, Philostratus's *Life of Apollonius of Tyana*, Inca and Hindu traditions, the stories of Romulus and Oedipus, and many others.[25] Girard's interpretation of these stories is that once the scapegoat has appeased the contending factions by uniting them against him and is safely dead, he can be recognized as a sacrifice for the good of the community, which then appreciates the reconciliation and peace brought about through his death. The return to calmer times after the sacrifice first appears to confirm the guilt of the innocent victim. But fear of the dead could—perhaps at the urging of an erudite elder—then lead to reverence for the victim in light of the better times that follow the sacrifice. Rituals devised by the wise shaman reinforce the restrictions to which the opposing factions have agreed, and the united community over time elevates the victim to the status of a god to complete the myth's inclusion in the cosmological order.[26]

As myths or superstitions solidified into traditions and then cultures, they produced the order that prepared the way for the agricultural societies that have been the focus of almost all systematic research. Agricultural society was carnivorous, even though the human esophagus does not seem naturally to pass meat. So why did people start eating meat? Girard suggests that the eating of animals and the domestication of animals both emerged from sacrificial ceremonies. So did kingship, most trade, and the state itself. Even today, British coronation ceremonies begin with pseudo-lethal threats against the new monarch and end with him or her accepting responsibility for bearing the burden and sins of the community—while the subjects accept the obligation to obey this now sacred leader.

With agricultural society, we bring the story of humanity up into recorded history, where Girard joins the historians in recognizing cosmological civilizations everywhere. Most historians give Athens the greatest importance in the development of the West, for its philosophers and its commercial empire. But Girard observes that the enduring dominance of the traditional Greek gods frustrated Athens's break from its mythical roots, and led to the scapegoating of Socrates when he questioned the status quo. The only real challenge to cosmological civilization came from Jerusalem and its founding revelation. While Jews had a tradition of sacrifice and taboo, they also questioned those traditions. Their God preferred mercy over sacrifice, and their prophets challenged sacred kingship. With Deutero-Isaiah came the positive image of the suffering servant.[27]

The final break from cosmological civilization—as conceded even by Rousseau, Machiavelli, and Nietzsche—was with Jesus, who followed in the Abrahamic tradition, although he died not as dumb sacrifice, but by exposing the evil in the scapegoat myth underlying the ancient religions. Jesus called not for sacrifice but for each individual to follow the Father's command of love by mimicking the Son, who did not seek death but refused to be intimidated by it. This God differed from others in allowing individuals the freedom to disobey him. A god who forced obedience, Girard says, would have been no better than Satan and no different from all earlier gods who relied upon human force.[28]

In the ancient world, nothing was more powerful than violence and its collective contagion, which is why archaic religion not only tried to limit it but even divinized it. Nietzsche drew a comparison between the story of Jesus and the primitive myth of the counterhero Dionysus, the god of collective violence, of the frenzied mob, of madness. In Euripides' play *The Bacchae*, Dionysus is put on trial as a false god by King Pentheus. But he warns the king that although slight in size he has godly power, while Pentheus is but mortal, like an ox not able to "kick against the pricks" or goads that drive it. Dionysus then manipulates the humble human pricks of the mob to tear the king apart. A similar phrase is attributed to Jesus in the Acts of the Apostles, when Saul on the road to Damascus has his dramatic encounter with an arresting presence that asks: "Saul, Saul, why do you persecute me? It is hard for you to kick against the goads."

Both Dionysus and Jesus were martyred, but they are fundamentally antithetical, said Nietzsche:

> Dionysus versus the "Crucified": there you have the antithesis. It is not a difference in regard to their martyrdom—it is a difference in the meaning of it. [To Dionysus] Life itself, its eternal fruitfulness

and recurrence, creates torment, destruction, the will to annihilation. In the other case, suffering—the "Crucified as the innocent one"—counts as an objection to this life, as a formula to its condemnation.[29]

Dionysus philosophically accepts life and death as they are, in a natural cycle of fruitfulness, torment, and destruction, while Jesus rejects earthly life with its natural violence and endless recurrence in favor of a peaceful order that leads to reward perhaps here but surely in the hereafter. Nietzsche viewed Christian pity for the victim as a slave morality of the weak arising from envy of the morally superior Dionysian supermen, or of the ancient gods who inflamed the mob to sacrifice Jesus and persecute his followers.[30]

Immersed in Roman and Germanic myth, Girard argued that the medieval Christian story was initially blind to the sacrificial problem and how the Resurrection was meant to overcome it. The biblical Letter to the Hebrews extolled the ancient sacrifice, rather than emphasizing the uniqueness of Jesus' self-sacrifice. This made the Christian message more attractive to those who still adhered to ancient myths, but undermined its central significance. The sacrificial emphasis continued even up to Thomas Aquinas, who relied heavily upon Aristotle, with his high regard for heroic sacrifice.[31]

In the post-Nietzschean secular world, we may have moved some distance "beyond good and evil," but no one today openly endorses sacrificing innocent scapegoats for the common good. Not even the most vicious ruler will publicly repeat the words of Caiaphas justifying the killing of an innocent to help the community. On the contrary, power seekers can easily exploit "victimhood," manipulating society's new awareness of and sympathy for victims as a way of scapegoating the once-privileged, anointing themselves the champions of the disadvantaged and riding their moral legitimacy to power.[32] Even Hitler claimed that he and his people were the victims of those he actually persecuted. Similarly, Arab anti-Semitism draws upon claims of victimhood at the hands of Jews in Israel. But as the world supposedly rationalizes, such manipulation perhaps becomes more difficult.[33]

With René Girard, we get a view into chapters of the human story discounted by most historians and philosophers: perhaps 3,000 to

6,000 years of agricultural evolution and 7,000 years of hunter-gatherer socialization before that. Looking at early beginnings puts into question the normal assumption, as in Aristotle, that humanity is naturally social. Agricultural man was social, but was he *naturally* so? Socialization clearly began long before the agricultural state, but at the very beginning? Girard's work is important not so much in its specifics as in its perspective on human time and on the power of acquisitive desire, the violence that results, and the necessity of finding a means to control it. Recognizing the extraordinary length of time this process required must affect how a serious person looks at every question of social organization.

Whatever internal constraints may have been inherent to humans, the desire for others' goods or mates—and the means to take them—must have dominated human communities for a very long period before city-states developed, and certainly before capitalism and the nation-state.

Divine-Right Nationalism

Primitive religion played a crucial role in creating social order and laying the foundation for the modern state. The Greek city-states remained within the pattern of cosmological civilization, but Judaism and especially Christianity made a decisive break from its all-encompassing social order, ushering in a world with greater freedom but also new disorders.

According to Rousseau, it was Christianity that introduced a fundamental new social division by separating religious authority from state power. In *The Social Contract*, he wrote:

> Jesus came to set up on earth a spiritual kingdom, which, by separating the theological from the political system, made the State no longer one and brought about the internal divisions that have never ceased to trouble Christian peoples. As the new idea of a kingdom of the other world could never have occurred to pagans, they always looked on the Christians as rebels who, while feigning to submit, were only waiting for the chance to make themselves independent and masters, and to usurp by guile the authority they pretended in

their weakness to respect. This was the cause of the persecutions. What the pagans had feared took place. Then everything changed its aspect: the humble Christians changed their language, and soon this so-called kingdom of the other world turned, under a visible leader, into the most violent of earthly despotisms. However, as there have always been a prince and civil laws, this double power and conflict of jurisdiction have made all good polity impossible in Christian States.[34]

As Rousseau saw it, the major cause of conflict in society was the competition for allegiance between the state and the institutionalized Christian Church with its own structure of authority and independent moral status.

Previously the state had a monopoly on power, but as early as the fourth century in the Roman Empire, a bishop like Ambrose of Milan could stand up to an emperor and escape with his life. When the seat of imperial power moved east to Constantinople, the church filled the power vacuum in Rome, whose bishop officially represented the city when it fell to Attila. Christian authorities had some influence over tribal chieftains and kings in the Dark Ages, but temporal powers also gained substantial control over religious institutions under feudalism. The Cluniac reforms beginning in the tenth century led to more independent monasteries and church institutions, and to an assertive papacy—particularly that of Gregory VII, who in 1077 induced the Holy Roman Emperor, Henry IV, to kneel in the snow and repent of his sins. Church influence over the state—but by no means control—would last for several centuries more, though it began to fade in the wake of the fourteenth-century plague that killed a third or more of the European population over a decade.[35]

Previously, the church had built a moral reputation for helping to mitigate the toll of plagues and pestilence, but it was unable to soften the devastating blow of the Black Death. As a consequence, the church lost moral authority.[36] The dramatic shrinking of the population had the effect of increasing per capita wealth and made survivors better able to bargain. But the surviving barons with their larger estates could bend the law in their favor. Feeling liberated from church moral strictures, great lords contended for dominance and for royal crowns.[37]

The Roman Church was also weakened by the papacy's residency in Avignon for nearly seven decades, followed by a schism from 1378 to 1417 with rival popes claiming legitimacy, and secular powers lining up behind one or another. But the church's moral authority was not directly confronted until 1517, when Martin Luther—the "indispensable firestarter" of the Reformation, to use Alec Ryrie's term—posted his 95 Theses on the door of the Castle Church in Wittenberg.[38] What had been a rivalry between church and state became a multidimensional competition.

Ryrie, an Anglican, explains that the pre-Reformation Catholic Church had historically been "flexible and creative, a walled garden with plenty of scope for novelty and variety and room to adapt to changing political, social and economic climates." That church had elastic boundaries, but they stretched only so far. "Those who wandered too far would be urged, and if necessary forced, to come back." Martin Luther was first and last a man who tolerated no boundaries. Driven by a passionate "love affair" with God, unbounded and unlimited, he disdained anything that got in the way.[39]

Luther "was not a systematic theologian," Ryrie observes. The iconic phrases so identified with him, "faith alone" and "Scripture alone," were essential ideals to him, but he would not enclose them in doctrine. He had "an intoxicating passion," even a "reckless" belief that Jesus loved the faithful absolutely and laid down his life for them as a pure gift that merely had to be accepted, without any institutional or sacramental intermediary. No Christian leader before him had so thoroughly rejected doctrine and left individuals so free to fashion their own morality.[40]

Luther's teaching of unmediated access to God seemed self-evidently true to many, and the message traveled widely, since Luther was literally the inventor of mass communication. The print medium had existed for more than half a century before he posted his 95 Theses, but it was all thick volumes, mostly in Latin. Luther instead produced pamphlets—about one every three weeks. He himself wrote 1,465 pamphlets, to which pro-Luther authors added another 800, while Catholic opponents produced only 300. The latter were mostly in Latin, whereas Luther wrote mainly in German, thus reaching a vastly broader audience.

The Catholic Church naturally tried to tame Luther through discussion and debate. (In fact, his theses were first meant to initiate debate in a traditional way.) But he would not budge, even when the church pronounced him a heretic and the Holy Roman Emperor, Charles V, condemned him as an outlaw. Luther was "saved by politics," in Ryrie's words. His powerful prince and protector, Frederick III, Elector of Saxony—for his own political reasons—deterred Catholics from exacting retribution.

But some of Luther's followers wanted to go further than he, demanding the elimination of infant baptism and holy images. Luther was appalled by these demands, which imposed nonbiblical doctrines on questions that should be decided by love alone. He was even more horrified when radicals invoked his idea of Christian freedom to demand the election of priests, the abolition of serfdom, the redistribution of church property, and even the abolition of private property. Prince Frederick cracked down with great severity on the peasants' rebellion of 1524–25, the largest insurgency in Europe before the French Revolution. Luther sided with the prince, because Christian freedom for him was inner liberation, not political anarchy.

"Fairly or not," says Ryrie, the disruption was blamed on Luther's teachings. It shook all Europe and helped alienate the more moderate advocates of reform, such as Erasmus. The rift in Christendom frustrated a united front against the Ottoman Turks, who were pounding on Europe's door. The Turks defeated the Hungarians at Mohács in 1526 and then laid siege to Vienna. The German Catholic princes and the Holy Roman Emperor could no longer ignore the divisions, so a temporary truce was made between Catholics and Lutherans in 1532. Then John Calvin made his break from the Catholic Church, and the further spread of Protestantism pushed Catholics to initiate their Counter-Reformation in response.

The Reformation became "fundamentally a struggle for the backing of secular governments," Ryrie observes.[41] The Augsburg Settlement of 1555, an effort to restore peace within the Holy Roman Empire, permitted Lutheranism in the principalities that had pretty much already adopted it, and de facto permitted Calvinism in cities. It was a liberalizing settlement, allowing for a kind of federalism that Johannes Althusius would describe in 1603: a division of power between his

Calvinist city of Emden, his Lutheran regional prince, and the Catholic emperor.[42] But as Calvinism grew more widespread, the Augsburg accommodation became insufficient to maintain order.

Religious and political conflict across much of Europe erupted in the Thirty Years' War of 1618–48. Order was finally restored with the Treaty of Westphalia, an international agreement setting out broader principles than the Augsburg Settlement. The treaty affirmed the legitimacy of both Protestant and Catholic monarchs as rulers of centralized nation-states, with the God-given power to determine even the religion of their subjects. Church and state were again united under divine-right kings, who gained a victory over feudal principalities.[43] From the time of the treaty's ratification, however, the following two centuries were littered with dynastic wars to decide who would exercise the new centralized power.[44]

A state-imposed religious creed, under a monarch claiming a divine right to rule, was a focus of growing opposition in a secularizing and liberalizing age. In Rousseau's view, it did not solve the problem of dual loyalty either. The solution he proposed was to unify power under the "general will" of the community. A set of rational, secular beliefs arising from the community itself would define public morality, "not exactly as religious dogmas, but as social sentiments without which a man cannot be a good citizen or a faithful subject." Only when morality and power were united in the general will of the citizenry could order be established and differing viewpoints tolerated. There would be freedom and tolerance, but within strict limits. The sovereign could banish, as "antisocial," those who questioned the civil faith, and "If anyone, after publicly recognizing these dogmas, behaves as if he does not believe them, let him be punished by death."[45]

Earlier divine-right monarchs, like Henry VII of England, had anticipated Rousseau by centuries in recognizing the problem of plural centers of power—political, clerical, and economic, as well as competing local powers. So they had set out to rationalize all of society's resources in their protective embrace, and to make subjects faithful only to the unitary ruler. Soon they placed themselves at the head of their established churches too. The divine-right monarchs were more practical than Rousseau. Not being interested in devising moral codes, they simply accepted the existing Protestant or Catholic morality, as taught by religious authorities, but set themselves on top as *pontifex maximus*, with a mute God as their only superior.

The potential revenue from growing capitalist markets drew the attention of monarchs as the means to build their treasuries and fund their wars, so they set about taxing all in highly regulated mercantilist economies.[46] At Blenheim in 1704, the more capitalist (as distinct from mercantilist) English shopkeepers defeated the "invincible" armies of Louis XIV, and within a century the French monarch was overthrown and executed. A claim of divine right was clearly an insufficient support for monarchy. Something else was required to redirect moral attachment to the nation-state, so every European state adopted some form of popular nationalism—a concept perfected by Napoleon.

Soon gone was the covenanted division of power that Althusius had described between city, regional prince, and emperor, each with historically agreed-upon rights and mutual responsibilities. The federal Holy Roman Empire itself dissolved in 1806. Finally, Germany would become a nation-state, as would Italy. Political unification, wars, markets, full treasuries, and demands for order and fatherland did help build a strong sense of nationalism across the continent. Prussians became German, Scots became British, Catalonians became Spanish, Lombards became Italian, Basques became French. Farther afield, Azeris became Iranian, Kurds became Iraqi, Ukrainians became Soviet, Alawites became Syrian.

Isaiah Berlin, the historian of political ideas, saw nationalism as fulfilling "a basic human need," a desire to "belong to a given community, to be connected with its members by indissoluble and impalpable ties of common language, historical memory, habit, tradition and feeling." He identified four principles of nationalism. First, all peoples belong to a culture with a unique way of life, which is not deeply valued or understood by outsiders. Second, a set of core beliefs and values are primordial, intrinsic to the nature of a people, and those beliefs and values trump all others. Third, if outside groups challenge the values of the people, their own way must prevail. Finally, the nation has the right to compel those who come into conflict with it or threaten its interests to yield.[47] These principles were developed to varying degrees in different states, but by 1850, as the historian John Lukacs noted, all the ancient European states had transformed into nation-states informed by such principles.[48]

The nation-state won the loyalty of mass populations—something that had been elusive for aristocracies—and rallied the passionate commitment of all classes, even to the ultimate commitment of risking lives in nationalistic wars. August 1914 confirmed nationalism to be "the new faith of the secular age," as Karen Armstrong put it. Under nationalism, "all differences of class, rank and language were flooded over at that moment by the rushing feeling of fraternity." What followed, however, was "a century of unprecedented slaughter and genocide," inspired not by religion but by the "equally commanding notion of the sacred" represented by the secular nation-state.[49]

Alexis de Tocqueville found the United States to have a sense of nationhood even though it had been multiethnic and multidenominational from colonial times and its federalism was as pluralistic as feudal Europe. To be sure, it was a different sort of nationalism—the result of the fact that the central government was mostly an ideal, demanding little and thus engendering little resentment. Yet as Tocqueville predicted, governance nationalized, and the U.S. central government demanded more uniformity among states, which led to a civil war. While it cost nearly a million lives, the Civil War resulted in a stronger cultural nationalism, later deepened by the First World War.[50]

Taming Capitalism

The nation-state proved expensive to maintain, so rulers had a major interest in appropriating the spoils of capitalism. It was Jean-Baptiste Colbert, the French controller general of finance under Louis XIV in the seventeenth century, who first thought systematically about capitalism as a concrete social institution. But it was for pragmatic purposes, since warfare, pomp and pageantry, and winning popular support were all costly. The Sun King needed revenue and this required control of economic activity.

Colbert's system is normally labeled mercantilism, which often is characterized as not true capitalism, but it certainly had historical importance in turning France and Prussia into major economic and political powers. By the eighteenth century, mercantilism had pretty much become the dominant economic system in Europe except for the British Empire, the Netherlands, and Venice. It rested on the belief that exports were the basis for power and should be encouraged, while imports were to be discouraged. A large, expert bureaucracy would manage the economy to enrich the monarch and provide subsidies to his supporters.[51]

Adam Smith, the Scottish economist, challenged this version of capitalism. The mercantilists seemed "to consider production, and not consumption, as the ultimate end and object of all industry and commerce," he wrote, and in their system "the interest of the consumer is almost constantly sacrificed to that of the producer." Home-country manufacturers are favored, while consumers are "obliged to pay that enhancement of price which this monopoly almost always occasions." The consumer must pay "first, the tax which is necessary for paying the bounty, and secondly, the still greater tax which necessarily arises from enhancement of the price of the commodity in the home market."[52]

In *The Wealth of Nations*, Smith proposed instead a true capitalism in which all "preference or restraint" was eliminated, allowing "the obvious and simple system of natural liberty" to establish itself "of its own accord." Smith continued,

Every man, as long as he does not violate the laws of justice, is left perfectly free to pursue his own interest his own way, and to bring both his industry and capital into competition with those of any other man, or order of men. The sovereign is completely discharged from a duty, in the attempting to perform which he must always be exposed to innumerable delusions, and for the proper performance of which no human wisdom or knowledge could ever be sufficient, the duty of superintending the industry of private people, and of directing it towards the employments most suitable to the interest of the society.[53]

This pluralist conception of capitalism greatly influenced Britain's economic policy and that of its colonies. But it was soon challenged by a third conception arising from Voltaire's search for a more rationalizing type of mercantilism. Voltaire admired Adam Smith but rejected his laissez faire views, seeking instead to tame the destructive freedom of capitalism for the public good, under the guidance of enlightened absolutism.[54] Also in that rationalist tradition, the influential English political economist David Ricardo offered a rationalized capitalism centered on "perfect competition" in a market of many buyers and sellers.[55]

Ricardo arguably had at least as much influence on modern economics as Smith because of his more mathematical and scientific conception, in comparison with Smith's open-ended, pluralist idea based on subjective human choice, which he called freedom. Marx himself used Ricardo's conception of capitalism, although perhaps because he considered it easier to dispute.[56] In the United States, the Sherman Antitrust Act of 1890 incorporated Ricardo's view of capitalism into antimonopoly law. It became the central ideal of trust-busting Teddy Roosevelt's nationalism, of Woodrow Wilson's progressivism, of Franklin Roosevelt's New Deal, and it carries over into progressive politics today. It even gained new appeal by being somewhat incorporated into President Trump's nationalist economic platform.[57]

All three conceptions of capitalism have some claim to the term, and all have influenced the practice and results of capitalism in actual history and in concrete nation-states. Various intellectuals, political factions, and social groups in the United States have supported either pragmatic or rationalizing or pluralist capitalism. All have been co-opted by nationalism at one time or another. There have been efforts to reconcile the three conceptions, but more often the basic principle of capitalism is held as an ideal without a clear account of its components and contours. Each represents an effort to tame purely laissez faire capitalism, with a pragmatic use of power, scientific rationalization, or fair laws with minimal coercion.

Different forms of capitalism have basically followed national interests ever since the Treaty of Westphalia made nation-states preeminent, but nationalism came under serious question in the early twentieth century. The most prominent American intellectuals of the day challenged the traditional nationalism that was characterized by pragmatic self-interest.[58] David Starr Jordan, the founding president of Stanford University (1891–1913), believed that just men could reform the world on idealistic terms rather than self-interest. Nicholas Murray Butler, president of Columbia University (1902–45) and a Nobel Peace Prize winner, taught that war was obsolescent. Elihu Root, who served as secretary of state (1905–9) and also won a Nobel Peace Prize, had unbounded faith in rational treaty-making. All these principles were incorporated into Woodrow Wilson's "Fourteen Points" proposal of 1918, offered as a rational yet idealistic way of achieving a more peaceful world order.[59]

The hopes attached to this vision were crushed by World War II. Wilson's idealism was soon written into the Universal Declaration of Human Rights, adopted in 1948 and affirming that "the equal and inalienable rights of all members of the human family are the foundation of freedom, justice and peace in the world." While its practical effect was limited, some Wilsonian idealism mixed with capitalist globalism was revived during the Cold War. The eventual demise of the Soviet Union suggested that the mission had been successfully accomplished, achieving the "end of history" and allowing a return to Westphalian normalcy.[60] But within a few years, a more aggressive internationalism reappeared, under Presidents Bill Clinton, George W. Bush, and to some degree Barack Obama, resulting in unpopular trade treaties, and military stalemates or losses in the Balkans, Afghanistan, Iraq, Syria, and elsewhere. Public dissatisfaction with this course led to the election of President Trump on a platform of America First nationalism.[61]

The Westphalian idea was that nation-states promoting their own interests in a rational way would result in a tolerably peaceful world order, but that no longer seemed to work. Wilsonian idealism did not appear to be universal among nations. Americans became more divided over foreign policy, among many other issues. Some even questioned the rationality of the nation-state.[62]

The Fall of World Order?

The most obvious failure for America's idealistic attempt to achieve world order occurred on September 11, 2001, when members of a small Islamic group hijacked airliners and smashed them into the World Trade Center in New York and the U.S. military headquarters in the Pentagon, besides attempting an attack on the Capitol that ended in a crash landing in rural Pennsylvania. Al-Qaeda's declaration of war against the United States a few years before had been dismissed as ludicrous, but the group had now destroyed the great icon of capitalism, the Twin Towers.

When the commander in chief of the leading world power subsequently declared a "war on terror," it was clear that something had changed radically since President Franklin Roosevelt had declared war after an equally dramatic attack. President George W. Bush initially targeted a non-state group, and then attacked Afghanistan only when it refused to surrender al-Qaeda's leader and expel his followers. Even though al-Qaeda clearly had moral and religious motives, President Bush studiously avoided any invidious reference to Islam or any sect thereof.

President Barack Obama followed the same pattern, asserting that the aggressors did "not speak for Islam," and identifying the enemy generally as those "embracing a perverted interpretation of Islam."[63] After the mass shooting by two Muslims in San Bernardino in December 2015, it was a *Washington Post* reporter who found one of the persons responsible, Tashfeen Malik, and revealed her identity. Her best friend in Pakistan, Abida Rani, disclosed that Malik had become radicalized while going "nearly every day" to a madrassa that "belongs to the Wahhabi branch of Sunni Islam."[64] Yet President Obama identified her merely as a "female terrorist," and described Islam as "a religion of peace, charity and justice."[65]

Most of the government, the media, and the public remained fixated on the amorphous term "terrorism" to denote the enemy. President Donald Trump was a bit more specific when he spoke of "fighting terrorist groups, such as ISIS [Islamic State of Iraq and Syria] and al-Qaida," but he too described a general war on terror.[66]

The neoconservative right criticized this vagueness as refusing to call the enemy by its name, "Islamic extremists."[67] While this term inserted a dose of specificity, Islam is not a unitary religion any more than is Christianity or Judaism. All have had deeply divergent sects almost from the start. The attack on 9/11 and others following it were motivated by a very specific tribal-based ideology, precisely Salafi Sunni Arab Islam, or the more radical Wahhabi Salafi Sunni Islam.[68]

Al-Qaeda's leader, Osama bin Laden, was a Wahhabi opponent of the Saudi Arabian government. Wahhabism is a puritanical belief system that idealizes the warrior Muhammad and his first associates, and rejects any later liberalization or modernization, which it regards as the cause of Islam's decline. Bin Laden made capitalism a particular target of his moralist attacks.[69] He and al-Qaeda followed Abdallah 'Azzam's Salafi teaching that every Muslim is obliged to defend Islamic lands against infidels. In 1996, bin Laden had issued a declaration of war against the United States for stationing troops in the Sunni holy land of Saudi Arabia, in violation of Salafi doctrine.

The United States had allied with Saddam Hussein, a self-professed Sunni, in the 1980s Iran-Iraq War. By later overthrowing Saddam in the 2003 war, it joined de facto with Iraqi Shiites and their ally, Iran. The U.S. joined with Sunni groups again in the 2011 Syrian civil war, and since 2018 it has taken the side of Saudi Arabia against Shiite Iran.[70] Sunni and Shia have regarded each other as heretical enemies since the seventh century, and their rivalry has deeply shaped the Middle East and a good deal of the world.[71]

Nation-states had not become irrelevant, of course. Al-Qaeda was harbored by a nation-state, and soon after its attack on the United States there were American and allied forces in Afghanistan, Iraq, Syria, and other nations. The involvement of Russia, Iran, and Lebanon's Hezbollah in the Syrian conflict was straight out of classical power politics.[72] Yet the whole time, the United States was reacting to religious and tribal threats rather than nation-state foes, and moreover it ended up straddling the fundamental Sunni/Shia divide, confusing all sides.[73] Nation-states remain the major support of the international order, but there is no question that religious, tribal, and factional loyalties have become serious challengers to nationalism and to capitalism.[74]

Civilizational Confusion

Almost three years into the Iraq War that began in 2003, a legal deposition obtained by the *New York Times* revealed that the FBI counterterrorism chief, Garry Bald, was unable to distinguish between the Sunni and Shia versions of Islam that defined the fault lines of that Middle Eastern conflict.[75] A subsequent survey by the *Congressional Quarterly* found that most top officials in Washington were confused on the Sunni/Shia division. One FBI official, for example, identified (Shia) Iran as a Sunni nation.[76]

Several years later, the director of national intelligence, James R. Clapper Jr., revealed a misunderstanding of the basic drivers of the conflict in Iraq. "I didn't see the collapse of the Iraqi security forces in the north coming," he admitted. "I didn't see that. We overestimated the ability and the will of our allies, the Iraqi army, to fight." But even then he referred to an "Iraqi army," as if the fighting force were a nation-state. In fact, the effective forces were primarily Shia militia, whose motives for fighting were sectarian and tribal.[77]

It is not too extreme to suggest that the sectarian and tribal divisions in Iraq and the Middle East generally were the only important things to know about the region. Most people there—and indeed in most of the world—prioritized family, tribe, and religion, rather than what many Westerners said they really valued: prosperity, freedom, rights, equality, free markets, democracy, and the like.[78] This reality should not have been difficult to understand. Before the decision to invade Iraq in 2003, I wrote:

> Since Iraq was drawn on the maps of a far-away colonial office in 1921, it has generated dozens of coups, eight Kurdish revolts, nine Shiite uprisings and three pogroms, all before Saddam imposed his terrible order on the local factions. He then allied his secular Arab revivalist Baathists with the minority Muslim Sunnis and Christian Chaldeans, who feared a united fundamentalism more than Hussein. Revolts since then have killed 100,000 Iraqi Kurds and 30,000 Shiites. Playing well with others is not a high priority in old Mesopotamia. While of pluralism there is aplenty, it is not the benign type required for a democracy.[79]

Muslim factions have been fighting for centuries. Even the most ruthless nation-state leaders, mainly Sunnis, cannot fully contain these conflicts. Many so-called nation-states are just collections of tribes that mostly view other tribes as evil. Iraq was 60 percent Arab Shia, living mostly in the south; 20 to 30 percent Arab Sunni, mostly in the west; 10 percent Kurd (mainly Sunni) in the north; with small groups of Turks and Christians. The Shia rallied to nation-state "democracy" in Iraq after the American invasion, but one of their leaders admitted to the author on a visit there that it was because they were the majority and could thus rule over their domestic enemies, the Sunnis and Kurds—which they will continue to do as long as majority rule prevails.[80] American officials often expressed pleasure when a new government was formed that included all three major groups, but the Shia actually ruled, the Sunni were resentful at best, and the Kurds kept threatening to break away.

Given the complexities of the situation, the emergent dangers and the shortage of solutions, it is not surprising that U.S. national leaders came to view Iraq, and foreign policy generally, mostly through the lens of their own parochial concerns.[81] When the Islamic State of Iraq and Syria (or, the Levant) killed 130 Parisian innocents in November 2015, President Obama first called it a "terrible and sickening setback," but then blamed Republicans, and particularly their reluctance to take in refugees from war-torn Syria. "I cannot think of a more potent recruitment tool for ISIL than some of the rhetoric that's coming out of here during the course of this debate" in Congress, he said. The president accused his political opponents of "hysteria" and an "exaggeration of risks" that caused "fear and panic" among Americans. He found it offensive and discriminatory that some Republicans had suggested favoring Christian refugees over Muslims.[82]

The Republican Speaker of the House, Paul Ryan, responded politically too, introducing legislation to delay the immigration of the 10,000 Syrian refugees proposed by the president until they were "certified" by the intelligence agencies as terrorism-free—though at the same time, heeding the president's criticism, he promised that there would be no "religious discrimination."[83] Both leaders spoke in terms of universal rights rather than considering the specific character of the refugee crisis, in which Christians and other minorities such as Yazidis and Jews were the ones being beheaded and crucified, and had no recent history of aggression against the United States. When the Islamic State took control of the city of Raqqa, they forced every Christian to either convert to Islam, or "face the sword," or pay an extra tax, dress modestly, and cease all public religious activity.[84]

President Obama suggested that technology could be used to identify terrorists aiming to enter the United States, but the FBI director, James Comey, testified to Congress that "a number of people who were of serious concern" had been able to slip through the screening of refugees from Iraq. Syrian refugees would be even harder to screen because U.S. soldiers had not been on the ground, collecting information about the local population, as they had in Iraq. "If we don't know much about somebody, there won't be anything in our data," Comey said. "I can't sit here and offer anybody an absolute assurance that there's no risk associated with this."[85] The one characteristic that was easiest to verify and most useful in identifying those less likely to terrorize Westerners was religion, yet political leaders of both parties put it off-limits as a criterion, deeming it unacceptably discriminatory.

But it was soon back to partisan politics following the murder of fourteen Americans in San Bernardino on December 2, 2015. President Obama blamed Republican gun-rights activists for refusing to enact tougher national laws, even though the attack took place in a state with strict gun laws. He urged Americans not to "turn against one another by letting this fight be defined as a war between America and Islam," although he also proposed that people on his no-fly list of terrorist suspects not be allowed to purchase firearms, a prohibition that would disproportionally affect Muslims. Republicans responded politically by calling for stricter limits on immigration and rejecting any additional gun-control laws.[86]

The politicizing of national security and foreign policy continued into the next administration. In early 2020, for example, President Trump claimed that the assassination of a top Iranian government official was necessitated by the weakness of his Democratic predecessor.[87]

This might all be good domestic politics. But applying parochial or partisan interests to complex international realities is not a way to advance nationalism or the national interest—nor to promote universal values such as peace, order, freedom, and democracy in an environment where tribal and sectarian loyalties are dominant.

The Return of the Tribes

There is confusion likewise about the dynamics of conflict in other regions of the world, such as Ukraine. Writing about the uprising against the government of Viktor Yanukovych, who was allied with Russia, in early 2014, the columnist Peggy Noonan asked, "Whose Side Are We On?" To most Americans, the answer was obvious. The people in the streets of Kiev were demonstrating for freedom against a powerful authoritarian president who was responding with force. We Americans were all with the demonstrators for freedom and democracy. We, like Ms. Noonan, knew in our hearts that they were the victims of an oppressive authoritarian regime.[88]

But Yanukovych had been elected only four years earlier, in what outside Western observers called "an impressive display of democracy."[89] In the face of the protests, he had agreed to European demands for early elections and a more representative cabinet. But the demonstrators rejected that proposal, instead forcing Yanukovych to flee. It might have seemed like a victory for the people, but was it really democratic to overthrow a fairly elected government?

The first parliamentary acts of the new government under Oleksandr Turchynov, the interim president, were to issue charges of mass murder against the previous leaders and to downgrade Russian from its status as a second official language. In response, Mikhail Dobkin, governor of the Russian-speaking eastern Kharkiv region, called the ski-masked Ukrainian-language demonstrators who helped overthrow Yanukovych "fascists" for the violence of their anti-Russian riots. The Ukrainian mayor of eastern Sevastopol was forced to resign under pressure from Russian speakers (and undercover Russian special forces), and his self-appointed replacement promised to resist the western Ukrainian government. Vladimir Putin recalled the Russian ambassador, cut aid, warned against oppressing the Russian-speaking majority of Ukraine's eastern regions, and then, under the guise of protecting his fleet, introduced troops into Russian-speaking Ukrainian Crimea and later simply incorporated it into Russia by force of arms.[90]

Putin's pretext was defending the interests of the Russian-speaking population, by joining them to their natural homeland. Voting in past elections, since independence, had been closely divided between primarily western Ukrainian speakers and eastern Russian speakers, decided by razor-thin margins. Were only the westerners in the right, or did all the Ukrainian people deserve sympathy?

In this situation, a Westphalian-thinking American nation-state was in fact reluctant to become engaged militarily with Russia, which was much stronger in the region and the only opponent with sufficient power to cause nuclear annihilation. So it responded with minor sanctions of the kind that tend to irritate but have limited results.[91]

There was a similar response in 2015 when the Islamic State advanced its forces from Syria into Iraq in an effort to create a "caliphate" encompassing both nations. President Obama responded in classic nation-state terms with airstrikes to blunt the advance. In doing so, he invoked the congressional authorization to use military force (AUMF) enacted in the wake of the al-Qaeda attack in September 2001. That AUMF empowered the president to use:

> all necessary and appropriate force against those nations, organizations, or persons he determines planned, authorized, committed, or aided the terrorist attacks that occurred on September 11, 2001, or harbored such organizations or persons, in order to prevent any future acts of international terrorism against the United States by such nations, organizations or persons.[92]

Rather than rallying to the president with wide nationalist support, as would have been typical in the past, politicians from both parties criticized his decade-old justification as inadequate and demanded that he submit to Congress a proposal for a new AUMF specifically covering ISIL, which of course had not been involved in the 9/11 attack. When the president did submit a draft, suddenly neither party was sure it wanted an AUMF. Criticism of the Obama proposal came from pundits on both sides of the spectrum. On the right, in *RealClearPolitics*, Lawrence Kudlow complained that Obama's plan was designed to prohibit "boots on the ground" and that it would "tie the hands of the U.S. military."[93] In the leftist *Huffington Post*, a contributor charged that "Obama's proposal asks Congress to rubber-stamp his endless war against anyone he wants, wherever he wants."[94]

The lack of patriotic support for the AUMF proposal was due in part to its vagueness. First, it did not revoke the 2001 AUMF, and it even claimed that a new authorization was not necessary for acting or even expanding operations. Why then was a new one proposed? Second, the new proposal did "not authorize the use of the United States Armed Forces in enduring offensive ground operations." But what did "enduring" mean? And what was offensive as opposed to defensive action?

Third, besides allowing military force against ISIL, it also authorized extended action against "individuals and organizations fighting for, on behalf of, or alongside ISIL or any closely-related successor." What was "closely-related"?

Following classic Westphalian national-interest doctrine, the former attorney general Michael Mukasey (and David Rivkin) argued that a new AUMF was not required because "Congress cannot restrain the president's core authority to wage war, even when congressionally-imposed restrictions are minor." Mukasey continued, "The Founders were careful to vest responsibility for waging war in a unitary executive, rather than a multi-member legislature." Of course, he knew the Constitution did give Congress the power to "raise and support armies," to "provide and maintain a navy," and to "declare war," and that the "careful" vesting of power in the president was simply as "commander in chief," without further specifics.[95]

Jack Goldsmith, a former assistant attorney general, countered: "Some of Congress's very first authorizations of force, in the quasi-war with France in the 1790s, authorized the President to use only limited military means (U.S. armed vessels) against limited targets (certain French armed vessels). The Supreme Court recognized these limits in *Bas v. Tingy.*" Indeed, as Goldsmith and Curtis Bradley argued, "most authorizations to use force in U.S. history have been of this limited or partial nature."[96]

But the Constitution was pretty much beside the point. U.S. tribal politics had again overridden nationalist and rationalist state interest. Members of Congress were preoccupied with parochial domestic concerns, and wary of local constituencies' opposition to military action and the resulting casualties, so they simply let the AUMF proposal die without final action. Another attempt at getting a new authorization under President Trump likewise failed.[97]

The world was full of complicated conflicts that weren't defined by obvious nation-state interests. The Middle East was an especially complex tangle of religions and tribes and weak states. Ukraine too was divided by culture and religion, and it bordered a dangerous Russia, with which it had a troubled history.[98] The same could be said about much of the world. Moreover, in deciding whether to ratify engagement in a significant new military operation, U.S. representative bodies mostly followed parochial political interests—like much of the rest of the world—rather than national interests supported by patriotic ideals.

Nation-States and Modern Tribes

If there is any remaining political belief about the modern world that is universal among Western intellectuals, political leaders, and even supporters of capitalism, it is the essentiality of the nation-state. The large nation-states we know today were cobbled together by force or by dynastic union—such as that between Ferdinand of Aragon and Isabella of Castile—pulling separate peoples into one unit.[99] In principle, the nation-state governments "managed everything, from consciences to the patterns of the silk fabrics of Lyons," as Joseph Schumpeter put it. The power of the rulers was "never really absolute," but they did more or less consolidate various local communities into a national order with a common market.[100]

War was the nation-state's greatest accomplishment. Previously, in the nine hundred years between the fall of the Western Roman Empire and the fourteenth-century dynastic wars, Europe had lost perhaps 2 million lives in warfare, half of those in the Crusades spanning more than two centuries. Then, the Hundred Years' War alone took more than 2 million lives, and the clearly nationalist Thirty Years' War vastly exceeded that toll, causing 7 million deaths in merely three decades. The Napoleonic wars resulted in almost 2 million dead in little more than a decade. World War I killed 15 million, and World War II took the lives of 55 million.[101] Then there was the atom bomb.

For a while following World War II, foreign policy still reinforced a sense of nationalism in the West, and not least in the United States. When the Soviet Union declared that its purpose was to make the world Communist and perhaps had the power to do so, Americans were generally united against it. But the anticommunist wars in Korea and Vietnam became unpopular, and the military draft was gone by 1973. When the Communist adversary had fallen and a new enemy attacked, the goal of foreign policy expanded again to promote freedom, equality, capitalism, and democracy. But the expensive and prolonged wars in Afghanistan, Iraq, and elsewhere proved more divisive than unifying.

Between the end of the Cold War and the beginning of the War on Terror, a prestigious international gabfest was organized in Salzburg, in 1996, to discuss "cross-cultural perspectives" on the major political divisions in modern nation-states. Political parties and movements designated as "conservative" at home or regarded as such by the European organizers of the Salzburg Seminar were invited to explain their views and to discuss what the different national versions of conservatism held in common. With representatives from across Europe to Turkey, and even from China, it was difficult to find any commonality on social or economic matters.

This author, as the U.S. representative, played by the rules and suggested that decentralizing power among pluralist local institutions might be one way to define conservatism. The French representative nearly swooned, furiously insisting that conservatism was precisely the opposite: love of the *patria,* the nation, and its embodiment in the national state. This point was immediately seconded by the Turkish representative. The Spanish, Italian, Belgian, and eastern European national representatives actually denounced localist movements as threats to the highest conservative ideal—national order. When I observed that subnational movements were restive throughout the world, even in Britain, the group of intellectuals found the possibility of localism in such a long-unified nation-state so preposterous they broke into laughter, and the Englishmen questioned my sanity.

By that time, parts of the former Soviet Union and its satellite states were dividing into nationalities. The Czechs and the Slovaks had accomplished an amicable separation. Two decades after the conference, 45 percent of Scotland's population voted to break a three-century union with England to become an independent country again. Millions of Catalans took to the street to demand independence from Spain, and the regional legislature of Catalonia voted to hold a referendum to exit. Basques threatened to do the same. Flemish nationalists in barely united Belgium pledged that if Scotland received European Union representation, they would demand it too. The Northern League demanded separation from Italy. Every so often, Bavaria has threatened to secede from Germany.

Indeed, Germany and Italy were not unified until the later nineteenth century. Many of the old principalities and cities retained their local customs and institutions, often nursing old grievances and developing new ones against remote and unresponsive national governments. To build national consensus, nation-states had necessarily discriminated against ethnic, cultural, and religious minorities, stirring up xenophobia and anti-Semitism (especially in France, Germany, and westernizing Russia). But ethnic, religious, and other minorities remain identifiable as important social forces. Even France today has restive Basques, Bretons, Savoyards, and others demanding local rights, or either partial or total independence.[102]

Some see a broad revival of nationalism in the British referendum to exit from the European Union; in the nationalist party victories in Italy; in the German Alternative party, countered by a shaky left-right coalition; in the strong showing for Marine Le Pen's rightist party in France; in the populist turns in Hungary, Poland, the Czech Republic, and Bulgaria; in the active threats of secession by Catalonia, Scotland, and others; perhaps even in the rise of outsider Emmanuel Macron in France; and in Donald Trump's defeating the Washington establishments of both political parties and questioning of international agreements on immigration, trade, and climate change.[103]

But are these trends and movements truly nationalist, or rather pluralist and regional? Brexit, the choice for a more independent Britain, was favored by majorities in England and Wales, but opposed by majorities in Scotland and Northern Ireland. In France, the support for Marine Le Pen came mainly from the northeast. The German Alternative party has been most successful in the former East Germany. Italian nationalists have represented the prosperous northern region. It is only the smaller, more culturally homogeneous countries such as Hungary that have a true "nationalist" commonality.[104]

As David Goodman has demonstrated, Western cultural nationalism is actually expiring. The original justification for the nation-state was to provide internal order and protection from foreign threats. Nationalism traditionally rallied the patriotic support of citizens in the common defense. But current data show that in the great majority of western European nation-states less than one-third of citizens say they are willing to fight (and die) for their country. On the other hand, most of the European countries once occupied by the Soviet Union show higher rates. Western European nations also have very low birth rates, portending difficulty in raising large military forces in the future; in fact, birth rates across Europe and into Russia are below replacement level. By contrast, Turkey and especially Israel manifest a high willingness both to fight and to procreate.[105]

America long considered itself an exception to the decline in patriotism. But only 44 percent of American respondents told pollsters in 2015 that they were willing to fight for their country.[106] The United States has not been immune to the secession impulse either, even after the Civil War was thought to have settled the matter. A poll taken during the George W. Bush administration found that 32 percent of liberals and 17 percent of conservatives thought state secession a good idea. In 2008, the total was 18 percent.[107] A Reuters/Ipsos poll taken in 2014, during the Obama administration, found that nearly 24 percent of Americans would like to see their state secede, including almost 30 percent of Republicans and 21 percent of Democrats.[108] Under President Trump, 35 percent of Republicans and 41 percent of Democrats would allow their state to secede.[109]

President Trump, of course, was elected in 2016 promising a revival of American nationalism.[110] Yet the nation became more politically divided than ever, as documented in an extensive study by a new international organization called More in Common, published in 2018. It concluded that the only way to make sense of the American populace was in terms of "hidden tribes": Progressive Activists (8 percent), Traditional Liberals (11 percent), Passive Liberals (15 percent), Politically Disengaged (26 percent), Moderates (15 percent), Traditional Conservatives (19 percent), and Devoted Conservatives (6 percent), with those at each end of the political spectrum holding the strongest and most consistent views.[111]

The tribes on the ends of the spectrum were divided on practically everything. Progressive Activists overwhelmingly believed that citizenship is not equal in America: 94 percent said that white people have advantages over others, 91 percent said that women have lesser rights, 86 percent believed that an individual's own efforts don't determine life outcomes, and 94 percent said that government must "take more responsibility to ensure everyone is provided for." Devoted Conservatives took precisely opposite positions: only 20 percent agreed that white people or men start out with significant advantages, a mere 2 percent said that individuals cannot control their own destinies, and 3 percent said that government should do more to provide for all. These two groups also diverged widely on the question of different roles for men and women, and on whether it is more important to raise children to be well-behaved or to be creative.

Looking at the whole spectrum, the study targeted four issues as especially polarizing: immigration, race, gender, and religion. When asked whether they thought immigration is good for the country or a burden, 56 percent of respondents said it is good and only a minority from the right considered it a burden. Overall, 41 percent believed that undocumented aliens take jobs from Americans, and 47 percent believed that the rights of immigrants are "more protected than the rights of American citizens." Progressive Activists were almost unanimous (95 percent) in disagreeing with that view, while 92 percent of Devoted Conservatives agreed. But most respondents did support a path to citizenship for undocumented aliens brought into the country as children.

Majorities in all of the political groups agreed that problems of racism are at least somewhat serious, and 60 percent overall saw white supremacist groups as a threat. Only 40 percent of Americans (but large majorities of Progressives and Liberals) said that many people do not take racism seriously enough, while 60 percent said that many people are too sensitive about racial matters.

While a majority of respondents believed that whites have social advantages, and 60 percent of Progressives supported racial preferences in college admissions, 72 percent of all respondents opposed racial preferences. While 51 percent of Americans believed that the police discriminate against blacks, 56 percent thought the Black Lives Matter movement had increased racial tensions.

A slight majority overall believed that changing attitudes toward sex and sexuality were making the country more tolerant, while the remainder thought these changes were damaging the country's moral foundation. At the same time, 55 percent believed that changing views on marriage were hurting family values. On questions of sexism and bias against LGBT people, Progressives and Liberals demanded more government action. Conservatives were concerned that claims of sexual harassment are too often unjustified. In fact, 51 percent overall agreed with the latter view, and 46 percent said that too many feminists are just out to "attack men."

Concerning religion, 83 percent said they believed in God and 63 percent said that religion was very important in their lives; just 21 percent said it was not at all important. Majorities in all groups said that people of different beliefs can be good Americans. But the question of religion did divide the political tribes: while majorities of Progressives and both Traditional and Passive Liberals said that America needs more science and reason rather than more faith and religion, majorities of the Disengaged and Moderates as well as Traditional and Devoted Conservatives said the opposite.

On questions directly related to nationalism, 73 percent said it is essential that Americans speak English—although merely 22 percent of strong Progressives considered even that necessary. Half of Americans said U.S. birth is central to American identity, while 38 percent believed that national identity is tied to a particular religion. Most important, only 53 percent told the pollsters that they were proud to be American. Among Devoted Conservatives, 90 percent were proud, but only 18 percent of Progressives and Liberals. Pretty much everyone said they believed in freedom, equality, and free speech, but seven in ten said they feared to speak freely about race, gender, immigration, and Islam, except "when I am with people who are like me."

A follow-up study by More in Common suggests that "Americans imagine themselves to be more divided than they really are." The study found that Democratic and Republican voters both tend to regard the other party's voters as being mostly on the far opposite end of the spectrum in their opinions, although many in fact are closer to the middle. This is called a "perception gap" between the image and the reality of partisan opponents. But perceptions matter too.

More education and following the news closely did not help close the perception gap, but were likely to accentuate it. Conservatives who followed internet news and talk radio, and likewise Progressives who followed the *New York Times* and the *Washington Post*, were more likely to have a stereotypically negative view of what their opponents believed than those who did not follow news media as closely. More highly educated people generally had a less accurate idea of what voters in the other party actually believe. Democrats with postgraduate degrees were three times as inaccurate in describing Republican beliefs as those with no high school degree.[112]

State control of education has not resulted in national consensus on many basic national ideals. According to Pew Research, key Western values like freedom of speech were held by minorities in most nations. Even in the United States, around one-fifth do not think it is very important to have freedom of religious practice, or regular contested elections.[113] A survey by the Cato Institute in 2020 found that "62% of Americans say the political climate these days prevents them from saying things they believe because others may find them offensive," with only "strong liberals" feeling free to do so.[114]

Everywhere, there were significant divisions by ethnicity, region, religion, ideology, and class. The one unifying belief among Americans seemed to be that the national government could not be trusted.[115]

National consensus has declined even in the United States, especially among the young.[116] Group identities have gained strength, and there is weakening commitment to crucial capitalist ideals like freedom, property rights, limited government, and the value of hard work.[117] At the same time, support for socialism is growing, also especially among the younger generation. All this has led to demands for more rationalized and moral forms of both nationalism and capitalism.

CHAPTER 3

RATIONALIZING NATION-STATE CAPITALISM

K arl Marx famously said that "working men have no country." He urged them to throw off their nationalist and capitalist institutional chains, and then use their new power "to centralize all instruments of production in the hands of the state, that is, of the proletariat,"[1] and establish class equality. A large proportion of his ten proposals for a working-class revolution would eventually be adopted by modernizing Western nation-states. It was the rationalizing states-man Otto von Bismarck who created the first welfare state, as a means to defeat pure socialism politically and manage his new German state.[2]

Marx insisted that only the elimination of property would produce true class equality, but the remarkable increase in wealth generated by pluralist capitalism convinced most Western liberal leaders that private property and markets needed to be preserved. At the same time, the pluralist freedom of capitalism had to be moderated by an ideal of equality and provisions for the general welfare, managed by experts in a centralized state. The liberal icon J. S. Mill promoted the major ratio-nalizing reforms that were codified in the British Poor Law of 1834. He argued that "no member of the community" ought to "be abandoned," and that "society can and therefor ought to insure every individual belonging to it against extreme want."[3] Late nineteenth-century British rationalistic progressivism, inspired by the philosophers T. H. Green and Leonard Hobhouse, began advocating the mitigation of all the inequalities associated with raw capitalism.[4]

Fighting Poverty

Alexis de Tocqueville remarked upon the unusually robust pluralism in the United States, with local and private organizations taking
responsibility for social welfare. This exceptional pluralism lasted into
the twentieth century. Then President Franklin Roosevelt, following
the lead of Woodrow Wilson, initiated comprehensive, rationalistic
national programs: regulation of banking, welfare relief, direct agricultural payments, and the Social Security Act of 1935. The next major
step toward rationalizing American capitalism was President Lyndon
Johnson's epic War on Poverty in the 1960s.[5] Yet on its fiftieth anniversary, President Obama found its goal to be largely unmet.[6]

Around the same time, a prominent progressive columnist, William
Galston, cited Ronald Reagan's quip that the United States had fought
a war on poverty and "poverty had won," but Galston maintained that
poverty had actually been fought "to a draw." Roughly 15 percent of
the population were below the official poverty line in 1983, the end of
the recession that began in the Carter era, and 15 percent were below
it after the Great Recession of 2008—which does suggest a stalemate.[7]
Another progressive columnist, Dana Milbank, disagreed, claiming that
"if you include all the financial assistance from anti-poverty programs,
the poverty rate drops to below 8 percent" in 2014.[8]

An empirical study by Bruce Meyer and James Sullivan in 2011
found the real poverty rate to be as low as 5 percent measured by consumption: large numbers of those officially counted as poor had computers, dishwashers, and central air conditioning.[9] A Census Bureau
report, published in 2014, found that while as many as 31.6 percent fell
below the official poverty line over a recent three-year period, only 3.5
percent were stuck there the whole three years.[10]

Further analysis by John Early, former associate director of the
Bureau of Labor Statistics, included all money income, the value of
benefits from government, and offsetting taxes paid, and found that the
average share of spendable household income for the poorest one-fifth
of the population rose by an incredible 493.2 percent—almost a sixfold
increase, mainly from government welfare transfers. The gain for the
second-lowest quintile was 97.7 percent. The middle group's income
share increased by a more modest 21.5 percent, while the income shares

of the two richest classes fell, by 9.2 percent and 31.9 percent respectively, mostly because of higher taxation.[11]

The bottom line was that "the average spendable income for the highest [20 percent] income group was only three times higher than that for the lowest group, with government redistribution eliminating 88.5 percent of the ratio between the highest and lowest" income categories. The spendable income gap between the highest and lowest quintiles was thus less than one-fifth the ratio of 16.2 to 1 highlighted in the Census Bureau estimate. Indeed, the middle-income group had only 20 percent more total income than the poorest group. As a result of government transfers, between 2009 and 2012 only 2.7 percent of the population lived in poverty for all this period. Sixty-one percent of households will have earnings in the top income quintile for at least two consecutive years during their lives, and about half of all households in the bottom quintile at a given time will rise into a higher income category within ten years.[12]

So was the War on Poverty won? Part of the problem in trying to evaluate the result is that it is difficult to calculate the initial poverty rate accurately. Another problem is that market forces undoubtedly had some influence in narrowing the gap between rich and poor. But the main question is whether economic equalization is the most important goal. As the *Washington Post* columnist Robert Samuelson noted, simple monetary gain was not in fact the aim of the War on Poverty, which President Johnson promised was for "most poor people [to] become more productive, independent and middle class."

This did not happen. The popular education program called Head Start, meant to change attitudes from dependency to self-motivation, showed gains for the poor at the beginning but they were dissipated as early as the third grade, with little or no sign of educational or earnings benefit thereafter.[13] A comprehensive study by Eric A. Hanushek and Laura M. Talpey of Stanford, Paul E. Peterson of Harvard, and Ludger Woessmann of the University of Munich, published in 2019, found that the billions of dollars of federal and state compensatory education spending since 1965 did not close the achievement gap between economic classes or racial and ethnic groups.[14]

Robert Rector, a welfare expert at the Heritage Foundation, calculated that the federal government spent $22 trillion (excluding

Social Security and Medicare)—more than $70,000 for every person in America—to wage the War on Poverty. While 82 percent of poor adults reported that they were never hungry at any time in the prior year for lack of money to buy food, and there was substantial redistribution of wealth, the most obvious result has been increased dependency. Fewer people sought employment, major parts of central cities became dysfunctional, and the black family was decimated—all of which was acknowledged by President Obama.[15]

Rector and his associate Rachel Sheffield documented that the War on Poverty especially "crippled marriage in the low-income community." As means-tested benefits expanded, welfare began to serve as a substitute for a major income earner, eroding marriage among lower-income Americans. At the outset, the welfare system directly penalized low-income couples who married by eliminating or reducing benefits. "As husbands left the home, the need for more welfare to support single mothers increased. The War on Poverty created a destructive feedback loop: welfare promoted the decline of marriage, which generated the need for more welfare."[16] Welfare directed to poor women allowed them to gain independence from wages earned by potentially abusive husbands, but transferred their dependency to government.[17]

The number of single-parent families has tripled since the War on Poverty began, increasing to 34 percent of households. Children from such households are more likely to drop out of school, get pregnant before age twenty, and be unemployed, thus perpetuating the cycle of poverty. In the 1960s, only 6 percent of men ages 25 to 54 were regularly without jobs, but in the early twenty-first century, with government jobs programs long in place, the figure was 17 percent.[18]

Early concluded that income redistribution significantly lowered the incentive of those in the bottom income quintile to seek employment. In 1975, the proportion of families in the lowest quintile with a prime-age head of household available but with no family member working was 24.8 percent higher than in the middle-income group. By 2015, this difference had risen to 37.1 percent. Compared with the lower-income quintile, the middle-income group had five times more families with two or more members employed in productive work.[19] After a century of rationalization, the paradox was that income and welfare equality for the poor increased but working and family attachments declined.

On the other hand, the U.S. poor were rich by international standards. The world median annual income in the early twenty-first century was $1,225 per person, according to Branko Milanovic, a World Bank economist. The income of the world's wealthiest 1 percent began at $34,000 per person (after taxes). Calculating from his data, the median U.S. income then was $34,500, which means that half of Americans were considered rich by global standards. Almost all of the world's 60 million "rich" were in the United States or Europe, and almost half of the world's 1 percenters lived in the United States. A family of four in the United States was considered poor on an annual income of $23,000, but that would place them among the world's income leaders.[20]

Nearly three-quarters of the officially poor in the United States have a car or a truck, and 31 percent have two or more; nearly two-thirds have cable or satellite television; two-thirds have at least one DVD player and a quarter have two or more; half have a personal computer and one in seven has two or more; half of poor families with children have a video game system such as an Xbox or PlayStation; 43 percent have Internet access; 40 percent have a wide-screen plasma or LCD TV; a quarter have high-quality digital video recorder systems; and 92 percent of poor households have a microwave.[21]

Even a transfer of $22 trillion to the poorest over half a century was not sufficient to overcome economic inequality in the United States. But consider what would be required to equalize income globally.[22] Once U.S. and European states had taxed away and then redistributed to the world all income earned by their populations above $1,225, the median that Milanovic found, other things would obviously have to change. With decreased tax revenue, defense spending would plummet and programs like Social Security and Medicare would have to be reduced drastically. High-end services and manufacturing could no longer be supported by the lower after-tax income, so investment would need to shift radically to agriculture and low-end manufacturing. Industrial waste would decline dramatically, but agriculture and primitive energy sources might actually increase pollution globally.

Imagine democratic political parties in the West proposing such a plan, telling the relatively poor in their own countries—including those

on welfare and much of the lower-income half of the U.S. population—
that they are too rich and must reduce their income to benefit the real
poor of the world, who live on only $1.25 a day or so. The result would
be economic equality, less overconsumption, simpler lifestyles, and
perhaps less waste—but who in the richer nations would agree to it?

Which Equality?

One can speak of moral equality before the Creator, or equality under
the law, or equal opportunity, or equality of results—but these are very
different principles. There is a tendency, however, to assume that people
mean the same thing in speaking of equality, when often they do not.
William Galston, a former aide to Bill Clinton, exemplified this assumption when he claimed: "Most liberals agree with most conservatives that
the objective is equal opportunity." In support of his progressive equality agenda, he quoted Paul Ryan, a top House Republican, saying that
"the condition of your birth does not determine the outcome of your
life."[23] But the two did not have the same idea of "equal opportunity,"
as Ryan spoke of moral rather than economic equality. Indeed, differences of understanding are often obscured in discussions of equality,
sometimes to keep civil peace. When the specifics come to the fore,
there is really not much agreement about what is required to promote
"true" equality.

For example, the columnist Harold Meyerson lauded President
Obama's commitment to equality and his new social programs for
achieving it, but objected to the president's support for a Pacific free-
trade compact. Meyerson claimed that the trade agreement would
reduce the number of U.S. manufacturing jobs and result in more
income inequality among Americans. "By now," Meyerson wrote, "even
the most ossified right-wing economists concede that globalization has
played a major role in the loss of American manufacturing jobs" and a
consequent "stagnation of U.S. wages and income."[24] In fact, most conservative economists did not agree that free trade increased inequality
or decreased net domestic employment, although the future Republican
president, Donald Trump, might have agreed in some measure.

Meyerson opposed free trade because he believed that poor foreigners would underbid American workers to win manufacturing

jobs, so blue-collar workers in the United States lost jobs and income while richer Americans benefited, thus increasing domestic inequality. But if poor foreigners then earn higher incomes, global inequality is reduced. For this reason, Galston and Ryan actually did agree that free trade helps the poor. Meyerson, however, supported higher income for Americans at the expense of the poor in other countries. This would put him at odds with Pope Francis, who promoted the moral ideal of an equality that is global.

The mathematical Gini index is widely considered the best measure of economic equality in a country, and shockingly it tends to show that countries with freer markets are generally more equal.[25] A front-page *Christian Science Monitor* story in 2016 announcing a World Bank study said, "Global poverty has fallen faster during the past 20 years than at any time in history," and reported that two of the top three explanations were market-related.[26] An earlier World Bank study suggested that additional welfare spending by national governments had actually slowed economic growth and exacerbated inequality.[27]

Equalizing Race and Ethnicity

The efforts that have been made thus far to reduce inequality under capitalism are widely seen as inadequate. The Civil Rights Act of 1964 was supposed to end racial discrimination in the workplace, and civil rights protection was later extended to other categories. On the fiftieth anniversary of the law's passage, Jacqueline A. Berrien, chair of the Equal Employment Opportunity Commission, summed up the net results of a half century of trying to rationalize capitalist employment toward greater equality:

> The EEOC receives nearly 100,000 charges of discrimination each year, with retaliation and racial discrimination remaining our greatest challenges. Moreover, too many women are paid less or shut-out of job opportunities; too many people are forced to choose between their jobs and religious beliefs; too many workers are segregated on the basis of national origin; too many persons with disabilities are excluded from jobs they are qualified to hold; too many older workers are screened out of job opportunities because of their age; and

too many LGBT employees suffer harassment in the workplace based
upon stereotypes.[28]

The vast majority of cases filed at the EEOC over the years have
been dismissed as being without merit. But the numbers of employment discrimination complaints filed were higher than in the earliest
desegregation years—around 30,000 cases at the time Berrien spoke.
The Civil Rights Division of the Department of Justice announced in
2013 that it had filed more employment discrimination cases than in
any previous year. Even before taking into account the addition of new
categories, such as LGBT, is it possible that employment discrimination
could really have become worse than before the civil rights laws were
passed a half century earlier?[29]

The Supreme Court declared racial segregation in public schools
unconstitutional in its *Brown v. Board of Education* decision back in
1954. Other national civil rights acts were passed in 1957 and 1960, but
the mass demonstrations in Birmingham, Alabama, headed by Martin
Luther King Jr., led to the major legislation in 1964 that set up much of
the modern civil rights apparatus, forbidding racial discrimination in
voting, employment, and public accommodations. Sex, ethnicity, and
religion were added to the protected categories. Feminists succeeded
in getting an Equal Pay Act in 1963, no-fault divorce in the 1970s, and
protection from sexual harassment in 1986.

The Violent Crime Control Act of 1994 imposed criminal penalties for "hate crimes" on the basis of the actual or perceived race, color,
religion, national origin, ethnicity, or gender of the victim. Same-sex
marriages were granted equal federal benefits with traditional marriages by Supreme Court decisions in 2013, and state laws allowing only
opposite-sex marriages were declared unconstitutional two years later.
In 2016, protection from discrimination was extended by executive
order to include "gender identity," though President Trump somewhat
modified the order.[30]

Public attitudes had actually changed before the major civil rights
laws could have had much effect. Gallup polls found in 1963 that a
plurality of Americans said they would vote for an African American
for president.[31] The Protestant mainline denominations had already
shifted from a central emphasis on theological doctrine to a "social

gospel without the gospel," as Joseph Bottum described it in *An Anxious Age*. Those mainline Protestant churches began to view bigotry, discrimination, and inequality as the major moral evils to be opposed, and assumed a duty to enlighten and reform those clinging to their prejudices.[32]

Yet despite the antidiscrimination laws and changes in public attitudes, 71 percent of blacks and 56 percent of whites recently told Pew pollsters that race relations in America were "generally bad," even before the new wave of protests in reaction to claims of police maltreatment of African Americans in 2020.[33] In 2018, while 57 percent expected white/black relations to improve, 39 percent said they will always be a problem. Majorities of whites thought blacks have an equal chance at getting jobs for which they are qualified, and equal housing opportunities, and that their children have an equal chance of success. Majorities of blacks disagreed, with most reporting that they have been subject to discrimination in the employment market.[34]

Until recently, it was widely assumed that antidiscrimination policy as a whole had been a great success. Consider women in the workplace: Women's average income is lower than men's, but when the data are controlled for type of work and length of time in the job, they show that women have mostly achieved equality of income with men. Never-married women actually out-earn single men.[35] On the problem of sexual harassment and assault in the workplace, the #MeToo movement has resulted in the firing of numerous high-level public figures.[36]

There is a large African American middle class today, though the ratio of black to white earned income has not increased greatly since the civil rights laws were implemented: in 1947 it was 52 percent, and by 2014 it had increased only to 57 percent. Black unemployment remained twice as high as for whites. Workforce participation rates were lower and welfare dependency higher among the black population. Education for blacks was often substandard, marriage was disincentivized and the poverty rate for unmarried blacks was an incredible 36 percent—compared with 22 percent for unmarried whites. But for married blacks, the poverty rate was only 7 percent.[37]

The earlier consensus on antidiscrimination laws began to dissipate, as whites perceived little economic gain resulting for minorities, but saw a reverse discrimination as adversely affecting themselves, while redistributive policies reduced their income share. When the former governor of Maryland, a white Democrat, was questioned about the phrase "black lives matter" and he responded that "all lives matter," he was forced to recant this simple statement of equality for all.[38] New categories of protected groups had less appeal to moral sympathy than did African Americans, and traditional values seemed to be under attack. Mainstream Protestantism was becoming too divided on cultural issues to continue providing its traditional moral authority, but no other moral authority was available to replace the mainstream religions.[39]

At the same time, African Americans saw only modest economic gains relative to whites, along with disintegrating family structures, and experienced continuing discrimination against themselves. They seemed elbowed aside by other groups demanding antidiscrimination protection, with grievances perceived as less compelling than what black Americans had suffered. The codifying of civil rights for black Americans set the precedent for all subsequent antidiscrimination policies, but now that group seemed to be just one among many. Some of the new protected groups were even perceived as posing threats to the values of most black Americans. Indeed, religious blacks provided the decisive margin in approving the legal proposition in California, in 2008, to limit marriage to one man and one woman.[40]

Public support for antidiscrimination policy was weakened by stories suggesting that it was sometimes used to gain an advantage. One prominent example was Rachel Dolezal, who claimed to be black despite being the daughter of two white parents. In this way, she had attained several race-related professional offices, including a university position and leadership of the NAACP of Spokane, Washington. When Dolezal became a subject of considerable publicity in 2015, a reporter asked Jelani Cobb, an Africana Studies scholar at the University of Connecticut, what he thought of Dolezal's claim to be black. Cobb replied that he did not find it troubling:

> Dolezal has been dressed precisely as we all are, in a fictive garb of race, whose determinations are as arbitrary as they are damaging. If blackness is a matter of African ancestry, then we should set about the task of excising a great deal of the canon of black history up to and including the current President.[41]

If even one's race is only a "fictive garb," or a matter of personal choice, it is difficult to avoid the conclusion that the whole antidiscrimination edifice has an entirely subjective foundation, and therefore its rules and strictures can be applied arbitrarily too.

Equalizing Marriage

The Proposition 8 referendum of 2008 in California to codify a traditional definition of marriage was effectively overturned a few years later by the Supreme Court in *Hollingsworth v. Perry*, decided in June 2013. Justice Elena Kagan had argued that not allowing same-sex couples to marry "stigmatized" them, and that this invalidated state laws limiting marriage to one man and one woman even if homosexuals were allowed to have "civil union" alternatives and were treated equally under all other laws—as by and large they were in California.[42] Then in 2015, in what President Obama called a judicial "thunderbolt," the Supreme Court issued its *Obergefell v. Hodges* decision requiring all states to accept "marriage equality" for all couple relationships.[43]

Chief Justice John Roberts, in his dissent, expressed a fear that the majority position would not allow a person to "exercise" a religion whose tenets were contrary to the ruling, but merely to "advocate" and "teach" it, and he argued that this annulled the free-exercise clause of the First Amendment.[44] The solicitor general, Donald Verrilli, had acknowledged in oral argument the possibility that if the logic of *Bob Jones University v. United States* (1983) were followed in light of a ruling for the plaintiffs in *Obergefell*, it would soon be possible to deny a religious institution the right to a tax deduction or a government grant if it took personnel or service actions favoring the traditional definition of marriage.

In an essay published soon after the passage of Proposition 8, Chai Feldblum, a law professor at Georgetown and an EEOC commissioner, was open enough to recognize a need to accommodate religious objections, but argued that such accommodations should not be allowed to "undermine the effectiveness of equality legislation for LGBT persons." Feldblum would agree to narrow exceptions for religious institutions themselves and for those in top leadership positions in religiously affiliated institutions open to the public, such as hospitals, as long as those institutions were "clearly and explicitly" defined by religious beliefs. But generally, once a religious person entered "a stream of commerce," that person "must be expected to adhere to a norm of nondiscrimination" regarding sexual orientation and gender identity.[45]

Enforcing nondiscrimination in actual court cases is a complicated matter. Consider the case of Elaine Huguenin, a professional photographer who politely declined to take pictures at a gay wedding in 2006, and was ordered to appear before the New Mexico Human Rights Commission, even though another photographer had been hired in her place. The commission ruled that the exemption for religion provided in the 1993 Religious Freedom Restoration Act did not apply to any business open to the public, and the state supreme court concurred in August 2013 (after the *Hollingsworth* decision came down). In a unanimous verdict, the court ruled: "When Elaine Photography refused to photograph a same-sex commitment ceremony, it violated the [New Mexico Human Rights Act] in the same way as if it had refused to photograph a wedding between people of different races." The cost to Ms. Huguenin was $6,637.94 in fines and attorneys' fees.[46] She appealed the case to the U.S. Supreme Court, which declined to hear it.

The Colorado Civil Rights Commission declared that Jack Phillips, owner of the Masterpiece Cakeshop in Lakewood, had illegally discriminated against a gay couple in 2012 by declining to create a wedding cake for them, saying it was against his religious beliefs. One commissioner proposed "reeducation" to cleanse the baker and his staff of their "despicable" discriminatory views.[47] The Colorado Court of Appeals upheld the commission's order, but the U.S. Supreme Court reversed it in 2018.[48] This hasn't made things easy for other religious objectors, though.

When Barronelle Stutzman, the Southern Baptist owner of Arlene's Flowers, was asked to provide floral arrangements for a gay wedding in Washington State in 2013, she held hands with the longtime customer, explained regretfully that her religious beliefs did not permit her to participate in such a function, and referred the customer to three other florists. Still, she was fined $1,000 for discrimination, and legal fees sent the business into bankruptcy. If she reentered the florist business, she would be required to perform the services she found religiously objectionable.[49] The case was eventually appealed to the U.S. Supreme Court, which in June 2018 instructed the state supreme court to reconsider it in light of the recent *Masterpiece Cakeshop* decision. The state court again ruled against Stutzman, and the case was again appealed to the U.S. Supreme Court.

In Oregon, Aaron and Melissa Klein asked to be excused from baking a cake for a lesbian wedding in 2013, saying it was against their Christian beliefs. They were threatened with a $135,000 fine for "emotional distress" inflicted on the gay couple. The Kleins raised much of that sum through the GoFundMe crowdfunding website, but then their page was shut down. GoFundMe said the Kleins were defending discrimination. The Oregon Supreme Court declined to hear the case, so the Kleins appealed to the U.S. Supreme Court, which had already ruled in favor of Jack Phillips in *Masterpiece Cakeshop*. The case was sent back to the state courts for a new hearing.

In *Obergefell*, Justice Anthony Kennedy repeated the "stigmatizes" criterion that Justice Kagan had advanced in *Hollingsworth* as the main justification for expanding the definition of marriage. But as John Safranek observed while the case was being argued, "Any political minority can suffer stigma." If states are required to treat all varieties of marriage equally, it means that "opposite-sex couples opposed to same-sex marriage would suffer the legal and political insult of having their marriages equated with ones they viewed as discrepant and inferior to theirs. There is no neutral marriage statute. Every marriage law will violate some citizens' equality, and in fact most legislation does."[50] While the majority opinion emphasized "equal dignity before the law," Justice Clarence Thomas, in his dissent, observed that "government cannot bestow dignity, and it cannot take it away."

Ryan T. Anderson, a Heritage Foundation analyst, has argued that marriage between a man and a woman is a "natural right," and that the exercise of such rights must be kept "within the limits of justice and the common good." His case for giving preference to the traditional family was very basic: "it takes two to make a baby."[51] Anderson proposed a constitutional amendment to prevent the government from denying tax exemptions, grants, contracts, or certifications to an individual, association, or business because of a religious belief that traditional marriage only is a natural law right.[52] He argues that a high bar should be set to limit freedom, including freedom of religion, and the bar should not be lowered to meet demands of gay people who are not actually being persecuted today.[53]

Courts had already needed to deal with the children of gay couples differently, especially in custody cases after divorce. When two biological parents sought custody, there was a natural equality in that both had participated equally in the child's creation. That was not the case with gay couples, so a second stream of law evolved, and this points to acceptable legal differences with regard to marriage. Some proponents of gay marriage—for example, James G. Dwyer, a law professor—believe there is a sufficient state interest in treating traditional marriage distinctively in law.[54]

On the other hand, equality might be accomplished by making marriage a purely private or religious act, without government involvement other than by offering a state contract as an alternative for those who might prefer it, and all would be enforceable in state courts as simple contracts. Conservative religious denominations would not need to function as state agents certifying civil marriages, and could even change their terminology, using the word "matrimony" to affirm the distinction between their religious sacrament and a purely civil contract. Yet such compromises seem remote.

Equality as Nondiscrimination

A major problem with antidiscrimination law as a way of reducing inequalities is that rather than working directly, as in granting economic benefits and subsidies, it pivots through the back door by forbidding something. But there is no clarity on the thing to be forbidden. While there is a vast literature on the subject of discrimination, there is surprisingly little scholarly appetite for defining exactly where discrimination improperly violates equality.

A small number of pluralists have questioned the basis and rationale for the general nondiscrimination agenda. Professor David Bernstein of George Mason University and the libertarian Cato Institute argued that every free act requires "discriminating" against one product, service, provider, friend, potential spouse, etc. in favor of another. To oppose all discrimination is to use the ideal of equality to trump the virtue of liberty, to forbid free choice.[55] Daniel Oliver, former chairman of the Federal Trade Commission, wondered, "why *shouldn't* people be allowed to discriminate on the basis of what is known as sexual orientation?" After all, we have freedom of association as a right, which implies the freedom to choose our associations. "In a country where the right to associate is guaranteed, and the right not to associate is protected, why should people be required to associate with people they don't want to associate with?" Such a requirement would be justified only to spare another person from bodily harm—for example, if you "run the only inn for fifty miles and a tornado is coming."[56]

Actually, common law recognized that by operating a public accommodation the owner gave license for all to enter. But since it was his property and others were there at his forbearance, a proprietor could deny access for any "good reason," unless doing so would endanger the other person. To make judicious choices was to be discriminating.

Perhaps the only recent serious review of the subject of discrimination from several perspectives is *Philosophical Foundations of Discrimination Law*, edited by Deborah Hellman and Sophia Moreau, and published in Britain. The commentators mostly just assumed that antidiscrimination is a virtuous goal in itself while the editors presented divergent philosophical principles as the higher principle from which antidiscrimination flows: Hellman identified it as "equality of respect," while Moreau pointed to "derivative freedom." Thus the discussion begins from a foundation of conflicting first principles.[57]

The various contributors to *Philosophical Foundations of Discrimination Law* seem to agree that the idea of generally forbidding discrimination by law arose relatively recently, and first in the United States out of the felt need to reverse the effects of two centuries of slavery and state-enforced segregation. Antidiscrimination laws, and later affirmation action policies, were initially viewed as exceptions to free choice and were expected to be in effect for a limited period of time, as Justice Sandra Day O'Connor argued in *Grutter v. Bollinger* (2003).

The most comprehensive analysis of the concept of discrimination in *Philosophical Foundations* is by Professor George Rutherglen, of the University of Virginia School of Law.[58] He began by identifying the U.S. Civil Rights Act of 1964 as the archetype for antidiscrimination statutes worldwide, but noted that the term "discrimination" rarely occurred in that law or related ones. When it did appear, it was in a catchall final example of prohibited action: "or to otherwise discriminate." It was simply assumed that everyone knew what the word meant. Rutherglen came to the paradoxical conclusion that no one understands "discrimination" because everyone does. Indeed, the other authors in the volume mostly assumed the term's meaning to be understood, and went straight to their remedies.

As Rutherglen demonstrated, the understanding of discrimination at the heart of all civil rights laws derived from the unique historical experience of African Americans, and it could be extended only by analogy to other categories: race in general, religion, sex, national origin, and so on. Since analogy is never precise, extensions from the foundational case could be applied reasonably or not, depending on whose "ox is being gored."

Rutherglen concluded that the term discrimination occupied an "uneasy middle ground" between abstract ideals such as equality and liberty, on the one hand, and the "intricacies of legal doctrine" and practical questions of enforcement and compliance, on the other. His greatest insight was that what is most prominent in antidiscrimination law is the many exceptions—for the size of a business, for "business necessity," for religious practices and institutions, for public order, for women's safety, for health protection, for other rights and freedoms, for the armed forces, for disability, for age, and more.

Different legal conclusions could be reached under doctrines of implicit bias, affirmative action, positive discrimination, or disparate impact. There is a question of whether discrimination can be used to overcome the effects of past discrimination. There are also differences according to "corrective justice" versus "distributive justice," as Michael Selmi of George Washington University added.[59] Rutherglen found that the antidiscrimination idea serves as a template for "intricate compromises" of social acts and a "catch-all" for a variety of circumstances, rather than a universal principle. It finally "dissolves" into a "myriad of rules and exceptions."

Consider this example of an acceptable exclusion from antidiscrimination law, according to the European Union:

A charity works with gay men and women that have suffered bullying at work and in education, or violence relating to their sexual orientation. They want to only employ gay counsellors as the NGO believes they would be better able to relate to the gay victims and provide advice. This would probably be a genuine occupational requirement and not be unlawful discrimination.[60]

Then why not extend a similar exemption to other organizations, including religious ones? The EU did not.

The United States did provide exceptions specifically for religion, most importantly in the Religious Freedom Restoration Act of 1993. But its application was later severely limited by the Supreme Court. A leading religious rights lawyer, Ed Whelan, conceded that discrimination in public accommodations should generally be illegal, with narrow religious exemptions. Richard A. Epstein, a New York University law professor, acknowledged the necessity of banning discrimination in "routine provisions of services," but would make an exception for "commercial activities that require personal participation in religious ceremonies or other activities against religious conscience."[61]

These may be reasonable exceptions, though many do not regard them as acceptable, and in the end they are just further exceptions in a sea of exceptions. When racial discrimination is illegal, and discrimination on the basis of sexual orientation is disapproved or forbidden, why is it still acceptable to disfavor less attractive people in employment—which is now more commonplace and at least as discriminatory? As Rutherglen concluded, antidiscrimination law might be made more coherent, but it cannot be made fully rational unless its myriad exceptions and uncertainties are confronted.

Politicizing Inequality

On August 4, 2014, Richard Cohen of the *Washington Post* asserted in his column that a "Southern strategy" had been "devised" by Republican strategists for Richard Nixon in 1968 as "an appeal to racism against African Americans." After this, Cohen claimed, there was "no turning back" from racism for the GOP, so that the Southern strategy "fouls our politics to this very day." As Cohen told the story, the Democratic Party expelled racists while Republicans "welcomed them," along with "creationists, gun nuts, anti-abortion zealots, immigrant haters of all sorts and homophobes." He described the current Republican Party, and thus a large part of the nation, as a collection of "bigots and fools" preaching racial intolerance.[62]

On the same day, it was reported that Congressman Mo Brooks of Alabama, from the other side of the political spectrum, had complained that the mainstream media treatment of Republicans who had concerns about proposals for immigration reform was "part of the war on whites that's being launched by the Democratic Party." The main weapon in this war was "claiming that whites hate everybody else." Brooks called it "part of the strategy that Barack Obama implemented in 2008, continued in 2012, where he divides us all on race, on sex, greed, envy, class warfare."[63]

Dana Milbank responded to the congressman's remarks in a *Post* column that ran under the headline "A welcome end to American whiteness." The progressive Milbank saw good news in a study finding that whites would no longer be the U.S. majority by 2043.[64]

Is all this media argument and blame merely talk without real-world effect? A few days after those commentaries were published, riots broke out in Ferguson, Missouri, in protest of the tragic shooting of Michael Brown, an eighteen-year-old African American, by Darren Wilson, a white police officer. The protesters claimed that racism was the motive for the shooting, and that the police were regularly targeting black men. They believed that a political party composed of white bigots and fools was waging an ideological war fueled by a half century of racist paranoia.

Where would they have gotten those ideas? Later that year, in *Time* magazine, Darlena Cunha defended rioting as "a necessary part of the evolution of society," and characterized the rioters as "change agents."[65] When a grand jury declined to indict Officer Wilson on criminal charges, and then the U.S. Department of Justice cleared him of civil rights violations, this was taken as evidence that the judicial system is racist too. The police shooting of George Floyd in Minneapolis in 2020 stirred up these issues all over again.[66]

But were the police really targeting black men and wantonly killing them? Heather Mac Donald of the Manhattan Institute later reported that in 2014 there had been 6,095 police homicides of blacks compared with 5,397 homicides of whites and Hispanics combined. While the black population is much smaller than whites and Hispanics together, it is important to take crime rates into account. Blacks represented 15 percent of the population in the nation's seventy-five largest cities, but were charged with 62 percent of robberies, 57 percent of murders, and 45 percent of assaults. The leading cause of death among black men between the ages of 15 and 34 was murder, and 90 percent of those murders were perpetrated by other black men. Moreover, black police officers were more likely to kill blacks than were white officers.[67]

Juan Williams, a centrist African American columnist, recognized an "imbalance in economic and political power among racial groups" as a cause of black alienation and a factor in the Ferguson riots. But he also addressed the reasons for white people's fear of young black men, and cast blame on "the drug dealers, the gangbangers, the corrupt unions defending bad schools, and the musicians and actors who glorify criminal behavior among black men," as well as the political and cultural forces that romanticize or excuse it.[68]

Michael Gerson, a former George W. Bush staffer and self-identified anti-Trump Republican moderate, drew a link between racism and libertarianism, criticizing Senator Rand Paul specifically. In his *Washington Post* column, Gerson noted the senator's "disturbing history" on race and claimed that he was insincere in attacking the "militarization of law enforcement" and the general "erosion of civil liberties and due process in searches, broad general warrants, pre-conviction forfeiture." Senator Paul, wrote Gerson, carried "discrediting baggage" in the form of his earlier "discomfort with federal civil rights law" along with his "belief in a minimal state."[69] In other words, libertarianism was a cover for indifference to the well-being of black Americans.

On the other hand, a black editorialist for the *Wall Street Journal*, Jason Riley, published a book around the same time titled *Please Stop Helping Us: How Liberals Make It Harder for Blacks to Succeed*. Riley blamed much of the problem of racial inequality on all the liberal help that suppresses incentives to work, learn, marry, and become socialized.[70]

In 2015, there were riots in Baltimore after Freddie Gray, a black man, died in police custody. It would be hard to call Baltimore a racist city when three of the last four mayors, every city councilman, most police commissioners, a plurality of policemen and every officer over major were African American. But one-fourth of Baltimore families were receiving welfare, and 51.8 percent of residents ages 16 to 64 were unemployed. The site of the riots was a declining neighborhood called Sandtown, which urban development experts, charities, and government leaders had targeted in the 1990s for a model sociological and economic transformation, spending $130 million on the small community. It wasn't just money invested, but human capital too, in numerous personal visits to the local community and educational outreach. And it all seemed to do little good.[71]

Racial questions would be divisive in the 2016 election, as would social issues, chiefly gay marriage. A poll in early 2015 showed Americans narrowly favoring gay marriage rights, though with certain exceptions for religious beliefs.[72] But the most important socially conservative organizations, led by James Dobson and Franklin Graham, took a strong public stand against it just as the Supreme Court began oral argument in *Obergefell*. They issued a Pledge in Solidarity to Defend Marriage, proclaiming:

> Experience and history have shown us that if the government redefines marriage to grant a legal equivalency to same-sex couples, that same government will then enforce such an action with the police power of the State. This will bring about an inevitable collision with religious freedom and conscience rights. The precedent established will leave no room for any limitation on what can constitute such a redefined notion of marriage or human sexuality. We cannot and will not allow this to occur on our watch.[73]

The signers pledged "obedience to our Creator when the State directly conflicts with higher law." The declaration invoked Martin Luther King's "civil disobedience," a refusal to obey laws that are "out of harmony with the moral law." As James Poulos noted, this appeal to conservative religious values would be seen as ideological and partisan, unlike King's appeal to universal conscience.[74] However, Poulos assumed that the goal was changing public opinion, rather than firing up supporters—which, come to think of it, was also a large part of King's success.

While *Obergefell* was being argued, a state representative in Texas, Cecil Bell, introduced a bill to prohibit state and local officials from using state funds "to issue, enforce, or recognize a marriage license …for a union other than a union between one man and one woman." The gay-rights group Equality Texas claimed the bill would "subvert any ruling this summer by the U.S. Supreme Court" and could at least tie up the issue in litigation for years. The end of the legislative session stalled the bill, but the sponsor claimed he had majority support and said he would try again the next session.[75] Popular opinion changed, however, and social-conservative efforts moved into courts and elections.

After the *Obergefell* decision came down, Kim Davis, the county clerk of Rowan County, Kentucky, had refused on religious grounds to issue marriage licenses to gay couples. She was jailed for contempt by a federal judge and served almost a week before being released on the promise that she would not overturn the licenses issued by her assistants under court order. Davis lost the battle, but she and her husband met with Pope Francis, who reportedly encouraged her to "stay strong."[76]

Obergefell was a big success for gay rights supporters, but opponents of gay marriage were invigorated too. Conservative state legislatures aimed to limit the effects of *Obergefell*—as they did with *Roe v. Wade*, expanding exceptions to the original interpretation of abortion rights.[77] In Alabama, a state senator introduced a bill to end the issuing of marriage licenses by state probate judges, who instead would simply make an official record of a marriage. A bill was eventually passed by the legislature, in May 2019.[78] The reaction played out more subtly in the 2016 presidential election, where Donald Trump won the support of the religious leaders and congregations who opposed *Obergefell*.

A racial difference in sense of well-being also had a part: a Washington Post–ABC News poll in the runup to the 2016 election found 46 percent of working-class nonwhites saying they were better off than previous generations, but only 30 percent of working-class whites saying the same.[79] That year, Trump won predominantly white and religious small towns and rural areas with 62 percent of the vote to Hillary Clinton's 34 percent, while she won minority-dominant cities by like margins. The two candidates split the suburbs more equally. Trump won among white men by a 32-point margin nationally, but by 48 points in rural areas (72 percent to 24 percent). Among white women, he won by a 10-point margin nationally (53 percent to 43 percent), but by 28 points in rural areas.[80]

Eight in ten self-identified white, born-again/evangelical Christians voted for Trump, while just 16 percent voted for Clinton. White Catholics supported Trump over Clinton by 23 points (60 percent to 37 percent). Hispanic Catholics, religious "nones," Jews, and other religions were strong Clinton supporters, although Trump did do better than Mitt Romney had done in 2012 among Hispanics, Asians, and other non-African-American minority voters. Weekly churchgoers backed Trump over Clinton, 56 percent to 40 percent, while those who attended more sporadically were closely divided, and those who did not attend at all backed Clinton by a 31-point margin. Similar divisions appeared in congressional elections.[81]

Despite these divisions along demographic lines, Fareed Zakaria, a columnist of Indian Muslim background, suggested that the election of Donald Trump in 2016 may surprisingly have been a victory for equality. How? Zakaria observed that many segments of American society are put at a great disadvantage by the "highly-efficient meritocracy" that has developed over the past few decades, centered in major cities. A merit-based society has many benefits, but people with lower skill sets cannot compete as equals. The urban meritocracy know little about the lives of the rural white poor whose struggles might be blamed on their own limitations—for refusal to pursue more education, or for poor work habits. While the minority-race poor often live closer by and can arouse their sympathy, urban elites tend to assume that lower-class whites in rural and small-town communities are lazy and bigoted. Conservative elites are isolated from the rural white poor just as liberals are, though at least they "demonstrate respect by identifying with them culturally, religiously and emotionally." But liberal elites' ignorance of poor whites and disdain for them created an opening for Trump to appeal to them.[82]

What seems clear is that division on various equality issues will continue. Equality will remain an ideal with broad public support, but not the only ideal that Americans value. Debate will continue over equality and its exceptions, and whether antidiscrimination mandates should be broadened or narrowed. But those rules and exceptions could be forced through by raw political and governmental power.

When Equality Worked

Is it possible to establish equality as the undisputed central principle of society? Some may value other ideals that compete with it. But could a modern nation be rationalized in such a way that equality really trumps all else?

To see how this might work—and actually has been done in practice—there is no better place to begin than *Secondhand Time: The Last of the Soviets*, by Svetlana Alexievich. The Belarusian author wrote her book in Russian, and it won her the Nobel Prize in Literature in 2015. In fact, it was the scores of former Soviet citizens she interviewed who basically wrote the book, but Alexievich had the genius to draw them out and explain their thinking with unprecedented clarity.[83] The Nobel committee hailed *Secondhand Time* as "a new kind of literary genre," describing it as "a history of emotions—a history of the soul." It truly is about emotions, especially the later chapters, but it is much deeper. Even the secular Scandinavians used the proper term, "soul."

Alexievich begins by describing herself, her family, and most of those she knows in the different nations of the former Soviet Union as *sovoks*—people who were nurtured from birth on socialism. That Communist ideology was internalized and it molded every person's "soul," although some of those she interviewed were more bound to the "Soviet ideal" than others. After the USSR officially dissolved on December 26, 1991, these individuals had differing views of the new regimes, but almost all tended to have fond memories of the old one. All had been shaped by the socialist policies of Joseph Stalin, and to some degree they all missed him.

Within this general set, Alexievich identifies basically two types of survivors of the Soviet era: those who emerged full, and those who came out empty. The full ones emerged with a moral code, remaining true-believing *sovoks*. The empty ones ended up with no moral structure at all.[84]

Those who came out of the Soviet system empty were the shellshocked casualties of its breakup. They simply could not cope with the new order. In the past, work had been (in memory, at least) pleasant and undemanding, but now it was hard to find, and stressful when it was available. Pay used to be automatic and relatively equal, but now it had to be earned. Wages were much lower in the old days, but there was time left for family around the kitchen table, and for reading books—almost all the interviewees mention books and kitchen tables. Hospitals were dirty, and it usually took months to get to the top of the waiting list, and some people never did, but all Soviet citizens had a right to be on the list. Everyone complained about the Soviet order, but only at home. Nothing worked well, but that was just how things were. Life was hard, but everyone was equal.

Then all those certainties were shattered. In the old system, said one woman, the workplace was where she and her friends did all their socializing; the work was an afterthought, and everyone received the same pay anyway. In the new capitalist system, money was front and center, and everyone was grubbing for jobs, food, and trifles. More freedom produced more goods, but also more work. Asked why he would not take freely available property, one man replied, "Who's going to break their back working the land?" The old stipends and pensions were meager but usually enough to live on. Even when people were released from the prison camps, they received pensions.

And there was pride. All Russian literature was about historic heroes and wars, and the pinnacle was winning the "great patriotic war," World War II. It proved to the world that the USSR was the best. The anniversary of the defeat of the Nazis is still the top annual holiday today. When the Soviet archives were opened under perestroika, people were shocked that Vladimir Lenin had not only personally encouraged murder but insisted that the murders be made public so all Russians would "tremble with fear," and yet most *sovoks* considered it unpatriotic to have made the crimes of Lenin and Stalin public knowledge.

When the controls were loosened, the resulting freedom tended to be met with fear rather than rejoicing. There was no instruction manual for going from Stalinism to free markets, and many mistakes were made. Moscow and a few other cities were more cosmopolitan than most of the old USSR, and the advantages of freedom accrued mostly to that city, and to ethnic Russians, criminal gangs, and oligarchs. Ordinary people lost their supposed guarantees. Inflation raged, and savings, pay, and pensions lost much of their value. The most hated man in a remarkable number of the interviews was Yegor Gaidar, the freedom-loving capitalist deputy of Boris Yeltsin. Gaidar was reviled even more than the brigands and oligarchs who gamed the system to accumulate great wealth. Almost all interviewees expressed disgust for the wild capitalist rush for cash and things. There was more sausage for all—many an interviewee mentions sausage. Their bellies were full, but their lives were empty.

The second group that Alexievich identifies came out full, possessing a moral code. They were still attached to the Soviet ideal and believed that its promised equality had been delivered. The Communist principle of equality was taught in every classroom from the earliest years to university, constantly on media, in art, theater, TV, cinema, and every other cultural medium. The teaching worked so well that everyone took the ideal to be the reality. Under the Soviet regime there had been no rich people to be envied and hated. Everyone was equal. Now there was greater plenty, but it was no balm. The interviewees scorned abundance as materialistic, and they felt offended that it was unequally distributed.

Shockingly, even those who were sent to the Gulag camps believed that the USSR was delivering on the promised equality—and still believe it today. One such person had formerly worked as an executioner in the camps. He tells of developing hand cramps from all the shooting he did—a repetitive stress injury so commonplace, he said, that the government had to devise a special therapy for it. He was later sent to war, as cannon fodder. After he miraculously survived, he was sent to the camps as a prisoner because he knew too much about what happened there. He endured ten years of incredibly brutal treatment, but ended up being thankful to Stalin for allowing him to survive and for the small pension that he got after his release! He praised Stalin for his true commitment to socialism, for rendering equal treatment to everyone, and hoped for a return to Stalinism.

He was not alone. Many innocents suffered frequent beatings, extreme cold and heat, backbreaking labor, and near-starvation in the Gulag, and then were so grateful to have survived that they were thankful ever after for being released. Decades later, they still loved Stalin for setting them free from the bondage he had imposed!

Those who say true equality is impossible are wrong. The Soviet Union, with its fabulous bill of rights, terrorized all equally. No one was exempt from persecution except the chief terrorists themselves—Lenin, Stalin, and their successor chiefs, who terrorized all below them equally. "Uncle Joe" killed his closest associates, right up to those in the Politburo. The highest official and the most innocent person in the USSR were as likely to be sent to the Gulag as the worst criminal.

While most of the interviewees were ordinary people, Alexievich did get one former high official of the Communist Party's Central Committee to speak. "N" (who requested anonymity) began his interview reflecting on Sergey Fedorovich Akhromeyev, whom he regarded as the only real hero to have emerged from the Soviet experience. Akhromeyev had been a marshal of the Soviet Union, chief of the General Staff, Order of the Patriotic War, hero of Afghanistan, and adviser to the president of the USSR, Mikhail Gorbachev. He was found dead in his Kremlin office on August 24, 1991, hanging from a window frame clad in full military regalia. He had written to Gorbachev to explain that he had supported the recent attempted military coup because he feared that the Soviet Union was doomed under perestroika. Five suicide notes in his own hand revealed that it took at least two efforts to get his suicide right. His notes say he wanted to show history that there was some "resistance to the death of such a great state," even though he expected the coup to fail.

Akhromeyev was the only idealist with the moral courage to oppose Gorbachev openly, said "N." As for the rest, "What kind of putsch was this, with no shots fired? The army beat a hasty retreat out of Moscow, like cowards." The Soviet Union could have lasted indefinitely, "N" insisted. It was a superpower with nuclear weapons and a society tough enough to weather nuclear war, unlike the spoiled inhabitants of the capitalist West. "The Russian soldier is not afraid to die," he said. The Soviet regime was invulnerable to threats from the outside and from below. Stalin had made a system that was vulnerable only from the top. Then Gorbachev came along with his misguided program of reform, which doomed the USSR. And he resigned without a fight! According to "N," this was not a surprise, since Gorbachev was the only Soviet leader with no experience of war or ethnic repression. He was soft.

The interviews in *Secondhand Time* suggest that violence works as a way of controlling a society, but it must be total. And it must make freedom look more frightening than oppression. Nazism terrorized its enemies but not its supporters equally, and it didn't last long, nor was it remembered favorably. Islamic State extremists are comparative pikers, killing a proportionately small number of their enemies (so far) but not their friends. The Roman emperors were ruthless, but normally one had to violate some law to call down the violence of the state. Only the Soviet state targeted everyone equally, and it was loved. Russians today still rate Stalin the "most outstanding" world leader.[85]

Alexievich concluded that "communism had an insane plan: to remake the 'old breed of man,' ancient Adam. And it really worked. Perhaps it was communism's only achievement." Communism created a populace that would love the regime no matter what.

By contrast, the modern capitalist world is unloved. There are few countries that still represent Smithian pluralist capitalism, and they are becoming more rationalized and less free.[86] Capitalist countries are seen by many of their inhabitants as hopelessly riven by inequalities of class, race, gender, political tribe, and so on. The United States has implemented rational plans for achieving social and economic equality, with great expenditures of wealth and talent. The result has been all kinds of exceptions, special treatment for one group or another, continuing inequality, constant grubbing for jobs, little time for family or books, and broadly based resentment at the unfairness of it all.

Creative Destruction of a Fragile World

Pope Francis's encyclical *Laudato Si'* ("Be Praised"), issued in 2015, presents an especially comprehensive and searing critique of modern capitalism.[87] It garnered wide acclaim in the international media, both from secular publications like the *New York Times* and from religious progressives like E. J. Dionne, while drawing a scathing response from free-market libertarians.[88]

Contrary to most media coverage, the document is not primarily about climate change, but something much more fundamental. It does refer to a "very solid scientific consensus" which "indicates that we are presently witnessing a disturbing warming of the climatic system." This warming "would appear" to be accompanied by more "extreme weather events" as well as a rising sea level. Acknowledging uncertainties about causation, the pope noted that "a number of scientific studies indicate that most global warming in recent decades is due to the great concentration of greenhouse gases... released mainly as a result of human activity." (§ 23) Francis worried that the "worst impact" of climate change "will probably be felt by developing countries in coming decades." (§ 25)

The pontiff's use of qualifiers like "indicate," "appear," "most," and "probably" demonstrates that this is not a polemic but a reasoned judgment. His focus was indeed more broadly on what he called threats to our "fragile world." (§ 78) Modern technology used in the service of "business interests and consumerism" is devastating our planet and wasting its resources. (§ 34) Damage to the natural environment and to the social environment are both caused by the same "evil": the belief that "human freedom is limitless." (§ 6) Reflecting its true moral tone, the encyclical is subtitled "On Care for Our Common Home."[89]

The pope described a zero-sum world with limited resources. The richer people of the world, especially in the developed countries, consume luxuries at the expense of the poor majority, mostly in the underdeveloped world, who will have increasing difficulty obtaining necessities.

> We all know that it is not possible to sustain the present level of consumption in developed countries and wealthier sectors of society, where the habit of wasting and discarding has reached unprecedented levels. The exploitation of the planet has already exceeded acceptable limits and we still have not solved the problem of [world] poverty. (§ 27)

The pope noted that many "opinion makers, communications media and centres of power, being located in affluent urban areas with substantial incomes, are far removed from the poor." They know little about the needs of the majority of the world's population. Francis rejected the idea of bringing the rest of the world up to the same level of prosperity as the affluent minority. That level of consumption could "never be universalized, since the planet could not even contain the waste products of such consumption." (§§ 49–40) So the pontiff was more concerned about unmanageable waste than about climate change.

To care for the planet, fundamental changes will be required in production and consumption, in government and in lifestyles. Even when global summits resulted in commitments to reduce greenhouse gas emissions, Francis was extremely critical, finding it "remarkable how weak international political responses have been." Richer nations have not been willing to curb the "harmful habits of consumption" that are spreading around the world, exhausting its resources. The pope singled out the increasing use of air conditioning as an example and said that stimulating demand for such appliances is "self-destructive." (§§ 54–55)

The early nineteenth-century English priest and academic Thomas Malthus was likewise concerned about exhausting the world's resources. He worried that natural growth in population would always outpace material production, finally leading to global poverty—although a "great law of necessity" would then decrease population naturally, so the people remaining could be provided for.[90] Malthus's modern proponents have thought population growth can be controlled by abortion and contraception, but these were morally unacceptable to the pope. The only logical and moral solution for him was a lower average standard of living around the world. But he would first redistribute wealth from the richer nations to the poorer, so the funds now spent on such luxuries as air conditioning would be reallocated to feed and house the world's poorest.

Laudato Si' sets out an agenda more sweeping than the demands of the most radical Green movements. The pope is fearless. There is no free lunch. If the Greens are right in their analysis of a climate crisis, and if infanticide is not an option, then the richer people of the world will have to sacrifice their standard of living and "leave behind the modern myth of unlimited material progress." (§ 78)

Francis does not mention it but, as noted earlier, 80 percent of the official poor in the United States tell their government that they have air conditioning and in fact are rich by world standards.[91] The poorer, less developed countries should not be punished with "burdensome commitments to reducing emissions comparable to those of the more industrialized countries." (§ 170) Poor countries "are bound to develop less polluting forms of energy production," but will need the help of richer countries in doing so. (§ 172)

The pope noted that a variety of material approaches to the global environmental problem have been presented, ranging from the libertarian "myth of progress that ecological problems will solve themselves" through technology, to the extreme-left view that the "presence of human beings on the planet should be reduced." Any "viable" solution would need to be somewhere between those extremes. The church "has no reason to offer a definitive opinion" on specific solutions, but would encourage "honest debate... while respecting divergent views." (§§ 60–61) It is not enough to rely on material approaches alone: "no branch of the sciences and no form of wisdom can be left out, and that includes religion and the language particular to it." Solving the global problem will require a synthesis of "faith and reason." (§ 63)

Most important, it is necessary to abandon "the idea of infinite or unlimited growth, which proves so attractive to economists, financiers and experts in technology." That idea "is based on the lie that there is an infinite supply of the earth's goods, and this leads to the planet being squeezed dry beyond every limit." (§ 106) This lie must first be confronted morally, by a change in thinking: "There needs to be a distinctive way of looking at things, a way of thinking, policies, an educational programme, a lifestyle and a spirituality which together generate resistance to the assault of the technocratic paradigm." (§ 111)

The pope made clear that we must not deny the "special value" of human beings in our concern to save the planet. "There can be no ecology without an adequate anthropology," he wrote, stressing the importance of steering between the extremes of anthropocentrism and biocentrism. "Human beings cannot be expected to feel responsibility for the world unless, at the same time, their unique capacities of knowledge, will, freedom and responsibility are recognized and valued." (§ 118) Technological progress should not replace human work, nor should financial assistance be a permanent solution to poverty. "Work is a necessity, part of the meaning of life on this earth, a path to growth, human development and personal fulfilment. Helping the poor financially must always be a provisional solution in the face of pressing needs. The broader objective should always be to allow them a dignified life through work." (§ 128)

Business leaders are criticized for selfishness, for a ruthless pursuit of profit, for insensitivity to the needs of their workers and the poor. Yet the necessity of employment leads Francis to make a rare concession to capitalism, saying "it is imperative to promote an economy which favours productive diversity and business creativity." This does not mean laissez faire, however. "To ensure economic freedom from which all can effectively benefit, restraints occasionally have to be imposed on those possessing greater resources and financial power." (§ 129) Even if there is "sustainable development," there must also be efforts at "containing growth by setting some reasonable limits and even retracing our steps before it is too late." (§ 193)

The pope espoused a concept of the common good based on "respect for the human person as such, endowed with basic and inalienable rights." The welfare of society as a whole requires honoring these individual rights, and developing "a variety of intermediate groups, applying the principle of subsidiarity," from the family up to the state and beyond. (§§ 156–57)

What Pope Francis envisioned is "one world with a common plan." (§ 164) Changes in lifestyle are not sufficient. "Enforceable international agreements are urgently needed, since local authorities are not always capable of effective intervention," Francis insisted. States "must be respectful of each other's sovereignty," he granted, "but must also lay down mutually agreed means of averting regional disasters which would eventually affect everyone." (§ 173)

Frustrated by the lack of progress, the pope quoted his immediate predecessor, Benedict XVI, who in turn cited John XXIII, on the "urgent need of a true world political authority." But Francis actually envisioned one with more extensive powers to "manage the global economy; to revive economies hit by the crisis; to avoid any deterioration of the present and the greater imbalances that would result; to bring about integral and timely disarmament, food security and peace; to guarantee the protection of the environment and to regulate migration…." (§ 175)

He again qualified his ideal by allowing that "there are no uniform recipes, because each country or region has its own problems and limitations." He conceded that "political realism may call for transitional measures and technologies, so long as these are accompanied by the gradual framing and acceptance of binding commitments." (§ 180) Decisions must be made "based on a comparison of the risks and benefits foreseen for the various possible alternatives." (§ 184) There must also be "greater attention to local cultures," involving "a dialogue between scientific-technical language and the language of the people." (§ 143)

As a pope, Francis must insist that the people of the world freely meet the highest moral standards, and be "selfless" in caring for creation. "Only by cultivating sound virtues will people be able to make a selfless ecological commitment. . . . There is a nobility in the duty to care for creation through little daily actions, and it is wonderful how education can bring about real changes in lifestyle." (§ 211) What is needed is a literal "conversion" into "a number of attitudes which together foster a spirit of generous care, full of tenderness." At the top of the list is "gratitude and gratuitousness, a recognition that the world is God's loving gift, and that we are called quietly to imitate his generosity in self-sacrifice and good works." (§ 220)

So it is not clear precisely how the pope's moral plan to rationalize capitalism and care for the planet should be achieved. It must be a "common plan," yet somehow respecting local cultures, and synthesizing religious and secular wisdom, state power and moral conversion.[92] What was clear to him was that the world is fragile and needs protection.

Pope Francis offered a fundamental critique of capitalism from an obviously well-intentioned perspective, but his general vision for addressing a global problem goes far beyond the reform of individual morality. It calls for governmental plans to restrain market capitalism—to limit the freedom of businesses to produce and of individuals to consume and make self-interested choices. This challenge to the morality of pluralist capitalism deserves a serious response, beginning with the question of how any such binding common plan could be implemented.

CHAPTER 4

THE EXPERT BUREAUCRACY SOLUTION

Since Pope Francis has advocated for a comprehensive plan to address important global problems, let us evaluate the basic institutions of government that would have to implement such a plan. The great sociologist Max Weber identified three basic types of government historically: cosmological governments such as monarchies, charismatic leadership such as by Napoleon or Juan Perón, and the legal-rational bureaucratic nation-state. In his *Theory of Social and Economic Organization* (1920), he argued that the third type was most appropriate in modern times:

> Experience tends universally to show that the purely bureaucratic type of administrative organization...is, from a purely technical point of view, capable of attaining the highest degree of efficiency and is in this sense formally the most rational known means of carrying out imperative control over human beings.[1]

If Weber was correct, then a major transformation of human society, as envisioned by Pope Francis, must rely upon the rationalized, bureaucratic administrative state.

Rationalizing Government

As early as the eighteenth century, France, Prussia, and other European countries, even Britain, had become expert-led bureaucratic states. The United States lagged in this trend, but American opinion leaders were impressed by the success of bureaucratic rationalization in Europe. The scholar who became president, Woodrow Wilson, took the lead in promoting expert administration.[2] By the time of Franklin Roosevelt's presidency, expert-controlled central bureaucracy had become the dominant force in U.S. policy-making. The political success of the New Deal made expert bureaucracy the undisputed governmental means for rationalizing capitalism.

The reader should be alerted that your author has some skin in this game. My Ph.D. in political science involved comprehensive study of Weber and Wilson, and experience in Weber-inspired public adminis-tration. This led to writing articles and books on the subject of public administration as a professor at the University of Maryland, and then to appointment as director of the U.S. Office of Personnel Management (OPM) in 1981, to head the U.S. civil service bureaucracy.

The civil service ideal as stated in law rests on six principles: 1) Recruiting, selecting, and advancing employees on the basis of their relative ability, knowledge, and skills. 2) Training to assure high-quality performance. 3) Retaining employees on the basis of their performance, correcting inadequate performance, and separating employees whose inadequate performance cannot be corrected. 4) Providing equitable and adequate compensation. 5) Assuring fair treatment in all aspects of administration without regard to political affiliation, race, color, national origin, sex, or religious creed. 6) Assuring that employees are protected against coercion for partisan purposes and are prohibited from using their official authority for the purpose of interfering with or affecting the result of an election or a nomination for office.[3]

Jimmy Carter, the Democratic governor of Georgia, ran for presi-dent in 1976 declaring his support for the New Deal programs and their Great Society expansion under Lyndon Johnson (Democrat) and Richard Nixon (Republican), but criticized the way they were being administered. The policies were not actually reducing poverty, increas-ing prosperity, or improving the environment, he argued, and to make them work required fundamental bureaucratic reform. He correctly charged that the operation of the bureaucratic system did not even meet the principles of the law, especially because almost all government employees were rated as "successful," and received the same pay regardless of performance. Even the worst were impossible to fire.[4]

President Carter fulfilled his campaign promise with the Civil Service Reform Act of 1978, which closely adhered to the original merit principles.[5] A new performance appraisal system was devised, generally with five rather than three rating categories. It specified individual goals related to agency missions that were to be used as the basis for promotion, benefits, and discipline. For midlevel managers and senior executives, superior work would lead to promotion, sizeable bonuses, and merit pay increases. But time ran out on Carter's administration before the system could be fully executed, so it was left to the following president—and to myself, previously given planning responsibility for personnel policies by John Sears, an early campaign manager—to implement the law.[6]

The normal opposition to change in a bureaucracy the size of the U.S. federal government made these reforms controversial, to say the least. There were aggressive public-sector union so-called job actions, slowdowns, and finally a strike by the Professional Air Traffic Controllers Organization. President Ronald Reagan's firing of the controllers for striking illegally won public support for reform, although efforts to expand merit pay to lower-level workers were blocked by Congress. Still, the best metrics suggested that the enacted reforms worked—for a short while. Yet by the end of the George H. W. Bush administration, there was little appetite for confronting the unions. Appraisals had returned to near-universal "satisfactory" performance ratings, manager merit pay was eliminated, and bonuses in many agencies were being divided between all executives rather than being paid to the best performers. The Carter-Reagan reforms were dissipated in a decade.[7]

Today, employee evaluation looks much like it did in the pre-Carter era. In 2015, for example, it was disclosed that the Veterans Health Administration executives who encouraged false reporting of waiting times for hospital admission were rated "outstanding" and received bonuses. The Senior Executive Association justified it by telling Congress that only outstanding performers would have been promoted to the Senior Executive Service in the first place.[8] The Government Accountability Office reported that pay raises for regular employees were automatic, mostly not based on performance. Within-grade quality steps and awards were supposed to be based on performance, but most employees were rated at similar appraisal levels, so this rule had little real effect.[9] President George H. W. Bush's OPM director, Kay Coles James, did require evaluations to be distributed across several rating categories, but 60 percent of employees and 89 percent of career executives were still rated in the top two categories.[10]

In 2013, by one estimate, only 9,513 of 2,054,175 federal employees were dismissed from their jobs, a rate of 0.46 percent, compared with 3.2 percent in the private sector.[11] The official GAO found that the federal government dismissal rate was only 0.15 percent, and those dismissed were generally low-level probationary personnel. Since the removal statistics include separations for abusive and even criminal behavior, the rate of separation for poor performance alone was actually much lower.[12] Managers who tried to remove poor performers had to face multiple levels of review in their own agencies. Employees could make additional forum-shopping appeals to the Merit System Protection Board, the Federal Labor Relations Authority, or the EEOC, and even up to the U.S. Court of Appeals. Understandably, few managers were willing to fight this byzantine appeals system.[13]

At the U.S. Census Bureau, its inspector general found that one employee manipulated the system over many years to collect $1.1 million in improper salary, and two years later the case was still unresolved. After a three-year investigation, the Trademark Office moved to suspend eighteen employees who had charged for eight hours a day but worked only two hours, and charged for overtime that was not actually performed, costing the government hundreds of thousands of dollars. After three years, no penalties had been imposed.[14]

Your author was actually present when the government abandoned the merit recruitment exams that had been the basis for creating a meritocratic bureaucracy. As the Carter presidency was winding down in 1980, the Department of Justice and OPM lawyers colluded in a "consent decree" with private lawyers friendly to civil rights activists to end civil service intelligence tests, claiming they were discriminatory because minorities scored lower, but not giving proof of any specific problems with the tests themselves. As head of the civil service agencies transition team, I was unable to convince our lawyers to contest the decree, so the court ordered an end to IQ exams for a decade. The ban has been extended several times in different ways and is still mostly in force.

Obviously one cannot run an institution—even a government, which doesn't need to be profitable—without some way of assessing applicants for jobs. So merit exams were replaced with a system called USAJobs, based on resumés, recommendations, and a self-assessment questionnaire. Of course, no one failed a "test" that consisted of assessing one's own skills and qualifications. But there was a suspicion that honest applicants were penalized for evaluating themselves more accurately, while clever insiders discovered the key words that enhanced their odds of being hired. Such "exams" did little to narrow down the search for good employees, so agencies that were serious about finding well-qualified people were overwhelmed by applications they could not adequately evaluate.

How did the poor bureaucrats deal with this reality? They selected people they knew. If they could not use a merit test, what else could they do? In fact, most vacancies in the mid to upper level of the federal government are filled through what were dubbed "name requests." It is a spoils system, but one favoring bureaucratic acquaintances rather than political allies. Nepotism is still illegal in Washington, but promoting a friend's relative or buddy if she will do likewise for you could get past the law. Who would discover the plan—or even really blame you?[15]

A few years ago, the GAO surveyed the work evaluation documents of 1.2 million federal employees and announced that 99.6 percent were rated fully successful performance or above. A mere 0.3 percent were rated minimally successful, and 0.1 percent unsuccessful.[16] J. David Cox Sr., president of the largest federal government union, the American Federation of Government Employees, claimed that this remarkable performance was achieved because government workers were in fact that good: "Could it be that the careful scrutiny and open competition among applicants for federal jobs produces a workforce of highly skilled, responsible, productive employees?" he asked. "If agencies take care to hire the best possible people and in turn almost all of those people do at least everything they are asked to do in a satisfactory manner, where is the problem?" William Dougan, president of the National

Federation of Federal Employees, was a bit more modest, saying he thought the percentages were "a little high."[17]

The union officials were more confident in current recruitment practices than was the agency in charge. In 2015, OPM announced that it was planning to introduce a merit exam called USAHire, which it had been quietly trying out in a few agencies for a dozen job descriptions. The tests had multiple-choice questions with only one correct answer. Some questions required essay replies, and they would be changed regularly to minimize cheating. Although OPM deserved high praise for this audacity, the implementation of USAHire had been limited for years because of a fear that some groups would not do as well as others, and the government would be forced once again to scrap tests that truly measure intelligence and ability.[18] Undeterred, President Trump ordered that USAHire be expanded to more agencies.[19]

There does not seem to be another President Carter on the horizon who would be able to rewrite the law and its regulations to return the old civil service principles to the management of the federal bureaucracy. Past experience suggests that such an effort would not go very deep or last very long, in any case.

What matters, of course, is not the institution itself, but the results. One measure of the federal bureaucracy's effectiveness is that faith in government has become seriously attenuated. In the 1970s, public trust of "the government in Washington" was usually in the 70 percent range.[20] In 2015, the Pew Research Center found that 22 percent of the American public were "angry" at the U.S. government, 57 percent were "frustrated," and only 18 percent were "basically content" with how it worked. Loss of confidence was bipartisan: 89 percent of Republicans and 72 percent of Democrats said they could seldom, if ever, trust the federal government. Six in ten believed the government needed "very major reform," up from 37 percent in 1997. Forty percent of Democrats and 75 percent of Republicans said that government is almost always wasteful and inefficient. Just 42 percent of Democrats and 30 percent of Republicans believed the government did well in lifting people out of poverty.[21] In 2018, Pew found that only 35 percent of the public had a favorable view of the federal government.[22]

Even in the world's most advanced and richest nation, the bureaucratic government could not realize its own rational principles or hold the confidence of the public. Such an institution would seem to be a weak instrument for carrying out the great task that Pope Francis had in mind. But it is necessary to look specifically at the policy results to make a full evaluation.

Keeping Government Simple

The challenge of running a national government efficiently is enormous. The U.S. bureaucracy operates in 25,000 separate installations worldwide, administering thousands of different programs, organized into 700 to 800 different management systems. In the larger agencies, top executives direct employees perhaps down through fifty levels to where the frontline work is done. Programs overlap across multiple agencies: there are 70 bureaus in six different agencies to feed hungry children; 105 programs in nine agencies to support math, reading, and science education; and 75 different work programs. The nation's top expert on public administration, Paul Light, concedes that this multilayered complexity makes it extraordinarily difficult to faithfully execute the nation's laws.[23]

The largest programs in the government are the popular entitlements such as Social Security, Medicare, Medicaid, military and civil service pensions, and the like. These programs pursue efficiency by keeping things simple, primarily sending checks to repeat customers. Even in these programs, losses can be significant, including about $80 billion annually in improper payments and waste in Medicare and Medicaid alone.[24] The sum sent out in checks to individuals has increased over the years from 56 percent of the total U.S. budget in 1994 to 70 percent in 2014. In that year, 49.2 percent of U.S. households were receiving federal benefits.[25] In addition, checks were regularly sent to 18 million private consultants and contractors.[26] A majority of government offices in Washington avoid market complexity by simply managing paper and leaving most actual operations to local and private resources.[27] Any more direct activity introduces enormous bureaucratic complication.[28]

Consider a relatively simple operation where the feds do act directly: setting dietary standards so citizens can adopt better eating habits. The recommendations of the Dietary Guidelines for

Americans bureau are produced by experts from the Department of Agriculture, the Department of Health and Human Services, the National Academy of Medicine, the Dietary Guidelines Advisory Committee, policy oversight officials, writing staffs, reviewers and technical assistants, and hundreds of outsiders who make recommendations through the government's comment process. The *Dietary Guidelines for Americans, 2015–2020* has 134 pages including numerous tables and charts, references to scores of outside studies, plenty of footnotes, and fourteen appendices.[29]

The problem is that the question of optimal diet is very complex but the government must make simple recommendations. One of the "Key Recommendations" in the 2010 edition of *Dietary Guidelines for Americans* was to limit cholesterol intake to 300 milligrams per day. In the 2015 edition, this warning was removed from the Key Recommendations—though deeper inside it is recommended that individuals "eat as little dietary cholesterol as possible." The Key Recommendations include limiting saturated fat to 10 percent of calorie intake, and to choose dairy products with little or no fat. But the guidelines also say that replacing fats with carbohydrates is not correlated with a lower risk of cardiovascular disease. After forty years of warning against saturated fat, the government now told Americans they could consume eggs, rich foods cooked in saturated fat, and even bacon!

A 2018 study by the American College of Cardiology and the American Hospital Association gave up on general guidelines for cholesterol and recommended only individualized decisions.[30] So what can the consumer conclude? Christopher Gardner, a professor of medicine at Stanford University, explained that knowing what saturated fats actually do in the human body would require studies of a kind that cannot be done in real life. Comprehensive studies would be too expensive, and subjects often "cheat" on what they are really eating. "But you can't have no advice, so this is the best advice from the data that is available and may not be useful," he explained. Gardner wondered whether there should be any government guidelines at all.[31]

Unfortunately, scientific studies often contradict each other. For example, a Columbia University study of an earlier guideline that recommended eating breakfast every morning found that doing so contributed to obesity.[32] Perhaps this disagreement among experts is one reason they are mostly ignored anyway. Americans eat too little of the vegetables, fruits, whole grains, seafood, and beans that the Dietary Guidelines recommend. They eat 70 percent more sugar than they are supposed to. They consume 90 percent more salt than the experts say they should.[33]

Another agency that makes recommendations for Americans' health and well-being is the Centers for Disease Control and Prevention. The CDC is one of the world's most highly regarded scientific institutions, and its protocols are accorded nearly religious reverence. The agency dramatically entered the public consciousness in 2014 when two nurses at a hospital in Dallas, Texas, contracted Ebola, which is often fatal, apparently after properly following CDC protocols in caring for a patient who had gotten the disease in Liberia.

How the Ebola virus had been transmitted to the nurses—and by extension outward into the nation—became the most pressing issue of the day. Several states implemented mandatory quarantine policies, going beyond the CDC's recommendation of voluntary quarantine. President Barack Obama opposed those state policies, saying, "We don't just react based on our fears. We react based on facts and judgment and making smart decisions." We use the "best science," which came from the CDC, and it pointed to relying on public education and only voluntary quarantine.[34]

The official educational document called "CDC Facts about Ebola in the U.S." had four lessons on the front cover, one of which was: "You CAN'T get Ebola through AIR." If one continued on to the question-and-answer section in the document, however, one would read: "Although coughing and sneezing are not common symptoms of Ebola, if a symptomatic patient with Ebola coughs or sneezes on someone, and saliva or mucus come into contact with that person's eyes, nose or mouth, these fluids may transmit the disease."[35] Don't sneezes come through the air?

When President Obama declared early on, "We know how to do it," to stop the spread of Ebola, he was mistaken. We did not know, which he later essentially admitted by appointing a "czar" to lead the Ebola response separately from the CDC bureaucracy. But the czar still needed that same bureaucracy to do the collection of data and make recommendations to him. Adding more complexity, the World Health

Organization announced that new cases of Ebola had ceased but could reoccur because it could hide from analysis.[36]

There have also been more inexcusable failures. The CDC was found to have inadvertently sent a live Ebola sample, rather than a killed virus, to a lab not equipped to handle it—thus threatening a technician's life. The agency had also mishandled samples of deadly anthrax and H5N1 (bird flu), in what was described as "a broad pattern of unsafe practices."[37] An Army laboratory in Utah shipped at least one live sample of anthrax to a Maryland lab, and eight others that were possibly still alive as far as South Korea.[38] (At least it was not to North Korea). At a National Institutes of Health building used by the Food and Drug Administration, a scientist found decades-old vials of supposedly controlled and suppressed smallpox in a storage room.[39]

A relatively small but critical agency, the National Institutes of Health (NIH) has been called "the crown jewel of government-run medical research." Its activity takes place in quiet labs, with an occasional announcement of a possible new cure for cancer or the like. The temple veil is occasionally lifted by an employee. In early 2015, a pharmacist at the NIH Clinical Center reported—not to her boss but to the FDA—that she had seen discoloration in a medicine vial, which turned out to be a fungal contamination. After an FDA inspection, the NIH suspended operations of its Pharmaceutical Development Section until corrective steps could be taken.[40]

Five additional inspections that same month found more compromises of sterile environments. Two pharmacy administrators were advised that they might face dismissal, but eventually they were only reassigned. When NIH informed Congress, the chairman of the Senate Committee on Health, Education, Labor, and Pensions proposed an outside review. A year later, this review found "a lack of compliance" not only in the NIH pharmacies but more broadly. Indeed, the entire Clinical Center "doesn't meet the standards you need to have when dealing with human lives," as the question of patients' safety had become "subservient to research demands." This finding resulted in reassignment of three senior hospital officials, but still no dismissals. Top management at NIH then extended the review to all of its labs, and the pharmacy unit was closed indefinitely.[41]

That should have been the end of the story. But a few months later,

NIH announced that a nurse had discovered "environmental mold" in a mouth-rinse solution being readied for a scientific experiment. Nine bottles were found to hold particles, three with mold. Fortunately, the experiment was stopped before the solution was administered. The mold was traced to the NIH Microbiology Section.[42]

NIH spent $24.5 billion on scientific research in 2016, handing out large sums of public money in grants. The Office of Research Integrity had only eight investigators to conduct oversight of how those grants were used. The office made only ten to fifteen findings of misconduct a year, and had not made a single finding of plagiarism for two years. Given the large number of research projects being supported with NIH funds, it is unlikely that there was so little misconduct and no plagiarism. The former office director had resigned earlier to protest the unit's "profoundly dysfunctional" second-level career leadership compared with the private sector.[43]

In 2013, the inspector general of Housing and Urban Development reported that 25,000 public housing tenants had exceeded the income ceiling for eligibility. In fact, one recipient of public housing earned $497,000 a year.[44]

What might seem to have been a relatively simple task was protecting government IT records. The Department of Homeland Security and the Office of Management and Budget were ordered by Congress to take on this responsibility, and in 1996 they created what was called the Einstein Program. Despite its name, the program unfortunately failed to stop large raids on government records in several agencies, including the CIA, the U.S. Senate, the IMF, and my own old agency, OPM, where 21 million confidential personnel records of government officials were hacked by Chinese spies in 2015.[45]

The Internal Revenue Service made headlines in 2012–13 when it emerged that Lois Lerner, head of the Exempt Organizations Unit, had led an effort to delay applications for tax-exempt status by organizations she disapproved of for political reasons, so that many of them could not operate in the run-up to the 2012 elections. Eventually, the Department of Justice concluded that there was "no evidence to support a criminal prosecution." The assistant attorney general, Peter Kadzik, however, conceded that the IRS "mishandled the processing of tax-exempt applications in a manner that disproportionately impacted applicants affiliated with the Tea Party and similar groups, leaving the appearance that the IRS's conduct was motivated by political, discriminatory, corrupt, or other inappropriate motive." But he chalked up this "appearance" of discriminatory or corrupt motives to "ineffective management," which "is not a crime."[46]

It was, however, ineffective management and probably worse. Lerner was revealed to have referred to certain applicants as "crazies" and "a-holes" in emails. But she was awarded $129,300 in bonuses before it all became public. In spite of a subpoena, 400 backup tapes were erased and 20,000 Lerner emails were destroyed somewhere within the bureaucracy, though the inspector general later recovered 700 tapes and a thousand Lerner emails.[47]

The Treasury Department's inspector general reported in 2015 that 1,580 IRS tax agents had willfully violated confidentiality laws over the past decade, and only 39 percent of those agents were fired.[48] That same year, it emerged that over 610,000 tax files had been hacked by outside organizations, potentially to become public.[49]

Was the answer to these management problems more government? When President Obama's health-care reforms were being criticized as inefficient, Paul Krugman of the New York Times said that the problem with Obamacare was that it did not go far enough toward full government control of health care. He argued that the total government control at the VA was preferable because it was "socialist."[50]

Unfortunately for him, Krugman wrote just before a VA waiting-list scandal. Only a fifth of veterans actually used its services, yet even with low demand, the VA for thirty years had been queuing veterans. The agency's budget had doubled under Obama, but demand for free services without market costing always exceeded supply. The only nonmarket solutions to prioritize resources were to issue vouchers for specified services within the VA or outside—an idea that VA unions and service associations have opposed—or to create waiting lists, which is what the VA managers did.[51] Funding was increased again under President Obama, and some reforms were adopted under President Trump, along with still more funding, but the basic structure remained in place.[52]

Government administrators are often in pretty much the same place as the secretary of energy was during the BP Horizon oil spill in 2010, when he announced that the federal government would "take over" if BP did not act more expeditiously to clean it up. He was publicly corrected by the Coast Guard commander in charge, who was standing next to him and stated that the government did not have the equipment or expertise to do so. Only the private capitalist BP company did.[53]

Even the government agency whose charge is protecting the environment has ended up being a polluter. In Colorado, the Gold King Mine, abandoned in 1923, was long a source of concern because of acid drainage into the Animas River. The state plugged all the mine's portals and installed drainage pipes, but it was an unstable situation. The Environmental Protection Agency was called in to fix the problem in 2014–15, but the EPA-directed contractor unintentionally destroyed the plug that was holding water inside the mine, sending 880,000 pounds of heavy metals such as lead and arsenic into a tributary of the Animas. About 3 million gallons of toxic mustard-tinted sludge spilled into a river system extending into New Mexico and Utah.[54]

Managing the Economy

What happens when the task is more complicated, such as managing the entire national economy? The champions of bureaucratic rationalization claim it is superior to pluralist capitalism because it can make the market work better than simply allowing it to seek its own levels. But let us see.

Consider the Federal Reserve System, which has been the top expert manager of the U.S. economy since President Wilson signed the Federal Reserve Act in December 1913. The Fed's prestige is so high that Morgan Stanley's chief global strategist, Ruchir Sharma, called it "the central bank of the world."[55] There is near-universal agreement among finance experts that the Fed's most powerful tool, manipulating the federal funds rate, can keep the U.S. financial system in balance, and through it the whole global system.

When Janet Yellen became chairman in 2016, after eight years of tepid economic growth, she aimed to reassure a nervous investor class that her agency was firmly in control. Yellen announced that the Fed did not need to resort to that ultimate rate tool, but would continue relying on its more modest "quantitative easing" bond-purchasing tool for an indefinite period.[56] At first the market shuddered, but it soon calmed down to the soothing voice of its overseer, who would scientifically control the fragile U.S. and world economies with the current program, keeping the big gun ready in case things got worse.

Brian S. Wesbury and Bob Stein of First Trust Advisors had routinely questioned the Fed's Delphic pronouncements, and explained that there was no longer any ultimate federal funds weapon: "In the past, the Fed manipulated the funds rate by making reserves 'scarce' or 'plentiful.' It withdrew reserves to push rates up and added reserves to push rates down—a simple 'supply and demand' calculation." But once the Fed injected $2.6 trillion in "excess reserves" into the economy to fight the 2007–8 recession, there could no longer be any scarcity. "As long as banks have excess reserves, they do not need to borrow reserves from other banks to meet their reserve requirements," so Fed manipulation becomes irrelevant. The federal funds rate was "an anachronism."[57]

All the Fed can do, the First Trust analysts later noted, is create or destroy bank reserves "by buying or selling bonds, to or from, banks." Rate hikes do not tighten money unless the Fed is shrinking bank reserves. Controlling the rate simply becomes a calming action to quiet the market's animal spirits. "The Federal Reserve has convinced itself, and many others, that it has ultimate control over the economy," but it does not.[58]

Sharma agreed that the Fed's years of "loose policies," building up vast reserves, simply immobilized it.[59] While the Fed eventually did start reducing assets, its proposals were vague, small-bored, and based on current conditions and pragmatic corrections rather than a comprehensive plan.[60] Sharma's research even demonstrated that the stock market was no longer being driven by private market trading but was following political signals from the Fed. He found that the S&P 500 had gained 699 points since January 2008, but "422 of those points came on the 70 Fed announcement days." This was a new form of over-rationalized capitalism. Between 1960 and 1980, Fed announcements had had little market effect, and between 1980 and 2007 those announcements had had only half as much influence as they would subsequently.[61]

When 60 percent of stock market gains have come on days the Fed made a policy announcement, the stock markets have been driven not by capitalist supply and demand, but by investor reactions to mostly misguided governmental calculations.

Earlier, in 2014, the Bank for International Settlements observed that it once appeared that "economic science had conquered the business cycle" after the Great Depression. There had been inflationary recessions in the 1970s and 1980s, but "long expansionist runs" from 1961 to 1969, and from 1991 to 2001. BIS noted, however, that even with multiple billions of dollars of stimulus spending, Federal Reserve policy since 2007 had failed to increase U.S. output, which was then 13 percent below where it would have been if previous growth rates had continued. The pattern was similar in other capitalist countries: output was 19 percent below earlier projections in Britain, 12 percent below in France, and 3 percent below in Germany.[62] In fact, analysts had ignored the fact that bank and treasury stimulus did not work in the 1930s, 1970s, or 1980s either. By 2014, the idea that central banks could actually control economies was seriously in question.[63]

As early as the financial crisis of 2007–8, it seemed clear that those in charge at Treasury and the Fed were operating each day by the seat of their pants, without any guiding plan, in the face of enormous market complexity.[64] Indeed, all three top managers of the crisis— Ben Bernanke, Timothy Geithner, and Henry Paulson—would later essentially admit as much in their book on the subject, *Firefighting: The Financial Crisis and Its Lessons.* As the title suggests, they conceded that they had simply been putting out fires on the spot, with no comprehensive, rationalized plan to direct the markets.[65]

Casey B. Mulligan, a professor of economics at the University of Chicago, argued that the unintended effects of firefighting based on instinct might even have made things worse. He maintained that the incentives to work, earn, and recover had been reduced by the stimulus programs, the six-year "emergency assistance" for the long-term unemployed, the expansion of the food stamp program, the mortgage assistance programs, and the like. The result was the lowest labor-force participation rate in thirty years and a historically slow recovery.[66]

By 2020, the former head of the Fed's research and statistics section, David Wilcox, conceded that "long-term interest rates are already too low by historical standards for [the Fed's] other tools to be fully capable of delivering the punch that will be needed to meet the next recession."[67] Economists from the San Francisco Fed and the University of California looked at all funds-rate increases in the United States, Germany, Japan, and Britain between 1955 and 2018, and concluded that economies are actually driven by "factors outside policymakers' control."[68]

In the face of the 2020 coronavirus work shutdowns, the Fed was encouraged by almost all experts to take bold action on the funds rate to stimulate the economy. It then took what the media called its "most dramatic action since 2008," followed the very next day by the largest drop in stock prices to that date. "The Federal Reserve is now effectively spent," wrote the chief economics commentator for the *Wall Street Journal*.[69] That only left the Fed offering more loans to private businesses, which was actually a less objectionable central-bank function. The other major response was a $2-trillion-plus stimulus that had positive aspects, but no serious economist could still believe that this historically high spending—swelling the already massive debt—would actually revive the economy even if people were able to go back to work, at least temporarily.[70]

Training for Productive Work

The federal government has recently spent a trillion dollars or so every year on social welfare, including $500 billion on Medicaid for one-quarter of the population, $270 billion on family and child welfare, $37 billion on unemployment benefits, and $50 billion on housing.[71] Indeed, one-fifth or more of the American population have received a means-tested government welfare benefit.[72] Yet, in recent years, one-third of Americans had not regularly participated in the labor force.[73] And we simply do not know how to run an economy when too many people aren't working, as the recent Covid-19 pandemic demonstrated.

In more normal times, one problem is that minimum-wage laws, intended to increase workers' pay, tend to reduce employment opportunities. One study found that "a 1% increase in [minimum] wages leads to a 0.3% to 1% decrease in the employment rate depending on whether wages increase citywide or in only one industry." Cities that increased their minimum wage to $15 per hour—Los Angeles, Seattle, and San Francisco—were estimated to see decreases in employment rates ranging from at least 1 percent to 3 percent or more (depending on how much of the city's labor force is affected). Only time will tell the specifics but history suggests there will be some negative effects.[74]

More women than men fall below the poverty line (14 percent versus 10 percent for men), especially during the childbearing years and in old age.[75] More women than men collect welfare benefits, though 40 percent of the women in family and child welfare programs do have jobs, as do a majority on Medicaid. Few regard this situation as ideal.[76]

Giving up welfare benefits in favor of work, unfortunately, carries an enormous cost in marginal tax rates. "Individuals who are deciding whether to enter or leave the labor force probably compare the total amount of income—after taxes and after transfers—that they would receive without working to the amount they would receive from working part time or full time," noted a study by the Congressional Budget Office in 2012. According to the study, the marginal tax rate associated with taking part-time work is 36 percent, while the marginal rate increase for accepting full-time work is 47 percent.[77] Another CBO study in 2017 confirmed the effect of taxation in reducing the incentive to choose work over welfare.[78] Work requirements for receiving welfare do increase employment rates, but what is legally defined as "work" for this purpose can include educational courses or community service, which do not necessarily lead to better-paying jobs with more productive or fulfilling work.

The standard answer is more training programs to provide the skills for better jobs. The first federal job training program began in 1933, and many more have followed, with limited success. Today, the federal government has forty-seven different job training programs spread across fifteen agencies, costing $18.9 billion in 2019. A study of their effectiveness by the Council of Economic Advisers concluded that "among the training programs with available data and rigorous impact studies, the evidence showed that most government training programs are not effective at securing higher paying jobs for participants."[79] Surely after so many previous attempts it would take great faith to believe that a new forty-eighth training program would finally succeed.

More formal higher education is another expert solution: data show that college education generally increases income. But it does not help everyone equally. Today, women earn a majority of bachelor's, master's, and doctoral degrees, and dominate in the expanding service sector and in well-paid government jobs.[80] Men, especially from the working class, are more likely than women to shun college. A consultant for Stanford University suggested that aversion to formal education starts early for boys, in elementary and secondary school. He noted that "boys have significantly lower literacy scores than girls, indicating weaknesses in crucial skills for getting through high school and succeeding in college." Boys also are less interested in the nonathletic extracurricular activities that admissions officers favor. The consultant found a "pervasive culture of anti-intellectualism for males," with an image of males as "crude, rude, childish risk-takers" becoming "ubiquitous in reality television, television commercials, sitcoms, music, and on the Web."[81]

On the other hand, men's participation in the labor market is 10 percent higher than that of women.[82] Yet between 1965 and 2015, the percentage of men not in the workforce more than doubled, from 3.3 percent to 11.7 percent. The drop in labor-force participation was smallest for those with graduate-level education and highest (16 percent) for men who lacked a high school degree.[83]

Boys generally find physical activity fun, which is why many prefer industrial work to service jobs.[84] The culture of classical antiquity looked down upon physical labor and manual work, for the most part. But the monastic movement beginning in the sixth century bestowed a new spiritual value upon physical work, saying it could be as important as prayer, so that even men from the warrior nobility might choose it. A moral culture honoring work was crucial to building a prosperous society. Without such a cultural work ethic, why choose to do physical labor, let alone a less-macho service job?

Historically, the most successful socializer of males, and a promoter of the work ethic, has been marriage. Never-married men have labor participation rates 20 percent lower than married men.[85] Today only half of U.S. adults are married, down almost 10 percent from thirty years ago, the lowest rates being among the less educated.[86] All the federal funds put into job training have not been able to offset the cultural dynamics that discourage work. Nor have government programs shown much success in providing productive and fulfilling work for those who want it.

Confronting Pandemics

Pestilence was commonplace in the premodern world. The devastating Black Death had particularly far-reaching consequences in the history of the West, but there were many other episodes of plague, among other diseases. Medical advances and economic growth have done much to reduce the severity of contagious disease. Still, the influenza (H1N1) pandemic of 1918 apparently infected about one-third of the world population and killed 50 million people in a year—more than the Black Death—including more than 600,000 Americans. Other pandemics followed: H2N2 flu in 1957, N3N2 in 1968, SARS coronavirus in 2002, H1N1 swine flu in 2009, and, of course, a pervasive H1N1 strain in 2019.[87]

Government health agencies such as the Centers for Disease Control, the National Institutes of Health, and the Food and Drug Administration were expected to have resources ready in case of a viral outbreak, and to provide guidance for the population. But basic medical supplies were insufficient. Developing and approving vaccines took a year or more. And in each case of viral outbreak, the experts emphasized worst-case possibilities, which the media hyped, and the politicians generally followed.[88]

Since there were no specific treatments available during the first months of each pandemic, the experts could only recommend general supportive care and individual isolation. In the severe Covid-19 outbreak in 2020, the CDC first simply recommended drinking water, covering sneezes and coughs, washing hands, cleaning surfaces, not sharing items even with family, and staying at home as much as possible. Strict social isolation was known to have negative effects on the sick, the mentally ill, the elderly, and other vulnerable groups.[89]

If Covid-19 symptoms appeared, one was to call rather than visit a doctor. The physician was to decide if the person should go to a hospital—since space and equipment in hospitals were assumed to be in short supply. Testing for the virus at first was imprecise and not widely available, while temperature checks were not conclusive. Testing always has measurement errors, and it can even spread disease.[90]

Very early in the pandemic, the elite CDC Atlanta laboratory spoiled a virus test, but the contamination was not reported outside the agency. One consequence was a delay in issuing travel advisories and other public warnings. A regular *Washington Post* columnist called the CDC's coronavirus website a "scandal" for its vagueness.[91]

CDC experts first said that closing schools was unnecessary since very few otherwise healthy children suffered severe symptoms from contracting the virus—and later reversed the advice. Although the coronavirus was far more dangerous for those with underlying conditions and especially for the elderly than for the general population, the CDC and the NIH recommended only broad restrictions rather than targeted ones. The health experts in Italy, the country with the most documented coronavirus cases initially, later found that 99 percent of those who died had a preexisting pathology and that the median age of a fatality was eighty years.[92]

All of the official recommendations and policies were, of course, based on the "best data." But a top Stanford University medical expert, John P. A. Ioannidis, wrote in March 2020:

> The data collected so far on how many people are infected and how the epidemic is evolving are utterly unreliable.... We don't know if we are failing to capture infections by a factor of three or 300. Three months after the outbreak...no countries have reliable data on the prevalence of the virus in a representative random sample of the general population.[93]

While other experts were predicting two million or more deaths in the United States, Ioannidis questioned such high numbers:

> The one situation where an entire, closed population was tested was the Diamond Princess cruise ship and its quarantine passengers. The case fatality rate there was 1.0%, but this was a largely elderly population, in which the death rate from Covid-19 is much higher.
>
> Projecting the Diamond Princess mortality rate onto the age structure of the U.S. population, the death rate among people infected with Covid-19 would be 0.125%. But since this estimate is based on extremely thin data—there were just seven deaths among the 700 infected passengers and crew—the real death rate could stretch from five times lower (0.025%) to five times higher (0.625%).

Adding other "sources of uncertainty, reasonable estimates for the case fatality ratio in the general U.S. population vary from 0.05% to 1%," Ioannidis wrote. By comparison, the CDC put the death rate from the 2009 H1N1 (swine flu) epidemic at 1 percent in the early months, but later reduced it to 0.02 percent.[94]

The Pew Research Center reported in mid-March that only 26 percent of Americans feared that the coronavirus would affect their own health while 70 percent feared the economic effects of shutting down businesses.[95] The practical president wanted to limit the duration of the severe restrictions keeping people from work, and to focus strong measures on cities and regions actually affected.[96] President Trump ultimately was forced to strengthen restrictions although as guidance only, leaving actual enforcement mostly to state governors and local government.[97]

In early April, Dr. Anthony S. Fauci—director of the National Institute of Allergy and Infectious Disease, and the most respected expert on the subject—explained confidentially to the White House task force overseeing the pandemic response that there were real limitations to the best available science as a basis for broad recommendations. He was quoted as saying, "I've looked at all the models. I've spent a lot of time on the models. They don't tell you anything. You can't really rely upon models."[98] Dr. Jeffrey Shaman, the epidemiologist whose data were the basis for the models, added separately, "We don't have a sense of what's going on in the here and now, and we don't know what people will do in the future . . . nor if it is seasonal as well."[99]

Closer to the front lines, Governor Andrew Cuomo of New York expressed his frustration with the expert advice and predictions in late May: "All the early national experts, here's my projection, here's my projection model, they were all wrong; they were all wrong." Cuomo continued, "We didn't know what social distancing would actually amount to, I get it. But we were all wrong. So I'm sort of out of the guessing business."[100]

Around the same time, a study of European containment efforts conducted by the Blavatnik School of Government at the University of Oxford found "little correlation between the severity of a nation's restrictions and whether it managed to curb excess fatalities."[101] In July, a *Lancet* journal published an article on Covid-19 responses around the world that concluded: "Government actions such as border closures, full lockdowns, and a high rate of COVID-19 testing were not associated with statistically significant reductions in the number of critical cases or overall mortality."[102]

In making recommendations, the national bureaucracy apparently had to rely on speculative models and experts' sense of what was happening at the moment. Those deciding policy, mostly at the state and local level, would then make pragmatic adjustments to the situation on the ground.

Managing Law Enforcement

The Federal Bureau of Investigation has been the iconic national governmental agency since its early days, when J. Edgar Hoover moved aggressively into local law enforcement to confront organized crime. The FBI's pioneering fingerprint laboratory came to symbolize its competence. But that reputation crashed in 2015 when mass errors in DNA analysis were discovered, undermining the validity of many thousands of convictions since 1999.[103] Errors were also discovered in various other forensic methods. For example, a study of court testimony based on hair analysis found problems in 95 percent of those cases, and another study co-funded by the FBI itself found that DNA tests were erroneous half the time.[104] What had happened to the icon of governmental competence?

The first principle of bureaucratic administration should be that private firms differ from government agencies. As Ludwig von Mises wrote in his classic *Bureaucracy*, private businesses can manage their bureaucracies down through multiple levels by tracking whether each lower unit is making a profit or not, and taking appropriate action on each. Government lacks a sophisticated internal signaling device for quality control—other than the personnel appraisal system, which is not very objective and rarely has serious consequences—so it has no meaningful way to judge whether a decision is rational.[105]

James Q. Wilson of UCLA concluded that bureaucratic institutions of government must therefore simplify their missions, as he explained in *Bureaucracy: What Government Agencies Do and Why They Do It.*

Simplification of responsibilities makes for easier management when the various levels and units have competing goals and differing management cultures, and when there is no reliable signaling from below to higher levels.[106] The problem with the modern FBI jumped out from its official mission statement: "The FBI is an intelligence-driven and threat-focused national security organization with both intelligence and law enforcement responsibilities. It is the principal investigative arm of the U.S. Department of Justice and a full member of the U.S. Intelligence Community." With competing missions in both intelligence and law enforcement—and with a large increase in size—the organization fell afoul of the imperative to simplify.

This mission complexity in the FBI is relatively recent. In testimony before the House Permanent Select Committee on Intelligence in 2011, Robert S. Mueller III, the FBI director at the time, explained how the bureau's priorities had changed after the 9/11 terror attacks, leading to a "paradigm shift" in mindset:

> The FBI significantly increased its intelligence capacity after the attacks of September 11, 2001, when the FBI elevated counterterrorism to its highest priority. Prior to the 9/11 attacks, the FBI's operations were heavily weighted towards its law enforcement mission; intelligence tools and authorities were primarily used for the counterintelligence mission. In the immediate aftermath of 9/11, the FBI quickly identified the need to enhance intelligence programs with improved analytical and information sharing capacities to detect and prevent future terrorist attacks.
>
> Protecting the United States against terrorism demanded a new framework for the way the FBI carries out its mission: a threat-based, intelligence-led approach....
>
> This new approach has driven significant changes in the Bureau's structure and management, resource allocation, hiring, training, recruitment, information technology systems, interagency collaboration, and information sharing, as well as a paradigm shift in the FBI's cultural mindset....[107]

A few years earlier, Judge Richard Posner had charged that the FBI was still putting too much emphasis on legal matters so as to win convictions in court, and thus failing in its intelligence and security mission. Posner called for the creation of a separate agency for intelligence only, like Britain's MI5. Mueller's deputy director of national security, John P. Mudd, replied that the FBI was "a national security agency; we are not solely a law enforcement agency." He described these two functions as complementary: "Under this new model, intelligence drives how we understand threats, how we prioritize and investigate these threats, and how we target our resources to address these threats."[108]

In fact, the FBI shifted emphasis so heavily to intelligence capacity that it reversed the problem described by Posner. Today, the legal ethos does not inhibit the intelligence work so much as the security emphasis overwhelms the legal mission. This was dramatically demonstrated when the FBI used the secret Foreign Intelligence Surveillance Court (or FISA Court) against American citizens, with the FBI arguably breaching the attorney-client privilege.[109]

Thomas J. Baker, who served as an FBI agent for thirty-three years, maintained that placing intelligence higher than the original law enforcement mission had changed the FBI fundamentally from a fact-gathering agency, where an agent would have to swear in court—and a "lack of candor" was a firing offense—to an intelligence agency dealing with "estimates and best guesses" not admissible in court. Intelligence is more centralized than law enforcement, and decisions are placed in more politically sensitive hands, like those of Peter Strzok, the FBI intelligence director who was found to have discussed the possibility of using FBI authority against a presidential candidate he opposed. He was fired after being exposed, but the problem went beyond one person. Baker concluded that politicization and polarization were endemic to the agency, with "no sense of the bright line that separates the legal from the extralegal."[110] One example is the FBI's wiretapping of Carter Page, a U.S citizen, and Trump campaign adviser, for a year without legal cause.[111]

The effects of the new "cultural mindset" at the FBI were on display in the devastating report issued in 2019 by its inspector general, Michael Horowitz, on FBI actions related to the 2016 elections. The IG investigation found that the information given by the FBI to the FISA Court as the basis to begin spying on Trump campaign officials contained seventeen "significant inaccuracies and omissions," and that the bureau had "failed to meet the basic obligation" to ensure that applications for surveillance were "scrupulously accurate." Many "basic and fundamental errors were made by three separate, hand-picked teams on one of the most sensitive FBI investigations that was briefed to the highest levels within the FBI," according to the IG. These errors represented "a failure of not only the operational team, but also of the managers and supervisors, including senior officials, in the chain of command."[112]

The IG established that the FBI "drew almost entirely" upon the controversial dossier from the British spy Christopher Steele, funded by the Democratic Party, to convince the FISA Court to initiate and sustain the investigation of the Trump presidential campaign. The FBI did not inform FISA of later information that was favorable to the campaign officials. The primary source for Steele told the FBI that the so-called facts attributed to him by Steele were "misstatements" and "hearsay." Also not reported to the court was that the so-called suspicious activities of Carter Page could be explained by the fact that he did work for the CIA.[113]

This was not the first time there were issues with the FBI's applications to the FISA Court. In 2002, the court reported that FBI agents had misled them in seventy-five cases, and in 2017 the court produced a dozen pages of errors the FBI had made in its surveillance applications.[114] Following the 2019 IG report, the FISA Court again requested explanations and changes in procedure by the FBI.[115]

In a follow-up study, the Office of the Inspector General (OIG) examined twenty-nine FBI applications to the FISA Court, all for surveillance of American citizens. The investigation looked into whether the FBI had adhered to its own established procedures for justifying a submission, called the Woods Procedures, which require keeping a file of supporting documentation for every assertion of fact made in the application. For some of the applications examined, no such file was found. "We could not review original Woods Files for 4 of the 29 selected FISA applications because the FBI has not been able to locate them and, in 3 of these instances, did not know if they ever existed," said the OIG report.[116] The applications that could be checked against the supporting documentation had an average of twenty errors each, indicating chronic problems in the submission process.[117] The OIG report concluded, "We do not have confidence that the FBI has executed its Woods Procedures in compliance with FBI policy, or that the process is working as it was intended to help achieve the 'scrupulously accurate' standard for FISA applications."[118]

In a fascinating interview with Jan Crawford on CBS News, the new attorney general who would oversee the FBI, William Barr, described a "Praetorian Guard mentality" in the bureau, conjuring up the ancient Roman military institution that was originally supposed to protect the emperor but gained enough independent power to pursue its own interests, even against the imperial interest. Barr said that FBI officials had likewise become "very arrogant" and had begun to "identify the national interest with their own political preferences." They came to see anyone with a different opinion as "an enemy of the state," and their arrogance could lead to subverting legitimate political control of their actions.[119]

Asked if a Praetorian Guard mentality existed at the FBI and the intelligence agencies during the 2016 election, Barr, whose Department of Justice has legal bureaucratic control over the FBI, replied that the question needed to be "carefully looked at because the use of foreign intelligence capabilities and counterintelligence capabilities against an American political campaign to me is unprecedented and it's a serious red line that's been crossed." He said that republican self-government is endangered if we allow law enforcement or intelligence agencies "to intrude into politics, and affect elections."[120]

The attorney general affixed the blame for the legal improprieties mainly to "a small group at the top" of the FBI career service, the "executives at the senior level," from "headquarters."

> I'm not suggesting that people did what they did necessarily because of conscious, nefarious motives. Sometimes people can convince themselves that what they're doing is in the higher interest, the better good. They don't realize that what they're doing is really antithetical to the democratic system that we have. They start viewing themselves as the guardians of the people, who are more informed and sensitive than everybody else. They can—in their own mind, they can have those kinds of motives. And sometimes they can look at evidence and facts through a biased prism that they themselves don't realize.[121]

Indeed, this insight undoubtedly applies more broadly to the Washington bureaucracy, within which many think of themselves as guardians of the uninformed and uncaring public, and feel empowered to act by their own lights rather than following the lead of elected officials or the political appointees nominated by elected presidents and approved by the Senate.[122]

Managing Homeland Security

Despite the FBI's shift of priority to intelligence, this mission too has suffered serious dysfunction. In 2016, after Omar Mateen had killed forty-nine people in an Orlando nightclub, the FBI director was forced to admit that Mateen had been under investigation for two years but agents had not used the time to search his electronic devices.[123] When Nikolas Cruz had killed seventeen people in a Florida school shooting in 2018, the FBI admitted that it had not followed up on a credible tip the month before.[124]

Thirty-one people were injured when an explosive device detonated in the Chelsea neighborhood of New York City in September 2016. It was placed there by Ahmad Khan Rahimi (aka Rahami), of Elizabeth, New Jersey. He had also planted a second bomb in Chelsea, which was found before it exploded, and another on a scheduled race route in Seaside Park, New Jersey, which detonated without causing injury. Several more explosive devices were discovered in a backpack at the train station in Elizabeth.

Two years earlier, New Jersey police had informed the FBI that someone named Mohammad Rahami told them his son was associating with dangerous people after returning from Pakistan, and was a "terrorist." The FBI Newark office interviewed the father, who then moderated his warning. The FBI found the son Ahmad's name on the Customs and Border Protection screening list for having visited the radical city of Quetta, but did not even interview him. Why not? You cannot make this up: Ahmad was in jail, charged with domestic assault, and therefore "not available."[125]

CBP's National Targeting Center twice flagged Ahmad Rahimi for questioning on account of his visits to Quetta, which one official called a "stronghold of militants." His name appeared on the Customs list confusingly labeled "primary questioning," even though no one is questioned upon initial flagging. Rahimi was not subjected to "secondary" questioning either, because no further negative information was forthcoming by the time of his return flight!

"We don't have any 'pre-watchlist'" with a wide-ranging list of possible suspects, one CBP official explained. "We'd be so gummed up with a backlog of reviews, it would be impossible to do our job. This is all risk-based modeling, and questioning by professionals, and you make your best judgment based on the information that you have." The information they did have on Rahimi said he was a military-aged Arab male returning from a militant stronghold.

Two days after the Chelsea explosion, Rahimi was apprehended in New Jersey, and police searched his room. They found a journal praising Osama bin Laden, the radical preacher Anwar al-Awlaki, and the Islamic State spokesman Abu Mohammad al-Adnani, who was quoted telling adherents to "attack the kuffar [non-Muslims] in their backyard." Incredibly, the CBP official said that even if Rahimi had been found to be carrying the journal with these statements while passing through customs, he would only "probably" have been subject to extra questioning.

The FBI boasted to the media about how quickly Rahimi had been caught. In fact, it was the local police in a small New Jersey town who apprehended him—after federal agencies supposedly had him on their radar for two years.

Most foreign intelligence, for obvious reasons, is kept secret for many years. But a top foreign correspondent for the *Washington Post*, Gregg Miller, recently obtained a secret CIA and National Security Agency record authored in 2004, about a joint operation with West German intelligence in the 1950s. A Swiss firm, secretly owned by the CIA and German intelligence, had developed a process that would give American intelligence "the means to crack much of the world's encrypted communications." However, conflict between the CIA and the NSA delayed obtaining control of the process for a dozen years and hindered full operationalization for many additional years.

Miller found that the NSA was "full of people who were technically brilliant but struggled to grasp the potential of the operation, impeded efforts to expand its scope and at times put the program's secrecy in jeopardy with sloppy tradecraft." The CIA, for its part, looked like "an overbearing elder, impatient with its more timid counterpart, dismissive of its intermittent objections." Miller quotes his source saying that CIA officials "made the rules as they went along," and that they were "much more inclined to ask forgiveness than permission."

Miller concluded, "The operation continued to be marred by frictions between the United States' preeminent but perpetually squabbling spy siblings. Their rivalry is widely acknowledged in Washington, and has been cited as a factor in intelligence failures including the lack of warning before the terrorist attacks of Sept. 11, 2001."[126]

The security and safety of air travel became an area of heightened concern after 9/11, and serious government agency failures have been demonstrated here too. When undercover federal agents tested the effectiveness of the Transportation Security Administration's system for preventing firearms and explosives from being carried onboard, over 90 percent of the test weapons passed through the screening undetected.[127] One Saturday in August 2015, a glitch in a new computer system installed by the Federal Aviation Administration to minimize air and runway gridlock—part of an effort to modernize air traffic control—resulted in canceling 476 flights along the East Coast.[128]

The Secret Service has the key responsibility of protecting the president, but officers are distracted by numerous other duties—or pleasures. In 2012, nearly a dozen agents were found to have solicited prostitutes while advancing a visit to Colombia by President Obama.[129] The previous year, more seriously, Oscar Ramiro Ortega-Hernandez was able to fire at least seven bullets from a semiautomatic rifle out of his car window into the White House family quarters and then speed away. It took Secret Service professionals four days to realize that gunfire had hit the residence, but this was actually discovered not by its professional agents but by a housekeeper who noticed broken glass and cement on the floor.[130]

White House security has been penetrated numerous times over the years, especially since the 1960s.[131] In 2014, Omar J. Gonzalez jumped the fence and was able to run into the White House past the guard who stood immediately inside the door, which was not locked as required. Gonzalez ran up a half flight of stairs to the family living quarters and then to the East Room. He was found to be carrying a folding knife in his pocket. The alarm near the front entrance had rung too softly to be heard, and the agent inside the door did not immediately alert others.[132]

Just a few days before, President Obama had been allowed on an elevator in Atlanta with a contractor whom the agents did not realize was armed. In March 2015, two top Secret Service agents drove to the White House after spending five hours at a bar and crashed into a barrier within inches of a suspicious package. Inspectors concluded that the agents were "more likely than not" impaired by alcohol. The House Oversight Committee discovered a total of 143 serious security breaches allowed by the Secret Service over the preceding ten years and six additional serious security breaches previously unreported by agency officials.[133] Later, President Trump's winter residence was breached too.[134]

When questioned by Congress in 2015 about the agency's serious failures, Secret Service bureaucrats responded by leaking derogatory information to the media about the chairman of the House Oversight Committee, Rep. Jason Chaffetz. It was later discovered that the Secret Service assistant director, Edward Lowery, actually told his associates, "Some information that [Chaffetz] might find embarrassing needs to get out." Inspectors found that forty-five agents had searched Chaffetz's confidential personnel file, and eighteen top officials, including the deputy director and the chief of staff, were aware of its contents, resulting in numerous negative media reports.

"It was a tactic to intimidate and embarrass me and frankly it is intimidating," said Congressman Chaffetz, a Republican. The ranking Democrat on the committee, Rep. Elijah Cummings, said he was "deeply troubled" by the actions of the Secret Service.[135] Forty-one career bureaucrats were disciplined, but the greatest penalty was suspension without pay for forty-five days. Most of those involved merely received a letter of reprimand and went right back to work.[136] Not long afterward, Chaffetz retired from Congress.

Can Congress Fix the Bureaucracy?

The Constitution vests all legislative power in Congress, giving the House of Representatives and the Senate sufficient authority to pass laws that might make the executive bureaucracy able to function more effectively. But most rule-making in the United States today is actually done by the bureaucracy itself rather than by the legislature.

While Congress passes about 23,000 pages of laws every year, the bureaucratic agencies issue about 80,000 pages of federal regulations annually, and the number has generally grown over time. Clyde Wayne Crews of the Competitive Enterprise Institute has quantified the growth. "While an utterly imperfect gauge," he writes, "the number of pages in the *Federal Register* is probably the most frequently cited measure of regulation's scope." Crews has also estimated the cost of compliance with all those regulations. For 2014, he found it to be approximately $1.88 trillion, or about 11 percent of GDP.[137] Crews characterized President Trump's serious efforts to cut regulations as exceeding some goals, but still of limited consequence in view of the remaining regulatory load on the economy.[138]

Regulatory compliance puts a great burden on the nation's economic competitiveness. The World Economic Forum's report on global competitiveness for 2014–15 ranked the United States 82nd out of 144 nations in "burden of government regulation," which meant that 81 countries had a lighter regulatory burden.[139]

Congress can override bureaucratic rules, but the process for doing so is cumbersome, as James Gattuso of the Heritage Foundation has documented.[140] Trying to address the problem, Congress passed the Congressional Review Act in 1996, with "expedited" procedures for resolutions of disapproval, to encourage up-or-down votes to reverse counterproductive bureaucratic regulations. However, congressional reluctance to overrule the president, along with fears of taking responsibility for controversial actions, resulted in only one such disapproval passing Congress in the next twenty years. In the early months of the Trump administration, Congress did nullify thirteen regulations issued late in the Obama administration but not yet fully implemented.[141]

The progressive Center for Effective Government opposed legislative disapproval resolutions in principle because this allowed Congress to "second-guess agency expertise and science on food safety, worker safety, air pollution, water contamination, and a host of other issues."[142] But even disregarding the fact (as we have documented) that bureaucratic expertise in these areas is often more in the promise than in performance, voting on such issues seems precisely what the Founders expected Congress to do.

A major obstacle to reining in the regulatory state is that Congress cannot see most of what the bureaucracy does. In 2010, Congress required that the Department of Defense conduct what would be the first-ever full audit of its operations, as a minimum basis for effective oversight. Eight years later, the audit was finally delivered at a cost of $400 million. Even then, Sen. Charles Grassley called the result "unacceptable" since it did not provide useable data. He asked, "How is it possible that the Pentagon is able to develop the most advanced weapons in the world but can't produce a workable, reliable accounting system?"[143]

Actually, most of the real work of the bureaucracy is performed through even more hidden means, resulting in what Crews labeled "regulatory dark matter," by analogy with mysterious dimensions of cosmology. Astrophysicists have concluded that less than 5 percent of the universe is constituted by visible matter like the "planets, the Milky Way, the multitudes of galaxies beyond our own." Most of the universe consists of "dark matter and dark energy," which is beyond our direct observation.

> Here on Earth, in the United States, there is also "regulatory dark matter" that is hard to detect, much less measure.
>
> Congress passes a few dozen public laws every year, but federal agencies issue several thousand "legislative rules" and regulations.... We have ordinary public laws on the one hand, and ordinary allegedly aboveboard, costed-out and commented-upon regulation on the other. But the [legal] requirement of publishing a notice of proposed rulemaking and allowing public comment does not apply to "interpretative rules, general statements of policy, or rules of agency organization, procedure, or practice."[144]

These informal rules include presidential and agency memoranda, guidance documents, interpretation of formal rules, notices, bulletins, directives, news releases, letters, blog posts, and even oral statements, which might contain explicit or veiled threats. All of this happens outside the purview of Congress or courts or other external observers.

Courts do hear cases that challenge bureaucratic regulations, but the Supreme Court's *Seminole Rock* ruling of 1945 set a precedent for courts to give deference to executive agency bureaucratic interpretations of law when statutes are ambiguous. Justice Clarence Thomas questioned the logic of this deference in *Perez v. Mortgage Bankers* (2015), arguing that because *Seminole Rock* "precludes judges from independently determining" the meaning of laws, it amounts to a "transfer of judicial power to the Executive Branch." The Court thus makes it harder for Congress to serve as a check on the executive branch and enforce the rule of law.[145] In *Department of Transportation v. Association of American Railroads* (2015), Justice Thomas argued that the current test for allowing a grant of rulemaking power to administrative agencies had become a "boundless" standard and was contrary to the constitutional allocation of legislative power.[146]

But Thomas's arguments for more judicial checks on executive agencies have generally been ignored, and the bureaucratization has continued pretty much unabated.

Arbitrary Law

Even as the executive branch churns out thousands of new regulations annually, and courts tend to give deference to agency rules, many people still believe that the government is in the pocket of capitalist corporations. With all their wealth, big businesses are presumably able to corrupt bureaucrats, legislators, and courts, to get rules crafted in their favor. It is believed that government turns a blind eye to corporate malfeasance, and that tougher legal enforcement against business fraud and corruption is needed to reform how government works.

The Corporate Fraud Task Force in the Department of Justice under President George W. Bush announced in 2007 that it was vigorously cracking down on business fraud:

In the last five years, the task force has yielded remarkable results with 1,236 total corporate fraud convictions to date, including: 214 chief executive officers and presidents; 53 chief financial officers; 23 corporate counsels or attorneys; and 129 vice presidents. Additionally, the Justice Department's Asset Forfeiture and Money Laundering Section has obtained more than one billion dollars in fraud-related forfeitures and has distributed that money to the victims of corporate fraud.[147]

The Obama administration's record was equally impressive. The chairman of the Securities and Exchange Commission, Mary Jo White, brought a record 807 enforcement cases in one year alone and extracted $4.2 billion in settlements from abusive stock traders.[148] In five years, the U.S. attorney supervising Wall Street, Preet Bharara, charged 107 stock traders with insider trading. *Time* magazine featured Bharara as the man who, far from coddling business, was "busting Wall Street," with a historically high 81 convictions and plea settlements for insider trading and conspiracy.[149]

Perhaps Bharara's top business conviction was against two portfolio managers, Anthony Chaisson and Todd Newman, who were accused of using confidential information in an insider trading scheme to amass a $72 million profit trading Dell Inc. and Nvidia Corporation stocks. The information had come to them indirectly through a network of investor relations representatives and analysts until it reached their employees. Judge Richard Sullivan assisted the prosecution in district court by instructing the jury that even if the original leaker had not acted for personal benefit, the investors could still be convicted if the jurors believed the trading was improper.[150]

The Second Circuit Court of Appeals, which oversees Wall Street market trading, overturned the convictions, and narrowed the range of actions that prosecutors could deem "insider trading." Judge Barrington Parker, writing for a unanimous court, noted that Chaisson and Newman were "three and four levels removed from the inside trader," and that "the Government presented no evidence that Newman and Chaisson knew that they were trading on information obtained by insiders in violation of those insiders' fiduciary duties." He pointed out that criminal intent was required for conviction, and that Bharara and Sullivan in their zeal had ignored two Supreme Court decisions affirming its necessity.[151] In December 2016, the Supreme Court loosened the restrictions on insider-trading prosecution imposed by the circuit court.[152]

Also among Bharara's prey was a well-known trader, Raj Rajaratnam, who had traded on information he did know was leaked by a corporate fiduciary. But Bharara could not resist expanding the case to his younger brother Renjan, who would later be acquitted by a New York jury. Nelson Obus, founding partner of Wynnefield Capital, would also be acquitted, although it cost him $12 million in legal expenses and consumed twelve years of his life.[153]

For showboats like Bharara, a wealthy celebrity is a tempting target, guaranteed to get headlines. Phil Mickelson may have been America's best-loved golfer, but his fame helped put him in the sights of an assistant U.S. attorney in Bharara's office, Brooke Cucinella.[154] In 2016, Mickelson was caught up in the prosecution of William "Billy" Walters, a well-known Las Vegas sports gambler who the SEC claimed made $40 million on illegal stock trades based on insider tips gained through his friendship with Thomas C. Davis, chairman of Dean Foods. The SEC, using its pet tool of promising easy treatment for some as a means of taking down others, pressured lower-level employees of the company to give up the big fish. By this means, the SEC bureaucrats pressured Davis to claim he had leaked information to Walters, who in turn had urged Mickelson to trade in the Dean stock.

The SEC alleged that Mickelson benefited from trades based on "nonpublic information" to which "he had no legitimate claim." But even though the violations of securities law were committed by others, the SEC argued it was still not proper for him to retain any of the profits, which totaled $931,000. The SEC and prosecutors in Bharara's office had no evidence that Mickelson knew there was anything improper about Walters's advice to buy Dean stock and had no evidence of intent to defraud anyone, so Mickelson could not be charged criminally or civilly. But he was still named a "relief defendant," even though that should require committing an unlawful act, which he had not done. Threats of further legal expenses forced Mickelson to repay all the money anyway. Basically he was punished for being rich and famous.[155]

The best-known celebrity conviction related to insider trading was that of Martha Stewart, the television personality and businesswoman. She was indicted in 2003 for insider trading, but eventually the trial court dismissed that charge as prosecutorial overreach. Then the ruthless prosecutors pursued amorphous "conspiracy" and "lying" charges, without an underlying charge, hoping to convince the jury that Stewart must have done something wrong. She was convicted and served five months in Alderson Prison Camp, followed by five months of electronic monitoring and five years of supervised probation. An appeals court upheld the conviction. Stewart endured court trials spanning two years, paid $225,000 in fines, and had her reputation sullied by what many regard as prosecutorial misconduct based on a plea agreement forced upon her financial adviser.[156] Today, few Americans trust jury trials, since 90 percent of defendants are intimidated into plea bargains by aggressive prosecutors.[157]

Picking on the rich is apparently good politics, but it creates bad law. Judge Parker, of the Second Circuit Court of Appeals, wrote that his court sits in the "financial capital of the world" but is expected to follow "amorphous" theories of stock trading that do not give traders "bright-line theory as to what they can and cannot do." This makes capitalists vulnerable to arbitrary prosecution with potentially severe punishment.[158] But it isn't only the rich who are affected, since a majority of Americans say they have money invested in stocks.[159] Thus they too could be subject to prosecutorial abuse in the name of controlling capitalist power.

Parker was asking in effect for a return to the rule of law. John Locke insisted on the necessity of "established, settled, known law," adopted by a representative legislature. James Madison warned that "It will be of little avail to the people that the laws are made by men of their own choice if the laws be so voluminous that they cannot be read or so incoherent they cannot be understood."[160] Today, not only are the laws voluminous and hard to understand, but too many rules are actually made in an insulated bureaucracy, where neither the legislature nor the courts can possibly know them all. And those bureaucratic rules are generally given deference by judges as valid interpretations of law.

Prosecuting the rich and famous is popular, so it can help advance a political career. Bharara was believed to have been second on President Obama's list for the position of U.S. attorney general when Eric Holder stepped down in 2014. He was also rumored that year to be contemplating a run for governor of New York. He was dismissed as U.S. attorney by President Trump, but quickly hired by CNN.[161]

Rich capitalists and celebrities are easily stereotyped by the media. Robert Weisberg, a Stanford University law professor, said that the rich and famous who are brought to trial seldom escape conviction. In 2005,

SEC employees testified to their inspector general that bureaucratic higher-ups instructed subordinates to alert them only to cases involving celebrities because they are much easier to convict than serious Ponzi operators such as Bernie Madoff.[162]

Even the courts are not free from bureaucratic games. Judge Ralph K. Winter, in a rare rebuke to a fellow judge, pointedly noted in the appeals hearing the "sheer coincidence that the judge who bought into the government's theory," Richard Sullivan, was the same judge assigned to other recent insider-trading trials. Winter asked whether prosecutors were "abating" courthouse rules that bar them from steering cases to judges they believe are sympathetic to their views.[163]

Well-publicized prosecutions teach capitalists that SEC and DOJ bureaucrats may have even more say in their business than the courts and the law. When Bharara targeted Anthony Chaisson, the FBI and the SEC raided the company he cofounded, Level Global, and agents melodramatically seized documents, data, and computers. Naturally, the media had been alerted in advance. Level Global's other cofounder, David Ganek, was never implicated in Chaisson's technology trades, but he was compelled to close the company's $4 billion fund and dismiss its eighty employees due to the "cloud of uncertainty" raised by the publicity surrounding the raid.[164]

Executive Leadership

If the bureaucratic agencies are tied up in red tape, or focused on advancing their own agendas, and if Congress and the courts cannot or will not restrain them, what could control bureaucrats in the public interest? The solution usually proposed by political scientists in modern times is to empower the leader of the bureaucracy—the president—to guide the executive branch more effectively and deliver the "best science" results.

Woodrow Wilson thought the way to remedy the inefficiencies arising from the separation of powers was through a stronger chief executive, supported by bureaucratic experts. In *Congressional Government*, Wilson argued that enacting any far-reaching national program would be extremely difficult for a legislature, since each member is swayed by local interests and will resist unpopular plans. Courts have made unpopular decisions, but they cannot control the details of governance. Cooperation between the president and Congress may be ideal, but partisan politics usually frustrates it.[165]

The president clearly does have significant independent power, and there have long been complaints that it is wielded excessively. When President George W. Bush attempted to replace regional U.S. attorneys in 2006, in the face of congressional opposition, Rep. Henry Waxman (Democrat) said that such claims of executive power against the will of the legislature must be umpired by the courts. Waxman said he was taught in law school that "the Constitution was what the Court said it was." President Bush replied, in effect, that if a court issued an order limiting his action in this regard, the order would be unconstitutional and he would not allow his attorney general to enforce it.[166]

In 1832, President Andrew Jackson simply refused to enforce a Supreme Court decision supporting Cherokee property rights in *Worcester v. Georgia*. President Lincoln ignored a circuit court decision, *Ex parte Merryman*, ruling that he could not suspend the judicial right of habeas corpus without authorization by Congress. But two years later Congress did give him that authority for the duration of the Civil War.

Still, a president must be careful how he uses his authority, since Congress can have the last word. Lincoln's successor, President Andrew Johnson, a Democrat, was impeached by Congress in 1868 for merely attempting to replace his own secretary of war. He was defying the recently passed Tenure of Office Act, under which the president needed the Senate's approval to remove a federal official. The Senate vote to convict Johnson fell short by one. But Republicans had a two-thirds majority in both the Senate and the House of Representatives, enabling the first Reconstruction Congress to nullify Johnson's veto power, uni-cameralize the legislature, and essentially run the bureaucracy too by meeting separately with cabinet members.[167]

Presidents have a long history of evading congressional checks and balances. Thomas Jefferson's purchase of Louisiana and Lincoln's proclamation abolishing slavery were done by executive order, which presidents argue does not require congressional approval. Franklin Roosevelt's creation of his Works Progress Administration during the Great Depression, Dwight Eisenhower's use of the military to enforce school desegregation in Little Rock, Lyndon Johnson and Richard

Nixon's policies of affirmative action to remedy employment discrimination, and Bill Clinton's bailout of Mexico all used executive orders. Ronald Reagan's centralization of agency regulatory approval in the White House is sometimes included among such examples.[168]

When President Obama was frustrated with opposition from the Republican majority in the House of Representatives, he announced in his 2014 State of the Union address that he would act directly by executive order to achieve his agenda without them, especially for health care.[169] Republicans and independents decried this as an unconstitutional power grab, a usurpation of authority granted to Congress by the Constitution. Democrats were mostly too embarrassed to defend what they so strongly opposed under George W. Bush. Obama went ahead and modified the Affordable Care Act by executive decree, delaying the law's employer mandate and allowing those who thought their new premiums were higher than their old private insurance to claim a "hardship" exemption.

That provided the precedent for President Trump to make major changes to the law by executive order too, which he did.[170] Republicans suddenly thought executive power was not such a bad idea. Indeed, what is done by executive order can be undone by it. President Reagan's order restricting the use of federal funds for abortion, called the Mexico City Policy, was rescinded by President Clinton, reinstated by President George W. Bush, revoked by President Obama, then reinstated again by President Trump.

President Obama entered office determined to overcome capitalist business resistance to his priority of protecting the environment. After the Paris Agreement to combat climate change was negotiated in 2015, the Obama administration pledged American commitment to the "long term, durable global framework to reduce global greenhouse gas emissions," based on "the best available science." An unprecedented 187 countries had submitted climate action targets as a basis for a comprehensive plan. "This new global framework lays the foundation for countries to work together to put the world on a path to keeping global temperature rise well below 2 degrees Celsius," as the scientific consensus demanded.[171]

Specific actions for meeting the global targets were mostly left for individual nations to determine. President Obama had already begun working toward U.S. climate goals when his Environmental Protection Agency reinterpreted a section of an old 1970 law into a new 645-page regulation. This rule mandated that the states reduce carbon emissions from their thousands of local fossil-fuel power plants by an average of 30 percent from 2005 levels by the year 2030; one state was required to reduce emissions by 72 percent.[172] Regulating this energy source, representing one-tenth of the gross domestic product, was second only to the Affordable Care Act (regulating one-sixth of the economy) in its ambitious scope.

The new energy mandate was controversial, having previously been voted down by the Democrat-controlled Senate, with particularly strong opposition by senators from energy-producing states. President Obama claimed that the total cost per year would be only $8.8 billion, but the U.S. Chamber of Commerce estimated it at $50 billion and noted that the government's figure rested on the assumption that alternative energy sources would be found at lower costs than the current technology could provide. The president risked alienating some of his core supporters, as the lowest 10 percent of income earners payed three times the share of their income for electricity as did the middle class.[173]

While the economic and political costs were high, the environmental benefit was modest. By the EPA's own accounting, even if every U.S. coal power plant closed immediately, it would reduce world temperature by only one-twentieth of a degree in a hundred years.[174] China was already emitting more carbon than the United States and would presumably increase coal energy production, and therefore world emissions, as soon as the U.S. stopped. But the science said the environmental danger was great, and doing something was better than nothing.

The public seemed generally supportive of such policies, as did most of the world's leaders. Gallup reported that even half of Republicans agreed with Obama's plan, at least in theory, if costs were kept low. But many considered the costs too high. In 2015, the Republican majority in Congress used the formerly neglected Congressional Review Act in an effort to undo the rule on carbon emissions, but President Obama vetoed this action. With a Republican president in 2017, Congress again used the CRA, this time successfully, to undo environmental regulations concerning coal-mining waste in streams and financial disclosure by energy companies.[175]

In France, President Emmanuel Macron enacted an even more comprehensive energy tax to protect the environment, but it was so violently opposed, especially by the country's poor, that it had to be abandoned.[176] As noted, Pope Francis considered executive actions to be insufficient throughout the developed world.

Beyond the limits imposed by Congress or the public or the actions of successors, chief executives are limited by their own executive apparatus. The top executive cannot easily break through the long-established layers where bureaucrats promote their own narrow agendas or those of outside interests. As Bruce Katz of the Brookings Institution observed, the Washington bureaucracy consists of "outdated agencies, many formed in the 1950s and 1960s, carrying out programs established in the 1970s and 1980s, through means more appropriate for a pre-Internet age." Moreover, federal funds go mostly into entitlements for the poor and the elderly, leaving a mere 17 percent for other domestic programs, "not sufficient to do big things"—even if the government could do them, one might add. For the problem isn't only about money. As the progressive Katz argued, "The source of the federal government's feebleness is structural." And he dramatically concluded: "As Washington fades into the background, the rest of the nation is engaging in a great experiment—can a country successfully invest in its future without the national government being a relevant player?"[177]

What Went Wrong?

"Federal failures have become so common that they are less of a shock to the public than an expectation," writes Paul Light, America's preeminent expert on public administration. "The question is no longer if government will fail every few months, but where. And the answer is 'anywhere at all.'" Light made these disturbing comments in a comprehensive analysis of the reasons for governmental failure, published by the progressive Brookings Institution under the title "A Cascade of Failures: Why Government Fails and How to Stop It."[178]

Professor Light began by identifying the major news stories from 2001 to the middle of 2014, using Pew Research Center's "News Interest

Index." Among these, he found forty-one "highly visible" government failures. Many of these examples of failure are all too familiar: missing the terrorist signals before 9/11, the botched response to Hurricane Katrina, the financial crisis of 2008, the Gulf oil spill in 2010, the Abu Ghraib prison abuse, the Boston Marathon bombing, the *Columbia* space shuttle disaster that killed seven astronauts, the I-35 bridge collapse in Minnesota, the Fort Hood massacre, the flu vaccine shortage in 2004, the Ponzi scheme of Bernard Madoff, among others. These big news stories during that time did include nine successes, and Light still believed that "the federal government creates many quiet successes every day."

The big government failures were split almost equally between inadequate oversight and incompetent operations. Professor Light was somewhat consoled that most of these failures "involved errors of omission rather than commission," and that some of the events had elements of success mixed in with failure. But in each case, the failures had serious consequences. While all of the failures had multiple causes, Light found poor policy definition and insufficient planning to be the leading causes of failure.

Professor Light identified the results of past decisions that help explain this cascade of failure: "the steady aging of the federal government's infrastructure and workforce; growing dependence on contractors; ever-thickening hierarchy; dwindling funds, staffing, and collateral capacity, such as information technology and accounting systems; increasing frustration with poorly drafted policy; presidential disengagement; and political posturing." He found fault in both political parties, although he viewed the Democrats as failing mostly by omission, in allowing government capacity to deteriorate, while he saw Republicans as acting directly to "undermine the bureaucracy" through budget cuts and other actions.

His proposed solutions, as he recognized, were all conventional and had been largely ignored over the years. He was hopeful that more funding and bureaucratic reforms could improve performance and restore public confidence, but acknowledged a general reluctance in the government to tackle the great challenges and to provide the necessary resources, regardless of party or ideology.

Even as a progressive and supporter of active government, Light conceded that many of the policy goals of the federal government were simply too vague and grandiose, making them "virtually impossible to deliver." Policies might be better designed, he suggested, but in the *Wall Street Journal* he wrote of "the kudzu that clogs government hierarchies and makes even routine decisions risky." Finally, he observed, "Americans know that the federal government already has too much to do. They also know that government could work better and cost less."[179] But in "A Cascade of Failures" he wondered: "It could be that bureaucracies are inherently vulnerable to failure regardless of funding, hierarchy, dependencies, and public angst toward big organizations of any kind."

Since welfare-state bureaucracies often fail in their basic responsibilities, how can we expect them to succeed in the more complex matter of socialist planning of whole economies, or in the kind of comprehensive global plan that Pope Francis envisioned? F. A. Hayek argued that the progressive idea that centralized, scientific planning of nation-states could solve humanity's manifold problems was based on analogies to physical science that vastly underestimated human complexity. In explaining the problem, he pointed to the complexity of a single human brain, in which the number of interneuronic connections in a few minutes might exceed the number of atoms in the solar system.[180] Decades later, many new discoveries have revealed previously unimagined complexity even in the simpler physical world.[181]

Hayek was especially skeptical of the notion that rationalizing experts in central government headquarters using inefficient bureaucracies, imperfect understanding of the facts, and inherently limited scientific methods could somehow perfect complex human nature— calling this presumption a modern "superstition" that would mystify future generations.[182]

PART II

Looking Forward

CHAPTER 5

SCIENTIFIC RATIONALIZATION

To understand more deeply the difficulties involved in trying to rationalize society and solve social problems through scientific administration, we must examine what science itself is. In particular, we need to look into the difference between what Friedrich Hayek characterized as "constructivist rationalism," an assumption that the methods of physical science can answer all social questions through abstract formulas, and a "critical rationalism" that takes better account of complexity and unpredictability in the physical world and especially in the social world.[1]

One admirer of Pope Francis's *Laudato Si'* suggested that much of the public has "something of a science problem," which makes it difficult to mitigate capitalist encroachment on the environment. Michael Gerson, an op-ed columnist for the *Washington Post* who served as a speechwriter for President George W. Bush, was concerned that ordinary people without scientific training are inclined to rely on their common-sense intuition and ignore the experts, or even outright reject the findings of the "best science," when it comes to how government should solve pressing societal problems. Customary common sense is a poor guide on problems like climate change, Gerson opined.

> Intuitions are useless here. The only possible answers come from science. And for nonscientists this requires a modicum of trust in the scientific enterprise. Even adjusting for the possibility of untoward

advocacy it seems clear that higher concentrations of carbon dioxide in the atmosphere have produced a modest amount of warming and are likely to produce more. This in turn is likely to produce higher sea levels, coastal flooding, shifting fisheries, lower crop yields and vanishing ecosystems.[2]

Gerson regards the effort to limit global warming as a "noble crusade," although he proposed a more pragmatic approach than the "strict global regulatory regime" demanded by many climate activists. He favors a plan to "make polluters pay" with a carbon tax, but one that supposedly would be rebated to taxpayers, compensating for higher costs. Even such a plan, however, would require more than "a modicum of trust" on the part of the public. Indeed, less dogmatic supporters of a carbon tax admit that it would be extremely unpopular, as it would double or even triple energy taxes.[3]

Gerson criticized the many so-called experts who operate outside their "actual expertise," but most interesting is his constructivist understanding of science itself. In explaining the limits of intuition on "matters intimately related to our survival—say on quantum motion or on the nature of black holes or the effects of radio frequency energy on the DNA in cells—our intuitions are pretty much useless." What Gerson missed is that the theories of quantum motion, black holes, and radio frequency energy are all based upon intuitions, only they are the intuitions of people who have some professional claim to be called scientists. Bertrand Russell and Alfred North Whitehead, both philosophers of science, maintained that all important scientific inferences are based on intuition.[4]

More important, some major scientific concepts—such as quantum mechanics, black holes, and DNA—are not fully understood by scientists themselves. As we shall see in more detail later, physicists are having difficulty reconciling the fundamental theories of quantum mechanics and general relativity. One attempt at doing so is with "superstring" theory, which posits unverifiable loops much smaller than elementary particles vibrating in six or seven dimensions of reality, empirically inaccessible in normal time and space. Common sense would surely question that as a description of reality, but so do many scientists.[5]

In criticizing popular skepticism about climate science, Gerson suggested that many people refuse to change their behavior concerning the environment because they suspect that "the vast majority" of climate scientists are "engaged in fraud or corruption." He called those suspicions "frankly conspiratorial," since the corruption "would need to encompass the national academies of more than two dozen countries including the United States."[6]

There certainly is no fraudulent conspiracy or corruption when a group of likeminded people agree on a goal they consider important and rally around the mission. But it is conspiratorial in the common understanding, and open to criticism, if the group suppresses alternative viewpoints. One example occurred about the same time that Gerson was writing, in 2014:

> On May 8, Lennart Bengtsson, a Swedish climate scientist and meteorologist, joined the advisory council of the Global Warming Policy Foundation, a group that questions the reliability of climate change and the costs of policies taken to address it. While Bengtsson maintains he'd always been a skeptic as any scientist ought to be, the foundation and climate-change skeptics proudly announced it as a defection from the scientific consensus.
>
> Just a week later, he says he's been forced to resign from the group. The abuse he's received from the climate-science community has made it impossible to carry on his academic work and made him fear for his own safety.[7]

Here the "climate-science community" was strongly united around the goal of limiting what was thought to be the devastating effects of climate change. Allowing any climate scientist to question the consensus was considered deleterious to the common purpose.

Scientific problems are generally best solved by communities of presumably competent individuals with shared interests. But communal knowledge always begins with individual persons, and therefore is subjective at bottom, as Michael Polanyi argued.[8] Thomas Kuhn, in *The Structure* of *Scientific Revolutions*, followed Polanyi, Hayek, and Karl Popper in explaining how normal science consists of people—communities of experts who adopt an investigative structure to explore problems in an effort to advance understanding, and their own interests.[9]

Is it "conspiracy theory" to recognize that opposing the consensus in one's field might have costs—that challenging the beliefs of the leading scientists or the community's reviewers in the scientific journals could affect one's future? "The stakes are so high. A single paper in *Lancet* and you get a chair and you get your money. It's your passport to success." The "conspiracy theorist" who said that was Richard Horton, himself the editor of the *Lancet*.[10] Scientists are not neutral about the subjects that are close to their hearts and central to their livelihoods. They are necessarily invested in the methods and paradigms of their disciplines.

A particular field may thrive with a successful method and paradigm. But a crisis of understanding may occur, leading some individuals to develop another paradigm or even create a new discipline, which might bring useful new understandings. The Ptolemaic cosmology gave way to the Copernican model. Alchemy was succeeded by chemistry. Eugenics was superseded by genetics. In climate science today, warnings about nuclear winter yielded to worries about global warming, which became climate change. Apparently no scientific consensus is forever—at least not so far.

No less an authority than Albert Einstein noted the paradox that as scientific understanding increases, new questions are constantly being opened, so the magnitude of what needs explaining increases at the same time.[11] There may be no ultimate answer, no perfect constructivist theory, no settled science. For this reason, even nonspecialists should not be reluctant to ask questions when experts claim that science is settled in their particular field of expertise.

The Pillars of Physics

In the master science of physics, the longstanding Newtonian mechanics finally began yielding to the theories of relativity and quantum mechanics in the twentieth century. Sabine Hossenfelder, a resident scholar at the Frankfurt Institute for Advanced Studies, offers us a short course on the basics of modern physics, and explains some problems and conundrums in the field today.[12]

The starting premise is that all matter is composed of fundamental elements that cannot be broken down into anything simpler. Particle physicists, working at the subatomic level, see their job as digging down to the core structure of matter. They have advanced a "standard model," consisting of twenty-five types of elementary particles that interact in ways described by the mathematical equations of quantum mechanics. In the vast reaches of the cosmos where particles cannot be detected, however, astronomic physicists postulate a "concordance model" in which matter interacts according to the general relativity equation, where the fundamentals are space, gravity, and time.

The search for nature's basic elements began with earth, air, fire, and water, but it was not until the discovery of elemental atoms that physics earned its scientific reputation. Subatomic physics then revealed the lower structure of electrons, protons, and neutrons. As a result of this real progress, the constant search has been for smaller and smaller particles, like quarks and bosons, to find the ultimate building blocks of matter. The elementary particles are presumed to compose the atoms and molecules that constitute the matter we can observe and measure directly. But subatomic particles cannot be observed directly, so scientists today are limited to measuring the results of their interactions.

Early observers of the physical world relied on light waves and microscopes to provide direct high-resolution measurement at the level of biological cells. X-rays got down to the level of atoms. Then, a new microscope beaming electrons rather than light was used as a way to explore subatomic structures inaccessible to light. In theory, accelerating electrons to high energy levels allows researchers to detect smaller and smaller particles, which explains the emphasis on building ever more powerful particle colliders. At this level, particles are hypothesized to dissolve into a wave function with properties of both particles and waves. The mathematical wave function allows physicists to calculate only probabilities of particle behavior.

Mathematics is the means for modeling systems of the simplest elements. The mathematical formulas that describe the observed outcomes of particle collisions are the basis of the standard model, which Hossenfelder calls "our best current knowledge about the elementary building blocks of matter." It was largely assembled by the late 1970s as a mathematical model based on quantum field theory, which combines quantum mechanics with Einstein's theory of special relativity.

One problem is that the data gathered from particle colliders are too massive even to store in the largest computers, much less analyze. Scientists must use algorithms to select subsets of data they find interesting for investigation. Another problem is that most important theories in science are enormously complex, requiring a great deal of time to test. It took a century to detect gravitational waves after the theory was proposed. Scientists do not have careers that long, so they naturally apply other criteria to confirm their theoretical models in the short run: continuity, aesthetic appeal, the beauty of the math. Measurement itself has arbitrary aspects, and many differences and contradictions are resolved only by approximation. It all seems to work within the limits of observation and mathematics, but there is no real controlling measure or theory.

Because of the relatively insignificant mass of the particles in the standard model, the effects of gravity can be ignored and only the rules of special relativity are considered necessary. But astrophysicists studying massive astronomical objects cannot ignore gravity, so they use the concordance model, which incorporates general relativity where space-time interacts with energy and matter by curving. Which implies that the universe expands.

In connecting the extremes of the subatomic and the cosmological, astrophysicists say that the visible matter relevant to atomic theory—the world we see and the subatomic realm beneath it—accounts only for an estimated 4.9 percent of the mass-energy content of the universe. They infer from indirect calculations that "dark matter," which does not emit or reflect light or any other detectable radiation and cannot be seen, represents 26.8 percent of the mass-energy budget of the universe. And more recent measurements of an increasing rate of expansion of the universe can be explained only by postulating the existence of "dark energy," which may constitute an incredible 68.3 percent of the mass-energy budget of the universe, leaving relatively little that can be seen and directly measured.

Hossenfelder notes that many top physicists are dissatisfied with one or both of the two basic theories, the standard model and the concordance model. Why is it necessary to postulate so much dark energy, which is mostly mysterious? Is there an unknown source of energy that does not include atoms or particles? Why does special relativity combine with the standard model, but not general relativity? Stephen Hawking was among those who have called the standard model "ugly," with so many parameters having no deeper explanation, and weak observational measurements. Steven Weinberg said he has spent a whole career in the field of particle of physics "without knowing what quantum mechanics is." Clearly there are many different interpretations of it.

More troubling is that the concordance model and the standard model seem to be incompatible. "As they are currently formulated, general relativity and quantum mechanics cannot both be right," writes Brian Greene, a particle physicist at Columbia University.[13] Hawking's superstring theory, as mentioned earlier, is one effort to reconcile the two "pillars of physics." Physicists in general have been seeking a "theory of everything" that explains the entire physical universe, both subatomic and cosmological. Is it possible that there are two accurate but incompatible theories of matter? Or are both models wrong?

In Hossenfelder's view, the central problem is that the models are basically all math, with little empirical confirmation because of the limits to measurement. They rest on mathematical standards of beauty, symmetry, naturalness, and elegance, organizing the complexity of the universe around aesthetic rather than empirical judgments. Indeed, Hossenfelder titled her book *Lost in Math: How Beauty Leads Physics Astray.*

Science Lost in Math

Hossenfelder describes herself as "one of some ten thousand researchers whose task is to improve our theories of particle physics." Most of her colleagues "make a career by studying things nobody has seen," such as other universes beyond our own, or wormholes in higher-reality space—things that are "practically untestable" empirically. She calls it a form of "magic." Intuitions are guided by "hidden rules" for naturalness, simplicity, beauty, and symmetry, which are all unconfirmable. Particle theory is mostly not even written down, but personally passed on to younger colleagues.

"I'm not sure anymore that what we do here, in the foundations of physics, is science," writes Hossenfelder, wondering if she has been wasting her time with it all. So she interviewed the top experts in the field to give her hope. Gian Francesco Giudice, manager of the Large Hadron Collider, the world's biggest and most important particle collider (sixteen miles long, and costing $6 billion to build), justified the mathematical rules for beauty and elegance as "universally recognized" between cultures. When Hossenfelder objected that they are not, he responded, "Most of the time it's a gut feeling, nothing that you can measure in mathematical terms: it is what one calls physical intuition." Discovering the rules of physics "requires not only rationality but subjective judgment," he said. It rests on a "sense of beauty" that is "hardwired in our brain." This "unreasonable" aspect of physics is what makes it "fun and exciting."

The empirical results from the particle accelerators have not been so exciting. When asked about the recent findings from the Large Hadron Collider, Giudice said candidly, "We are so confused." In the decade since the collider began operating in 2008, researchers had detected only one new particle, the Higgs boson, which had been predicted in the 1960s, but nothing more to advance particle theory empirically. Even the Higgs discovery turned out to be frustrating because it is the one particle assigned a very large mass, which should have been possible to measure, but it has not been successfully done.

Yet even without empirical measurements to support it, a model that shows mathematical symmetry is considered so beautiful that whatever derives from it is accepted as true. Anthony Zee simply said that "beauty means symmetry" for a physicist like himself. Hossenfelder traces the ideals of symmetry and beauty back to the origins of physics, to scientists of faith like Isaac Newton who were guided by a religious sense of beauty. This ideal was apparently translated into presumably more rational approaches to physics.

The theory of "supersymmetry" predicts many particles beyond the twenty-five in the standard model, and this prediction has guided generations of physicists who bet their careers on finding them. The known particles are divided into fermions (e.g. electrons) and bosons (e.g. protons and neutrons), all exhibiting different affinity properties. Each fermion and boson should have an opposite partner particle, but increasingly powerful colliders have not been able to find them empirically in almost two decades.

The search for possible particles in the early colliders was organized around the idea of symmetric "multiplets" of smaller entities that Murray Gell-Mann called "quarks," in a theory that won him a Nobel Prize. Symmetries have been identified to unify the electromagnetic force with the weak nuclear force and with the strong nuclear force, and have also been postulated for special and general relativity. But over time, many of these highly awarded symmetric theories have fallen by the wayside. As Hossenfelder summarized the matter, "aesthetic criteria work until they don't."

Hossenfelder notes the interesting fact that while there is constant interaction among particles in the subatomic world, atoms in the aggregate tend to behave alike. The differences lie in how they interact, which is governed by the electron shell. The nucleus seems merely to come along for the ride. Similarly, neutrons and protons seem mostly unaffected by their composition in quarks and gluons. For practical purposes, one can basically ignore the smaller components of atoms. In fact, the science of chemistry was built around the interaction of atoms, with no knowledge whatsoever of subatomic structure. Similarly, the science of astronomy was built around the interaction of large bodies orbiting in space, with no need for subatomic science at all.

While useful sciences have arisen from more limited knowledge, some physicists find the current theoretical models to be unhelpful and constraining. "That we have both relativity and quantum mechanics is an incredible constraint," says Nima Arkani-Hamed of the Niels Bohr Institute. It seems impossible to get beyond the models. The idea of symmetry itself is a constraint.

As the empirical is lost in mathematics, the idea of multiple universes has gained ground. Leonard Susskind finds it "exciting to think that the universe may be much bigger, richer and [more] full of variety than we ever expected." Steven Weinberg said it is "wild speculation" but still "a logical possibility" that there may have been many big bangs, perhaps an unlimited number, each producing different "constants of nature." But Paul Davies calls the multiverse theory "simply naive deism dressed up in scientific language." Hossenfelder is in the middle, describing the multiverse as "the most controversial idea in physics," but also a way to open up new thinking. Yet the multiverse certainly removes physics still further from the realm of empirical observation.

The speed of light sets a limit on what can be seen far away in space at any given time, since it is believed that nothing can travel in space faster than light. But space itself may expand faster. As the universe expands, it is plausible that some matter could never be observed even in theory, because the light it emits or reflects can never reach Earth, nor could we ever reach those objects ourselves. That is especially so in the multiverse scenario. So the idea that science deals only with empirically testable phenomena might logically be abandoned.

Stephen Hawking favored the multiverse theory, but his co-author George Ellis was concerned that today's physicists, including his famous associate, have given up on empirical testing as the scientific means to validate theory, simply because the theories "are such good ideas." Ellis thinks this is going "backwards by a thousand years," undermining the very idea of science as testable against reality. He doesn't reject the multiverse theory out of hand, but argues that it cannot simply be accepted as established science without some empirical basis, or at least a convincing philosophical justification. He cites David Hume on the necessary philosophical limitation of science to dealing with matter, which puts such questions as the existence of God outside its bounds. Hawking, with classic constructivist rationalism, maintained that the nonexistence of God can be scientifically demonstrated, because science requires matter for existence and God has none. In Ellis's view, this way of thinking undermines the legitimacy of the scientific enterprise.

Hossenfelder worries that unmooring physics both from empirical testing and from secular philosophy further amplifies the social dimension of scientific knowledge. "In science experts only cater to other experts and we judge each other's projects," she notes. Theories are advanced only with the approval of highly specialized colleagues, all dependent on publication to succeed in their field, so that research findings may be heavily influenced by current sexy theories such as supersymmetry or string theory. Testing of new theories is expensive and time-consuming, so unpopular theories do not get tested. Inventing new particles is one avenue to success, but since confirmation takes years, it requires an accommodating academic audience. The need to secure research funding means that "almost all scientists today have an undisclosed conflict of interest between funding and honesty," according to Hossenfelder.

In other fields of science, attention has recently been focused on problems of reproducibility in scientific experiments, raising questions about possible bias. Hossenfelder observes that theoretical physicists have not really tried to avoid bias in their instrument, the brain, given the large role played by aesthetic judgments of beauty, elegance, symmetry, and even faith. The standard model seems to be incompatible with general relativity, yet so much faith has been invested in the model that discarding it is "unthinkable" to particle physicists. Few theoreticians either in that area or in astrophysics are going back to the drawing board to resolve the contradictions.

Hossenfelder ends, "Physics isn't math. It's choosing the right math." To understand the quantum behavior of space and time, she argues, "it is necessary to overhaul gravity or quantum physics or both to describe actual nature consistently."

The Science Hope

The hope that science can save the world remains strong.[14] But when the foundational science of physics presents two incompatible basic theories of the material world, this represents a serious test for constructivist rationalism.

In particle physics, millions of dollars went into searching for the light sterile neutrino, a hypothetical particle postulated from earlier experiments that was widely expected to help fill gaps in the standard model. Researchers collected data for two years, and concluded "at approximately the 99% confidence level" that the postulated light sterile neutrino does not exist.[15] On the cosmological side of the house, astronomers were astounded when researchers using data from the Las Cumbres Observatory Global Telescope Network discovered something previously thought impossible: a supernova that had not simply faded away, as in the normally observed death of a star, but exploded again fifty years later, and then was observed to grow brighter and dimmer several times over a period of three years.[16]

In the more complex field of biology, evidence emerged that the long-held conception of the basic structure of the animal cell might not be universally valid, as a eukaryote without a mitochondrial organelle had been discovered.[17] Genetic science had long held that 98 percent of human DNA is noncoding "junk," until the recent ENCODE project concluded that at least 80 percent is active.[18] Scientists had long used a map of the brain cortex originally developed more than a century ago, which began with 50 distinct brain areas and gradually grew to 83. Then suddenly, new technology revealed 180 distinct areas. Researchers emphasize that much remains unknown about how they function.[19]

The Human Genome Project, launched three decades ago, promised to unlock the genetic code for every inherited human trait, revealing "the basic set of inheritable 'instructions' for the development and function of a human being." Over a period of thirteen years, scientists meticulously sequenced all three billion base pairs in the human genome and set out to create "linkage maps" to track inherited traits through generations. This "incredibly detailed blueprint for building every human cell" was expected to provide "immense new powers to treat, prevent and cure disease" through the manipulation of genes.[20] In addition to accelerated breakthroughs in medical science, many foresaw the potential to design superior human beings.

Some observers, however, were concerned about how those new powers would be employed, especially after a Chinese scientist was reported to have manipulated genes to produce "designer babies." In April 2019, members of the American Society of Gene and Cell Therapy, along with other scientists and bioethicists, called for a global moratorium on gene editing of human embryos that will result in births—that is, experimentation that alters genes passed to future generations, with possible deleterious effects.[21]

But the NPR science correspondent Richard Harris deserves a Pulitzer Prize for going beyond the National Institutes of Health press releases and interviewing geneticists on the actual possibilities for gene editing. "The reality is that biologists probably couldn't produce designer babies even if they wanted to," he concluded. "It turns out that the genetics underlying desirable traits such as athleticism, intelligence and beauty are so complicated it may not ever be possible to make targeted changes." There is not really a "gene for" this or that specific trait. Instead, many genes acting in concert are responsible for a given trait.

While many scientists have accepted this for years, some have thought they might be able to identify a small number of genes that together would account for common diseases.[22] This goal has turned out to be more elusive than anticipated.

Matthew Keller of the University of Colorado told Harris, "When we look at the 20 most studied genes investigated for schizophrenia, we find basically no evidence that any of those are associated at levels greater than we'd expect due to chance." The same is true of depression, he added. "You could have done just as well by throwing a dart at the genome." Jonathan Pritchard of Stanford University said pretty much the same for height: "It quickly became clear there's huge numbers of variants that affect height. We have estimated that it's probably something like 100,000 variants across the genome, so most of the genome affects height by a small amount." Some common traits might even involve *all* of our genes.

Ewan Birney of the European Bioinformatics Institute was a bit more hopeful that further research might reveal a clearer picture of genetic mechanisms. He stressed that manipulating genes to create a "designer baby" is different from repairing a broken gene that results in a genetic disease, although he noted that other options carry less risk than genetic manipulation. Birney emphasized how much is still unknown in genetics: "If anybody thinks we can understand how to change genomes to improve things, they don't have an appreciation for the lack of knowledge that we have."[23]

Many constructivist scientists will acknowledge gaps in our understanding of human genetics, but say that some doctrines are beyond questioning by anyone claiming to be scientific. The main example is the Darwinian doctrine of evolution by natural selection. Hardly anyone today doubts that evolution has had some part in the development of living organisms; it is even accepted by the world's largest religion. But philosophical constructivists are inclined to insist that natural selection alone can account for the origins and development of life forms, including intelligent life. To raise any doubt on the matter is seen as imposing religious dogma upon science.

Karl Popper, a critical rationalist and philosophical skeptic, observed that the theory of natural selection was often formulated "in such a way that it amounts to a tautology," which could not be disproved, and that even some Darwinians had called it a tautology. Popper himself had in the past called it "almost tautological," but observed that it "may be so formulated that it is far from tautological" and is actually testable. In such a formulation, he said, the theory of natural selection "turns out to be not strictly universally true. There seem to be exceptions, as with so many biological theories; and considering the random character of the variations on which natural selection operates, the occurrence of exceptions is not surprising."[24]

Stephen Meyer of the Discovery Institute hit number seven on the *New York Times* bestseller list with a dry academic book, *Darwin's Doubt: The Explosive Origin of Animal Life and the Case for Intelligent Design*. Meyer notes that Charles Darwin himself had one great doubt about his materialistic theory of natural selection by random mutation as the sole explanation for life on earth. Specifically, it could not account for what is called the Cambrian explosion as recently as 530 million years ago (out of an estimated 3 billion total years with life on earth), when many unique animal forms appear suddenly in the fossil record. How could they have evolved so rapidly, and why had no fossils been found showing a progression of forms up to that stage?[25]

Much of Meyer's book analyzes subsequent efforts to fill in the gaps in the fossil record with examples that supposedly illustrate a gradual growth in complexity. But Meyer demonstrates that science is not much closer to finding the missing data than it was 150 years ago when Darwin predicted that it would soon be conclusive. The likelihood of finding the missing links appears increasingly remote.

More important, Meyer stresses that the development of new body parts is now believed to require new DNA and epigenetic biological information being placed into informational sequences that already were enormously complex, where a single missed sequencing could be lethal to the organism. But a gradual transition into a different form of organism depends on viability at every stage. The odds against this happening are exceedingly high, and the evidence we do have appears to refute the possibility.

Darwin's contemporary and supporter Ernst Haeckel believed (as Darwin presumably did also) that a cell is a mere lump of matter. Today much more is known about the incredible complexity of biological cells. We have just begun to understand how DNA might work, and how some epigenetic information may be just as important as DNA.[26] Constructivist evolutionary scientists had been compelled (very reluctantly) to accept the Big Bang and the influence of extraterrestrial events like comet explosions on earthly life, both of which were fiercely opposed by the keepers of the Darwinian flame in the 1960s but are almost universally accepted today.[27]

Settled Science

Besides evolution, the area where science is most often claimed to be truly settled is, of course, global warming.[28] Or is it "climate change," when clearly no skeptic doubts that climate changes? Why the shift in terminology? Perhaps because the world's leading experts at the Intergovernmental Panel on Climate Change originally, in 2007, reported their central forecast for long-term warming to be 3 degrees Celsius. Since then, IPCC reports have given only ranges rather than a single estimate, and the panel has reduced its minimum expected warming from 1.5°C down to 1.0°C.[29]

NASA's Goddard Institute for Space Studies and the National Oceanic and Atmospheric Administration announced in early 2017 that the previous year was the hottest on record. James Hansen of NASA had earlier warned that annual rankings could be "misleading because the difference in temperature between one year and another is often less than the uncertainty in the global average." But the margin of error could be unhelpful to the constructivist narrative, so it wasn't mentioned this time. Holman Jenkins of the *Wall Street Journal* looked at NASA's data and found that 2015 and 2016 (two El Niño years) were tied for warmest year on record, while ten years—1998, 2003, 2005, 2006, 2007, 2009, 2010, 2012, 2013, 2014—were all in a tie for second place, very close behind.[30]

Judith Curry, a climatologist at the School of Earth and Atmospheric Sciences at the Georgia Institute of Technology, noted that IPCC models have forecast that surface temperatures will increase 0.2°C each decade of the twenty-first century. But during the first fifteen years of the century, temperatures increased by only 0.05°C. And the models cannot explain why 40 percent of the surface temperature increase since 1900 took place between 1910 and 1945, while that period contributed only about 10 percent of the total increase in atmospheric CO_2 concentration in the same period.[31]

Indeed, a report by the International Energy Agency in 2016 noted that the world economy had grown by 3 percent the preceding year without causing an overall increase in manmade carbon dioxide emissions. This "decoupling of global emissions and economic growth" was attributed to better technologies in energy production and advances in other economic sectors.[32]

Professors David Victor of UC San Diego and Charles Kennel of the Scripps Institution of Oceanography wrote in the journal *Nature* that there was little scientific basis for the IPCC goal of keeping global warming to a 2°C limit, "but it offered a simple focal point and was familiar from earlier discussions, including those by the IPCC, EU and Group of 8 (G8)." The goal "sounded bold and perhaps feasible," but Victor and Kennel considered it "unattainable" and "wrongheaded." They noted that the average global surface temperature had "barely risen in the past 16 years," up to 2014, while the oceans "are taking up 93 percent of the extra energy being added to the climate system."[33] Mark Maslin, a professor of climatology at University College London, said the 2°C goal "is not a sensible, rational target because the models give you a range of possibilities, not a single answer." He claimed that the number "emerged from a political agenda, not a scientific analysis."[34]

Looking into the data behind the IPCC executive summaries that so impressed Pope Francis and much of the world actually reveals all kinds of empirical qualifications. Between 2009 and 2019, the Heartland Institute published five large volumes on the subject, totaling over 4,000 pages, with at least 10,000 citations from peer-reviewed literature. The project was led by seven Ph.D. climate scientists and supported by scores of policy specialists in meteorology and related sciences. In sum, the researchers found that science has not conclusively proved that manmade carbon dioxide emissions will lead to disruptive climate change or even that a significant degree of warming will actually take place.[35] One may challenge these conclusions on one point or another, but the science hardly appears to be settled.

Why might some people insist on the absolute scientific truth of what others find questionable? In *The End of Doom: Environmental Renewal in the Twenty-first Century*, Ronald Bailey argues that "what

people believe about scientific issues is chiefly determined by their cultural values." Bailey draws upon numerous studies by the Yale Cultural
Cognition Project, derived from Aaron Wildavsky's theory of cultural
commitments, which "holds that individuals can be expected to form
perceptions of risk that reflect and reinforce values that they share with
others." Starting with Wildavsky's typology,[36] Bailey notes that "people
are adept at seeking out information that confirms their values."[37] In his
analysis, today's social and political leaders and cultural influencers are
predominantly from a low-risk cultural category that is anxious about
change and demands that it be controlled. Anxiety is not primarily
based upon evidence but upon preexisting cultural beliefs.

Bailey maintains that various common fears in the modern world—
population explosion, exhaustion of energy supplies, harm from bio-
tech crops, cancer epidemics, global warming, mass extinction—have
limited scientific foundations and mostly reflect assumptions rooted
in low risk tolerance. He admits that his own assumptions reflect a
more optimistic bias, but he also presents a large empirical basis for
his conclusion that population growth is slowing, that energy supplies
have been increasing with new technologies, that health is improving
globally, that incomes are rising, that global warming might be moder-
ating, and that the environment has benefited from economic growth.

On the other hand, Stephen Chu, secretary of energy in the Obama
administration and winner of a Nobel Prize in Physics, was asked in
2010 about temperature measurements showing no global increase
since 1998, and replied, "We don't fully understand the plateau that's
happened in the last decade." But he believed that the numbers would
eventually show a warming trend again.[38] It is not clear whether Chu's
answer was a constructivist statement of faith or a critical-rationalist
qualification. Temperatures did rise two years later, but no real-world
science can accurately predict how long such a trend will last.

Applied Science and Bureaucratic Administration

Bailey may be overly optimistic, and Chu overconfident in the consensus, but all constructivist "consensus science" relies on abstract models
that cannot precisely represent the complexities of the real world, as
critical rationalists emphasize. This is particularly true in dynamic

systems that include an element of human choice, with its greater unpredictability, and with necessary trade-offs.

Take the relatively simple matter of reducing carbon dioxide emissions from automobiles. The Department of Energy's Office of Energy Efficiency and Renewable Energy currently says,

> Highway vehicles release about 1.7 billion tons of greenhouse gases (GHGs) into the atmosphere each year—mostly in the form of carbon dioxide (CO_2)—contributing to global climate change. Each gallon of gasoline you burn creates 20 pounds of GHG. That's roughly 6 to 9 tons of GHG each year for a typical vehicle.[39]

Under the Clean Air Act of 1963, the Environmental Protection Agency issues science-based rules to reduce carbon dioxide in the atmosphere, in part with increasingly stringent tailpipe emission standards that automakers are required to meet. In 2007, the EPA boasted that "emissions from a new car purchased today are well over 90 percent cleaner than a new vehicle purchased in 1970."[40]

The EPA's very rational procedure is to publish its test protocols in detail well in advance, so that auto manufacturers will have time for the technological adjustments needed to meet new requirements. But this also allows them to game the standards by designing cars that perform better in the tests than on the road.

In 2015, the EPA charged that Volkswagen was using a "defeater device" with sensors for pressure, temperature, engine speed, and other measures to determine when a car was being put through the government's emissions test. If the sensors detected that a test was being performed, the engine would automatically adjust performance enhancers to reduce the car's emissions. A study by West Virginia University and the International Council on Clean Transportation found that nitrogen oxide emissions in real-world tests from two VW models were five to thirty-five times higher than the Clean Air Act regulations allowed.[41]

VW's actions might be put down to capitalist duplicity. But the EPA conceded that devices that might sense when tests are being performed could be legal, and even helpful if they protect the auto and driver in an emergency when more power might be required. The devices must be disclosed to testers, however, which VW did not. What the EPA does not admit directly is that defeater devices can be legally used if they aid in the goal of reducing carbon dioxide emissions—which VW claimed was its real objective.[42]

The fundamental problem is that making engines more fuel-efficient, and therefore reducing carbon dioxide emissions, results in the release of more nitrogen oxide, which the EPA also lists as a greenhouse gas. But the EPA nevertheless required VW to spend thousands of dollars per vehicle to change the devices—which could reduce carbon dioxide but also increase nitrogen oxide emissions. This is real-world scientific complexity, where trade-offs are the rule.

Drivers can also cheat the emissions tests that many states require for vehicle registration. They can purchase a device that improves overall efficiency but results in more pollution, and then unplug the device when the test is performed. It is an updated version of adjusting carburetors to fill the engine with more air and less gasoline just before testing: the engine will generate less power but burn the fuel completely, so the exhaust is rated cleaner in any test.

It is not clear that state-mandated emissions tests have much environmental benefit anyway. Gary Bishop, a research engineer at the University of Denver, said that emissions testing "costs lots of money but it does absolutely nothing to clean up the air." His study of Tulsa, Oklahoma, which never had mandatory emissions tests, found no more pollution there than in places that did require them.[43]

Trade-offs are often overlooked in matters of health and safety. Two decades ago, for example, airbags were made mandatory for all automobiles sold in the United States. Over the years, the National Highway Traffic Safety Administration began to acknowledge research finding that airbags could deploy when they should not, firing at 200 miles per hour and possibly causing death, especially of children, or they might fail to deploy when they should. Some adjustments were made, but NHTSA refused to reconsider its mandate or allow owners to remove airbags even though a series of studies in the *Journal of Trauma*, reported by the National Institutes of Health, found that airbags provided little protection beyond ordinary seat belts.[44]

Critical-rationalist philosophers of science emphasize that nature's complexity makes trade-offs unavoidable. Much of what happens in the real world disappears in constructivist probability estimates and measurement error.[45] In the world we actually live in, things can change unpredictably, and people can outwit the planners. But some people are confident they might get around that problem by scientizing the human mind.

Science and the Human Mind

The ultimate human science is psychiatry, the science of the brain. *Our Necessary Shadow: The Nature and Meaning of Psychiatry*, by Dr. Tom Burns, is a definitive insider study of the discipline. Dr. Burns, a professor of social psychiatry at the University of Oxford, does not shy away from the ambiguities of his profession. He believes that "psychiatry is a major force for good," but acknowledges that its claim to know the mind is inherently controversial. "Psychiatry touches directly on that which is most human in us, the central core of our being—our identity or, if you wish, our 'soul'." About this, we cannot be neutral.[46]

Psychiatry is a hybrid of "guided empathy" and detached cure. It has no unifying theory, according to Burns. There are only experimental approaches that may work for individual patients, who then are encouraged to take responsibility for themselves. Psychiatry is a medical science in that its practitioners can prescribe medicines in the form of brain-altering chemicals, and can recommend and sometimes perform surgery. Psychiatrists can even have legal power to decide when it is necessary for civil authorities to impose compulsory treatment. They typically practice psychoanalysis (and psychological therapy generally), which can also be done by psychologists who are not medical doctors.

Psychiatry treats mental diseases, normally classified as psychoses or neuroses. The former—schizophrenia, manic depression (bipolar disorder), and paranoia—are more severe and were once labeled madness. These conditions are impossible to ignore, since patients demonstrate extremely disturbed behavior and loss of contact with reality. The term neurosis, referring to milder conditions, is often considered less scientific due to its overdiagnosis in cases of depression, but a neurotic patient acknowledges there is a problem, while the psychotic one may not.

Mental illness has been recognized since ancient times, when treatment was entrusted to shamans, witch doctors, priests, hypnotists, and con men, many with wild theories and exaggerated promises of cure, but some with treatment methods that actually helped. Medieval society basically left matters to families and religious groups. The Bethlem (aka Bedlam) Royal Hospital in London started admitting mental patients in 1403. In the sixteenth century, in what Michael Foucault called the "great confinement," public officials began requiring incarceration of those considered dangerous madmen, informally and locally, but often locking up the merely unruly. In the seventeenth century, officials launched more extensive efforts to control dangerous madmen in scientific institutions. In the eighteenth century, the physician Franz Mesmer challenged ideas about demonic possession and developed a more rational method of treating mental illness that combined religion with hypnotic suggestion, which would influence Christian Science spiritualism as well as psychoanalysis.[47] The constructivist scientific thinking of the nineteenth century brought the first professorship of psychiatry, established in Berlin in 1864, but also a wave of asylum confinement that was later widely criticized.[48]

Sigmund Freud replaced hypnosis with psychoanalysis, trying out different approaches, as did other pioneers such as Alfred Adler and C. G. Jung. They began with different theories and used dissimilar treatments. All had their proponents and achieved some popularity, especially among intellectuals. All seemed to get equally poor results.

The first breakthrough in developing psychiatric drugs as an alternative came serendipitously through the use of malaria parasites to cure late-stage syphilis in 1917. Later came efforts to use insulin coma treatment for schizophrenia. Electroconvulsive therapy to treat schizophrenia and depression was first tried in 1938. The first brain surgery was performed in 1939. Burns recognizes the possible abuses in various forms of treatment that have been applied, including lobotomy and electrically induced epileptic fits, while considering them sometimes useful. He stresses the very real difficulties and suffering they are meant to alleviate, and says they often are the only relief for very disturbed individuals, not to mention their families.

"In the last hundred years," Burns wrote, "American psychiatry has lurched wildly from a broadly biological understanding to an almost exclusively psychoanalytic one between 1940 and 1970, and now back to an unapologetically and aggressive biological discipline." Burns, himself a surgeon, is generally supportive of this seemingly more objective orientation, and argues for a firmer scientific grounding of psychiatry in biology but would not eliminate psychoanalysis. He does acknowledge that so-called "multiple personality" was invented by psychotherapists through the power of suggestion. He notes that "recovered memory" has routinely been induced by both doctors and social workers, and he questions whether it is scientific at all, but finds some legitimacy in the idea of dissociation.

Burns tries to distinguish between "illness" and "disorder," psychosis and neurosis, biology and mind, psychiatry and psychoanalysis, to give his discipline a stronger scientific framework. He finally comes to the critical-rationalist conclusion that one can make abstract or legal distinctions, but not strictly scientific, medical ones. While there are psychological differences between psychosis and neurosis, Burns is reluctant to call the former a medical condition but the latter only a problem that simply requires nonmedical counseling.

Psychiatric treatment is now dominated by chemical medications, mostly classified as antidepressants, antipsychotics, sedatives, or mood stabilizers. Chemical treatment and surgery may seem more scientific than psychotherapy, but Burns calls the latter not merely an "add-on." It is essential to psychiatry, because there are always social problems that need to be resolved for full recovery. In fact, Burns observes that the success of sympathetic counseling by Quakers and nuns inspired early psychoanalysis and that treating people with patience and decency is still the main secret of success.

Medicalizing all of life's ordinary difficulties is the tendency today, and Burns is concerned that his discipline as a result is deviating too far from a scientific endeavor. Why not Prozac all the time for everyone? he asks rhetorically. It would calm us all down; but at what cost to our humanity? Indeed, he wonders whether psychiatry itself is "making us sicker." He notes that in the United States, 10 percent of ten-year-old boys are diagnosed with attention deficit hyperactivity disorder, which "surely cannot make sense" scientifically. In many cases, the diagnosis is a way to stop boys from being boys. He questions whether alcoholism is a medical condition, although psychoanalysis can be helpful in treating it.

The fact that the psychiatric profession has expanded the number of mental illnesses from 106 in 1952 to 297 in 1994—nearly a threefold increase—has undermined its claim to be a scientific discipline. "Of course," Burns said, "this does not mean that there are really hundreds more disorders," or that practicing psychiatrists recognize all of these categories.

Despite increasing public criticism of overdiagnosis, the demand for psychoanalysis remains high, especially if a third party or government bears the cost. People want a shoulder for comfort. Burns is concerned that the "extreme dependency" of the patient in such a setting, along with the profession's skepticism toward customary moral taboos, is an open door to abuse. A few therapists even justify sexual relationships with patients.

Mentally ill patients are often not able to make decisions for themselves, which must be turned over to families. Mothers tend to persevere in pursuing treatment for their children, but many family members simply want the situation resolved, no matter the nature of the treatment or the risk to the patient.

Burns is fair to critics such as Foucault, Erving Goffman, and Thomas Szasz who question whether psychiatry can be considered a science at all. But he says they have no solutions for the problem of mental illness and the harm it does to human beings. In his more pragmatic critical-rationalist view, psychiatry is essentially the "practical response to the reality of mental illness." Its limited success must be balanced against the real anguish of patients, relatives, and friends.

Psychiatric Science and Society

The mind is so complex, Burns argues, there are no cut-and-dried solutions to mental illness, but society often demands "scientific" remedies. Legislators and judges require psychological evaluations to help them keep order and make distinctions between "mad and bad." It is governmental officials more than psychiatrists who insist on medical support for confinement decisions.

"Compulsory community treatment" orders lasting a year or two are standard, but expensive. Burns's own study of compulsory treatment orders in Britain found they had "absolutely no effect" on recovery, but he believes that compulsory care is "an inevitable feature of psychiatry," since few will accept Thomas Szasz's idea of treating the mentally ill the same way as ordinary criminals when they cross the law. Whether the patient improves or not, additional treatment is routinely presented as the scientific solution.

The history of the profession has an even darker side in conjunction with state power. Between 1939 and 1945 in Nazi Germany, the Committee for the Scientific Treatment of Severe, Genetically Determined Illnesses ordered that 200,000 people be medically euthanized after being diagnosed by psychiatrists as incurably mentally ill. "How could such a terrible thing happen and why was there no effective opposition from psychiatry? For there was none," Burns observes. After all, it was "scientific." The only opposition came from some families and the church. The Soviet Union routinely classified political dissidents as mentally ill, so certified by psychiatrists.

It wasn't only tyrannical regimes that turned psychiatry against undesirable elements. In the heyday of eugenics, forward-thinking Sweden scientifically sterilized over 60,000 people, for example. Burns adds:

> [T]he systematic extermination of the mentally ill was a terrible consequence of more long-standing eugenic ideas which had been gaining strength in Europe, the UK and the USA for decades. "Social Darwinism," and a moral panic that the unfit were "breeding" faster than the educated and able, had become a preoccupation at the turn of the twentieth century. It is never that far from the surface, even now.

In the United States, the poet Ezra Pound was confined to Saint Elizabeths psychiatric hospital in Washington, D.C., ostensibly for being "unbalanced" but mainly for being an open fascist sympathizer in his speeches and writings. The "enhanced interrogation" undertaken by the U.S. government after the terror attacks of September 11, 2001 was supervised by psychiatric physicians.

Ewen Cameron, who asserted that the brain is simply a complex computer, was elected to the presidencies of the American, Canadian, and international psychiatric associations, as well as other professional organizations, in the 1950s and 1960s. He worked with the CIA and the

Canadian government in an experiment to repair the minds of usually unwilling patients by applying electroshock therapy twice a day (versus the norm of three times a week). The aim was to break all "incorrect" brain pathways and create the "correct" patterns. The project was carried out at McGill University in Montreal from 1957 to 1964, with no opposition from the academy, the profession, or the government.[49]

Burns mentions how the questionable science of "recovered memory" led to an "epidemic" of child-abuse accusations—basically a witch hunt—against teachers by government welfare officials in the United States during the 1980s and 1990s. He is also concerned about the dangers of "commercial and social pressures" to control atypical behavior. Still, he is "relatively hopeful that psychiatry is unlikely to be such an obviously unwitting tool of state oppression again. We have learned our lesson and the profession is now more open and international."

Two "errors" of early psychiatry cited by Burns suggest to the author the constant temptation to bend science to public opinion or elite fashion or government agendas. Burns notes that until 1973, homosexuality was listed as a mental illness by the psychiatric establishment. That characterization was then quickly dropped, with no scientific evidence to support the change. Today, homosexual behavior is considered normal, and discrimination against homosexuals is often illegal. Some professionals even regard "homophobia" as a contributing factor in mental illness, and some states forbid psychiatric treatment to "reverse" homosexuality. A scientific profession that once viewed homosexuality as a mental disorder now considers it an identity requiring legal protection.[50]

Burns was disturbed by the way that early psychiatry treated women as mentally different from men and somewhat inferior. He acknowledges that two-thirds of psychotherapy patients today are women, even though mental illness seems to be equally distributed between the sexes, and he is careful not to judge either the women or the therapists as responsible: "Who is influencing whom can be debated." He is ambivalent as to whether there actually is a mental difference between the sexes. But one suspects that any "correct" consensus statement on the subject by the American Psychiatric Association today would raise little dissent from the profession.

Burns makes very clear that psychiatry has no single scientific view of human nature, and no overarching theory. It is empirical, relative to given situations. So what is there to keep it within any limits? Burns recognizes that his critically rationalist understanding of a difference between mind and brain is a minority opinion in psychiatry, in the academy, and among experts in government. Darwin's view that there is no "mind" at all, only an evolved animal brain, is still extremely influential. This makes it difficult to ignore Burns's warning that social Darwinism and eugenics are "never far from the surface" of psychiatry.

As Burns demonstrates, it has not really become much easier to understand the human mind scientifically, to distinguish "mad" from "bad," to cure mental illness, or to limit the use of compulsory treatment. Patients who are unwilling or not fully informed are inherent to the discipline. Neuropsychology and genetic research promise further refinement of diagnoses in the future, though Burns confesses to being "unaware of any philosophical breakthrough in understanding the mystery of consciousness and identity" in recent years. Can a constructivist science that does not distinguish between brain and mind resist fads or political manipulation, or even understand humanity?

Androids and Children

Advances in information technology have fueled the consummate constructivist notion that the human brain can be understood simply as a form of computer. If machines appear to "think" as humans do, perhaps there is no essential difference between artificial intelligence and the human mind. Maybe humans can be programmed in much the same way that computers are, and brains can meld with machines as the ultimate rationalized being.

Ray Kurzweil, chief engineer of Google, has predicted that computers will think for themselves and link with human brains by the year 2040, and that humans will become immortal soon thereafter. He sees the Android phone as the precursor to artificial intelligence devices he is developing that will be able to learn—that will be more truly *android*, or manlike. Neural interface technology will allow humans to connect their brains directly to an AI machine, augmenting the individual's own intelligence. Everyone will become wise and quick-witted through the machine's inexhaustible data. Kurzweil promises that the human individual will still be in control. But how could that be if the machine acts instantly and authoritatively while the human would otherwise take time to analyze information? Which is making the decisions?[51]

Kurzweil's other top Google exploratory project, named Calico, uses genetic research toward the goal of ending human aging by controlling the harmful effects of mitochondrial DNA in the cells. Mitochondria are a form of bacteria captured and safely consumed by living cells eons ago, but they retain separate DNA, which is prone to error in the process of cell replication, causing health problems. Much of this effect was neutralized as cells over time moved parts of the mitochondrial genetic code into the cell nucleus, but that process came to an evolutionary halt. Kurzweil's aim is to finish what evolution left incomplete. This biological engineering in combination with connecting minds to computers is intended to produce unfathomably smarter and basically immortal beings.[52] The historian Yuval Noah Harari wonders if this is inevitable and any effort to stop it not worthwhile.[53] But would the products of this engineering still be human, or merely android?

Educational reform is the more traditional way of using constructivist scientific models to mold the human character. The most influential American educationist of the twentieth century, John Dewey, declared education the chief tool for inculcating morality as the means to strengthen democracy. Dewey's work was an inspiration for the Common Core State Standards Initiative launched in 2008.[54]

Common Core grew out of an earlier effort at empowering the federal government to enforce educational standards that began at the National Education Summit of 1989. Hosted by the National Governors Association under the chairmanship of Bill Clinton, then governor of Arkansas, the summit brought together the nation's top education experts to discuss ideas for reform. Businessmen were invited to a follow-up summit sponsored by the progressive Business Roundtable, which proposed a more "businesslike" approach to meeting educational goals, and created an organization called Achieve Inc. for that purpose. Senator Ted Kennedy then teamed up with President George W. Bush in a bipartisan effort to pass the No Child Left Behind Act of 2002, with the goal of making all American K–12 students "proficient in math and reading by 2014." States were required to test all their students in reading and math each year from grade 3 through grade 8, and then once in high school. Results had to be reported to the U.S. Department of Education, and significant failures by schools to meet their state's standards would result in various penalties, up to loss of federal Title I funds.

Unsatisfied with the results by 2008, two education experts—Gene Wilhoit, head of the Council of State Education Officials, and David Coleman, who was associated with the McGraw-Hill textbook firm—persuaded the Bill and Melinda Gates Foundation of the need for a fully funded program to implement national standards and assessments in education.[55] As Lyndsey Layton reported in the *Washington Post*, the Gates Foundation "didn't just bankroll the development of what became known as the Common Core State Standards" for reading and math. The foundation also spent more than $200 million to build political support across the country and persuade state governments to commit to the standards. The foundation spread money to the big teachers' unions, to business organizations including the U.S. Chamber of Commerce, to the National Governors Association (NGA), to Achieve Inc., and to Wilhoit and Coleman. While the NGA staff described Common Core as state-sponsored, the governors were clearly not the driving force, which Wilhoit later conceded.[56]

Common Core gained critical mass in 2009 when the Obama administration included it in the Race to the Top legislation, which allowed states that adopted the standards to escape the penalties of No Child Left Behind and to become eligible for new federal grants. Not surprisingly, forty-two states took the bait. In 2010, President Obama proposed that all federal education grants be conditioned on adopting the standards and offered another $4 billion in federal incentives, and the regulations for implementing the policy went forward.

When Common Core standards and assessments were issued some education experts were not so supportive. Ze'ev Wurman, a former senior adviser to the U.S. and California departments of education, was concerned that the math standards excluded algebra II and geometry content that was "a prerequisite at almost every four-year state college." Sandra Stotsky, professor of education at the University of Arkansas, described the "college-readiness" standards for English and reading as "empty skill sets," without any specific readings listed. A Fordham Institute study concluded that "the standards do not ultimately provide sufficient clarity and detail to guide teachers and curriculum and assessment developers effectively."[57]

Layton's main criticism of Common Core was that Gates's own company, Microsoft, profited by selling the project's software to the nation's 15,000 school districts.[58] Yet perhaps more concerning is the technology-based thinking behind the whole agenda. The unstated constructivist assumption is that children all have very similar capabilities, and similar ways of thinking and responding—more like cuddly androids than human individuals. In this view, every child's mind can be programmed with the same one-size-fits-all formula.

Opposition to the Common Core standards and testing regime, from students and parents and teachers, became so intense that five years after most states had adopted them, Congress passed a bill to roll back the federal standards in K–12 schooling. This was a time when Republicans and Democrats could agree on almost nothing, but the effort to standardize and centralize education succeeded in angering almost every constituency. Lamar Alexander, the Republican chairman of the Senate education committee, said the new law, called the Every Student Succeeds Act, would "ban the federal government from mandating any sort of education standards, Common Core or otherwise." The legislation "would lessen federal control in the education system" and "allow states to create their own accountability systems and determine how much standardized tests should account for student and faculty evaluations." Cosponsors included Patty Murray, the leading Democrat on the committee, and supporters ranged from Elizabeth Warren to Rand Paul. President Obama signed the legislation.[59]

Robert Luddy, an educational entrepreneur, has remarked that standardization "potentially sucks the life out of other great ideas." The reforms of public education that appear to be effective are the charter school movement, scholarships that enable students to transfer to better-performing private schools, and homeschooling. All of these permit some choice on the part of parents and students, rather than imposing a single plan that supposedly fits all but actually benefits few.[60]

An equally questionable "advance" to rationalize education is the heavy use of computers in the classroom. A study of forty countries by the Organization for Economic Cooperation and Development found that student performance improved somewhat with moderate use of computers, but extensive use resulted in poorer educational outcomes.[61] It may be that heavy use of computers interferes with developing creativity and adaptability. Human critical thinking skills and habits are simply not the same as a computer's information processing.

In *Our Robots, Ourselves: Robotics and the Myths of Autonomy*, Professor David A. Mindell of MIT uses the example of the Mars exploratory rovers to illustrate the limits of robots in processing information. Letting the rover move around and experiment by itself randomly is much less costly than sending a human to Mars to supervise it. Sending a signal from Earth to Mars takes twenty minutes, so the rover was designed to act without specific human direction in the interim, but it cannot do things quickly: "The rover can autonomously plan a route around a series of rocks or obstacles using imagery it gathers from its camera. Yet to do that it stops every ten seconds to look at the terrain for twenty seconds." The rover can travel faster when humans plan the route. Many experts think that having a human supervisor on Mars might be more efficient than an android alone, even at a tremendously higher cost.[62]

Programming computers can run into unexpected difficulties and failures too. In 2019, Google introduced a "quantum" computer to supersede all others, but was forced to admit it was not only more expensive but had much higher error rates too.[63]

The constructivist conception of science excludes such things as social custom and moral tradition, and this exclusion can have serious consequences. Soon after becoming chairman of the ruling Chinese Communist Party, Xi Jinping announced a new social credit system to rate the reputation of every individual in China. A mass electronic database of financial transactions and online interactions, searchable by fingerprints and other biometric data, would be used to assign every citizen a "social credit score" as a means of monitoring and controlling the population.[64] Is that where extreme constructivist rationalism, with its narrow focus on efficient technology and on brains rather than minds, will lead us all? Some observers, like Professor Harari, think it might be.

Can We Finally Retire Scientific Superstition?

As we have seen, critical-rationalist scientists and philosophers from Einstein to Popper, Hayek, and others reject the constructivist belief that the more we know, the more we understand. They have concluded that in the "best science," the more that is learned, the more questions and mysteries open up, and not only in the realms of the extremely small or the unfathomably distant.[65] Science itself is an inherently complex and unfolding process at its best. Research is done by people with their inherent human limitations. Good scientists have conflicts of interest, as Sabine Hossenfelder noted. They can be too attached to a theory, or may persuade themselves they have seen what they wanted to find.

A study published by the journal *PLoS ONE* found that 2 percent of scientists admitted they had actually fabricated, falsified, or modified data in their studies, and 33 percent admitted to using "questionable research practices."[66] In 2012, a Japanese anesthesiologist was forced to retract a whopping 183 research papers when it was proved that he had fabricated most of his results.[67] An internal investigation by the Harvard Medical School in 2018 found that fraudulent data from one of its top cardiologists had appeared in thirty-one refereed scientific publications.[68]

More fundamentally, research published in scientific journals has been notoriously difficult to replicate. Professor John Ioannidis of the Stanford University School of Medicine raised the issue publicly in 2005 with an article titled "Why Most Published Research Findings Are False."[69] He argued that scientists do not usually falsify their results intentionally, but rather "fool themselves" into finding something new that can be published, when most experiments merely confirm what is already assumed to be known, and thus do not merit being published. But publication is the basis for honors and promotions.[70]

In 2011 and 2012, two pharmaceutical companies attempted to replicate multiple academic studies on drug safety and efficacy, but for various reasons could not do so.[71] The Center for Open Science conducted a Reproducibility Project with 270 researchers trying to replicate findings from one hundred studies published in three prestigious psychology journals. Taking great care to follow the original experimental designs closely, they replicated the results in only thirty-five cases.[72] Subsequently, the Center for Open Science along with Science Exchange launched a project to replicate twenty-nine cancer studies from leading scientific laboratories. Among the first five replication studies published, three either failed or were inconclusive.[73] Then, scientists tried to replicate twenty-one social science studies published in *Science* and *Nature*, and could reproduce the results in only thirteen.[74]

Popper convinced Hayek and others that physical sciences and social sciences have similar difficulties in trying to explain complex realities.[75] But social sciences are widely understood to have even more complexity and certainly more unpredictability. So it was revealing to hear the confession of surprise that Alan Greenspan, former head of the Federal Reserve, made in his memoir, *The Map and the Territory: Risk, Human Nature, and the Future of Forecasting*, concerning the 2008 economic crash:

> It all fell apart, in the sense that not a single major forecaster of note or institution caught it. The Federal Reserve has got the most elaborate econometric model, which incorporates all the newfangled models of how the world works—and it missed it completely. I was actually flabbergasted. It upended my view of how the world worked.[76]

That this epitome of Washington expertise and establishment power actually believed that big data and little models could forecast the future takes one's breath away. Greenspan supposedly believed in capitalist markets, if presumably the rationalized Ricardian kind, and he must have known that all econometric models are understood to be only probabilistic. The models can be useful as rough guides for action and to keep naive investors calm, but if they were perfectly predictive, how had the Fed missed all earlier recessions too? Would better data help? Greenspan admitted that he had just then realized that irrational fears influence how people behave. Was he truly so obsessed with data and constructively rational models that he was unaware of something so fundamentally human?

While the material world remains a mystery in many, many ways, human behavior is a true black hole. The best psychiatrists acknowledge that there is no unified science for understanding the brain. What about millions or billions of minds, and innumerable interactions between them? How can anyone at the Federal Reserve or the White House comprehend this complexity, much less control it in a rational manner? If one begins with such a constructive rationalist assumption, one is likely to be "flabbergasted" quite often. Interestingly, several years later Greenspan admitted that the Fed could not even calculate inflation accurately, which would undermine all historical data models used to make economic predictions.[77]

Generations of economists, including to some degree Adam Smith, were obviously guided by the dominant Newtonian scientific understanding of the universe as a kind of machine operating by strict laws, as George Gilder explains in *Knowledge and Power*.[78] They constructed a science of system and order, where nothing could be unexpected. Science "came to mean the elimination of surprise," writes Gilder, acknowledging that this did provide a basis for centuries of progress. It also led to new and deeper mysteries.

James Watson, the ultimate constructivist, claimed that genetic biology had been shown to be "nothing but chemistry and physics," but the science of DNA has turned out to be more complicated than he thought. Bill Gates described the human genome as a software program "far more complex than any we could build," but a software program still. Gilder notes that beneath even chemistry and physics there is *information* in the DNA codes defining the amino acids that build proteins, which in turn form life. Those proteins cannot logically create the information in the codes. "Information and matter are complementary but intrinsically distinct," Gilder emphasizes. So where could the information come from?

He suggests that we can best understand the world through a new "science of information," which accommodates randomness and disorder. It offers a model of information entropy based on the logic of physical entropy. Order, regularity, and low entropy characterize one end of the scale; at the other, we find complexity, disorder, apparent randomness, entropy. The world we know is at the low-entropy end, and it is much, much smaller than the other end we do not know.

Here's the kicker: all information is "surprise." It must be conveyed through a low-entropy channel, but information itself is associated with entropy or disorder. Entropy is "a measure of freedom of choice." And human choice is intrinsic to investigating and explaining the world.

At the heart of science is "the free human being who conceives and defines it." The GeneChip can search millions of bits of information, but without intelligent choice there is no logic to the search. When physics bumps up against seemingly insoluble mysteries, the human mind postulates "dark energy" and "parallel universes" out of whole cloth to explain the inexplicable. Constructivist reductionism must throw up its hands in despair. Only a mysterious human consciousness can pierce through the disorder and come to new understandings.

The science of information is a blow to reductive determinism and constructivist rationalism, since it centers on freedom, which "defies every logical and mathematical system." Human choice cannot be quantified or even explained. Gilder cites critical-rationalist Nobel laureates on the futility of trying to reduce the human mind to material flux in the brain, as a simple product of Darwinian evolution. He calls this notion more ideological than scientific. Constructivist science fails at explaining complex human beings, especially mind and consciousness. Gilder ends with the same analogy of the human brain's neural complexity to the number of atoms in the universe that Hayek used many years earlier.

Professor Thomas Nagel of New York University, the very model of academic objectivity, wrote a fine book titled *Mind and Cosmos: Why the Materialist Neo-Darwinian Conception of Nature Is Almost Certainly False*, seeking a third, more critical way between constructivist materialism and traditional religion to explain the human mind. He rejects the idea of a divine creator, but concedes that pure materialism cannot account for consciousness. His argument is an intellectual tour de force that even a theist could admire as courageous and open-minded.[79]

Nagel demonstrates his critical rationalism by being open to phenomena beyond monist constructivism, but his openness has limits. Reviewing a book by Alvin Plantinga, a Christian philosopher, Nagel admitted that he earnestly desires an alternative that could explain what the materialist worldview did not, but one that was not a traditional theism either. "If I ever found myself flooded with the conviction that what the Nicene Creed says is true, the most likely explanation would be that I was losing my mind, not that I was being granted the gift of faith," he wrote. And he continued, "I want atheism to be true and am made uneasy by the fact that some of the most intelligent and well-informed people I know are religious believers. It isn't just that I don't believe in God and, naturally, hope that I'm right in my belief. It's that I hope there is no God! I don't want there to be a God; I don't want the universe to be like that."[80] Nagel became an object of scorn in the academy for even trying to challenge its dominant materialist assumptions, but has courageously and even passionately persevered.

Professor Plantinga, safely at Notre Dame University, has probably written the most radical argument on the limits of scientific explanation and of the materialist hypothesis that constructivist science rests upon. In *Where the Conflict Really Lies: Science, Religion, and Naturalism*, he challenges the idea that traditional religion is fundamentally opposed to science, and argues that the conflicts between the two realms are only superficial. Plantinga does not deny that evolution can account for much of the change that has occurred in biological organisms, but he argues that it cannot explain all human complexity. He quotes James Shapiro, a materialist molecular biologist, writing that "there are no detailed Darwinian accounts for the evolution of any fundamental biochemical or cellular system, only a variety of wishful speculations."[81] We have noted many wishful speculations in other branches of science too.

Certainly Nagel and Plantinga have their different beliefs, which some would characterize as superstitions in one case or the other. That is not a problem, at least for a critical rationalist, as long as those beliefs do not compromise the search for truth. The problem is when zealots are convinced that they know definitively what is rational or scientific, and what is good for everyone else, and when they insist on enforcing their ideas in the concrete world, no matter how much others might object. As Albert Camus noted, "The welfare of humanity is always the alibi of tyrants."[82]

Constructivist scientific materialism is, of course, an axiom, not empirical fact. Indeed, it might even be called a religious assumption, as the mischievous agnostic philosopher John Gray explained it.[83] Hayek's term "superstition" might fit as well. Any serious rational debate on empirical matters should fit within these epistemological limits.

The scientific picture of the world today is obviously far more complex than the Newtonian conception of strict, comprehensible laws that guided the nineteenth-century Prussian professors and our own progressive intellectuals in their plans for scientific public administration based on rational abstractions. As we have seen here, a constructivist science appears to have limits in explaining the physical world, and even more so a world including human choice. Here, a critical-rationalist perspective is much more useful.

As Hayek explained, critical rationalism "is a view of mind and society which provides an appropriate place for the role which tradition and custom play" in the development of science and societies. It "makes us see much to which those brought up on the crude forms of rationalism are often blind."[84]

CHAPTER 6

MORALIZING CAPITALISM

The enormous complexity of the world, the limits of science, and the social importance of tradition and custom impose serious limits on the prospects for engineering a rational economy and perfecting society. As Friedrich Hayek concluded, the capitalist market, registering the innumerable free choices of individuals with diverse needs, is a more efficient means of processing social information and bringing a kind of rational order to economic and social activity where the state cannot. In fact, the efficiency of capitalism is conceded even by many of its critics, including Karl Marx. It is the morality of capitalism—and ultimately its viability—that must be addressed here.

The Moral Basis for Individual Freedom

Let us return to the critique of capitalism and its "limitless" individual freedom presented by Pope Francis in *Laudato Si'*.[1] The pope's challenge to capitalism is not primarily based on empirical efficiency or state competency or political ideology, but on what he regards as its inherent immorality. Indeed, Pope Francis sees capitalism as inimical to a virtuous human life.

Concerning humanity and the natural world, Francis begins with the fundamental teaching on moral action in the book of Genesis. Here, God gave humans the mandate "to have dominion" over the earth, but with the corresponding obligation to care for and protect it. All human activity must remain within that boundary. (§§ 65–67) But the pope offers no biblical support for his central proposition regarding the planet's fragility. He frequently refers to and quotes his namesake, St. Francis of Assisi, on living in harmony with all creation, but these references say nothing very conclusive about the fragility of the planet.

The pope asserts that "Jesus lived in full harmony with creation," providing an example for all to live gently in accord with a fragile nature. As evidence he presents the passage where people see Jesus walking on water (Matthew 8:27) and exclaim, "What sort of man is this, that even the winds and the sea obey him?" (§ 98) Yet this seems to be a case of dramatically dominating nature rather than living in harmony with it.

Regarding the teaching on "dominion," the pope remarks that "we Christians have at times incorrectly interpreted the Scriptures," but now "we must forcefully reject the notion that our being created in God's image and given dominion over the earth justifies absolute domination over other creatures." (§ 67) While he urges reading the words about dominion in context with other biblical passages, he does not quote the preceding sentence: "Be fruitful, multiply, fill the earth and subdue it." Indeed, some biblical translations use "be masters of" or "rule over" instead of "have dominion over" the earth. Yes, Christians (and others) in the past have felt a need to "subdue" and "be masters of" a natural world that is often hostile, while also recognizing an obligation to "till and keep it" as good stewards. There is little evidence that Christians, or many other capitalists in the past, actually believed in exercising "absolute dominion" over creation, although they have generally envisioned a hierarchy of nature with humans placed decisively on the top.

Cardinal Joseph Ratzinger, who would become Pope Benedict XVI, responded to what he considered inaccurate claims about the Christian attitude toward nature in a set of homilies on the Genesis creation narrative. Christians had been accused of following a utilitarian "ideology of progress" that undermined the ancient cosmological worldview in which all material and spiritual existence was in a natural balance:

> What we had previously celebrated—namely, that through faith in creation the world has been demythologized and made reasonable; that sun, moon, and stars are no longer strange and powerful divinities

but merely lights; that animals and plants have lost their mystic qualities: all this has become an accusation against Christianity.

Christianity is said to have transformed all the powers of the
universe, which were once our brothers and sisters, into utilitarian
objects for human beings, and in so doing it has led them to misuse
plants and animals and in fact all the world's powers for the sake of an
ideology of progress that thinks only of itself and cares only for itself.

What can be said in reply to this? The Creator's directive to humankind means that it is supposed to look after the world as God's creation, and to do so in accordance with the rhythm and the logic of
creation. The sense of the directive is described in the next chapter of
Genesis with the words "to till it and keep it." (Genesis 2:15)[2]

While "keep it" and "look after" suggest some fragility, "till it" implies
a need for taming, as do dominion, subdue, and master. All are part of
the "rhythm" and "logic" of God's creation.

The Christian Church had long ago left behind the idea of a geocentric universe literally revolving around humankind. When Copernicus
developed a mathematical analysis demonstrating that a heliocentric
universe was more plausible than the Ptolemaic geocentric one, Pope
Clement VII asked him to publish his findings, which caused little
controversy for the next several decades. But in 1616, Pope Paul V
demanded that Galileo recant his argument for a heliocentric world.
The Ptolemaic model of an ordered geocentric world had long prevailed
in large part because it provided reliable scientific tools for naviga-
tion—leading to the discovery of whole new continents—right up
until the development of the satellite-based Global Positioning System.
Ptolemaic astronomers succeeded in keeping Galileo's works on the
church's Index of Forbidden Books, until it was removed in 1741 by
Pope Benedict XIV. In 1939, Pope Pius XII praised Galileo as one of the
"most audacious heroes of research...not afraid of the stumbling blocks
and the risks on the way."[3] By their own church doctrine, popes are not
infallible on empirical or scientific matters, and their pronouncements
can be superseded by new intellectual insights.

Cardinal Ratzinger had in fact argued that intellectual trends
had gone too far in decentering humanity. He challenged what he
saw as an antihuman understanding of the world that started with

Jean-Jacques Rousseau, reviving ancient myths that deified nature. He also criticized the progressive materialism pioneered by Claude Levi-Strauss and B. F. Skinner, with its hostility to the idea of an objective human nature and a God-given moral freedom.

> Reaction and resentment against technology, which is already noticeable in Rousseau, has long since become a resentment against humans, who are seen as the disease of nature. This being that emerges out of nature's exact objectivity and straightforwardness is responsible for disturbing the beautiful balance of nature. Humans are diseased by their mind and its consequence, freedom. Mind and freedom are the sickness of nature. Human beings, the world, should be delivered from them if there is to be redemption. To restore the balance, humans must be healed of being human. In ethnology, this is the thrust of Levi-Strauss's thinking; in psychology, of Skinner's.[4]

Christian morality taught the reverse, Ratzinger argued: that the human person is at the center of creation. Human beings are distinguished from other creatures by mind and moral freedom. They were given dominion over nature, for good or evil. The human individual is always tempted toward sinful exploitation of the world and other people. Humans have freedom to choose disobedience or to act ethically and responsibly. This freedom cannot morally be taken away by others without informed consent—which was John Locke's moral justification for a free capitalist society.

That argument for the morality of individual freedom is the fundamental moral defense of free-market capitalism. In his magnum opus, *Truth and Tolerance*, Cardinal Ratzinger even put freedom on the same moral plane as truth, while recognizing problems in how it has often been practiced. Given that the Ten Commandments "are the answer to the inner demands of our nature, then they are not at the opposite pole to our freedom but are the concrete form it takes." Freedom and truth are therefore complementary, not opposites.[5]

Other popes have taken a different view of capitalist freedom. Pope Paul VI, in *Populorum Progressio* (1967), criticized the principle of free trade as "inadequate" and called for a more equitable sharing of wealth globally.[6] His concern for the poor of the world was a traditional

Christian value, but the functioning of international markets is not really a theological issue. The pope's advocacy of more international aid from the rich nations to the poorer ones ignored the reality that such aid is often squandered or funneled into the bank accounts of dictators' pals.[7] And, of course, there is the critique by Pope Francis.

Pope John Paul II presented a more balanced perspective on capitalism and freedom in *Centesimus Annus* (1991), acknowledging the benefits of free markets while advocating some restraints. He emphasized the need for compassion toward the poor, but criticized the regulatory "excesses and abuses" of the welfare state. The principle of solidarity must be balanced with that of subsidiarity, in which the "free exercise of economic activity...will lead to abundant opportunities for employment and sources of wealth."[8]

For John Paul II and Benedict XVI, familiarity with the evils of communism and Nazism served to illuminate the virtues of freedom and the benefits of market capitalism, when disciplined by moral obligations. Their successor Pope Francis came from the very different formative experience of Argentina's crony capitalism. In a book published in 2013, the same year he became pope, Jorge Mario Bergoglio declared that "wild capitalism" must be condemned with "the same vigor" as communism.[9] It hardly seems possible he was speaking literally, given that communism killed 100 million or so of the people forced to live under it. He was more likely being dramatic to make a point.[10] But clearly he viewed capitalism as bad for humanity as well as for the planet.

Crony Capitalism and the State

Today the public face of capitalism, particularly as promoted by intellectuals, has often been closest to the portrayal drawn by Pope Francis and Charles Dickens: an image of Scrooge and his cronies. In 2016, not long after *Laudato Si'* was issued, Charles G. Koch, the chief executive officer of his family business, Koch Industries, wrote an op-ed for the *Washington Post*. A widely known supporter of free-market capitalism, Koch said that he agreed with the view of Bernie Sanders, the democratic socialist presidential candidate, that the American "political and economic system...is often rigged to help the privileged few."[11]

The online version of his article was absolutely swamped with comments. Almost all agreed on the evils of rigged capitalism, but many attacked Koch himself as hypocritical, charging that he was himself a rigging capitalist—even though he had presented examples of Koch Industries opposing government subsidies that could have been advantageous. The modern media and intellectual climate had made it difficult for him to crack this capitalist stereotype by distinguishing between the free market and crony capitalism.

Libertarian capitalists like Koch have vigorously denied the charge that their view of freedom is selfishly centered only on business, or that they support "limitless" freedom. They deny that they want no market regulation at all, but do insist that the market itself acts as a discipline on business, and that government action can distort the market. When Pope Francis inveighed against a policy of "saving the banks at any cost, making the public pay the price,"[12] he was actually describing the "crony capitalism" that free-market advocates criticized during the 2007–8 economic crisis, and that Adam Smith warned against. To the extent that today's capitalism is actually cronyism, no supporter of Smith would disagree with the pope's critique.[13]

When the government rigs policies to benefit particular corporations, pluralists do not view it as market capitalism, but rather as mercantilism at best. In 2016, when the *Wall Street Journal* ran a front-page story about how the big insurance companies that had lobbied for President Barack Obama's transformational health-care plan were now losing money on their new policies, what libertarian did not think "it serves them right"? Those companies had supported a program that forced people to purchase their product under threat of tax penalties.[14]

For many people, the Obamacare tax penalties for refusing to purchase insurance were lower than the premiums on the new government-mandated policies. Individuals were thus incentivized to delay buying "insurance" until perhaps they became ill. Many people—not being as foolish as the government officials who wrote the rules—therefore did not buy insurance unless they really needed benefits, which resulted in large premium losses for the insurance companies, just as critics of the policy had predicted. The loss for the health businesses was the direct market result of old-fashioned individual self-interest on the part of consumers.[15]

Why did the government fix the rules as it had? The majority of the public appeared to agree with the view of political science professors that the corporate lobbyists must have corrupted the public officials, to win special benefits for themselves and their businesses. As a political scientist myself, I once shared that assumption, until I gained unrestricted access to one of the nation's top legislators. I then saw unmistakable evidence that he was manipulating the lobbyists more than vice versa. The lobbyists were mostly on defense, trying to stop the latest bright idea from ruining their business rather than proposing some plan of their own—although many businesses did seek bailouts when they were in trouble, especially automobile companies, in the name of their employees, of course. On the other hand, the government had to compel some of the nation's biggest banks—against the will of two major banks in particular—to sell their shares and accept bailouts during the 2008 mortgage crisis.[16]

Bureaucrats lacking knowledge of an industry they are regulating often go to the dominant firms in the industry to learn how best to regulate it. This turns out to be a great advantage for the industry leaders, but is initiated by the regulators.[17] Most of the novel ideas for regulating the market come from politicians playing the good fairy, or from intellectuals trying to perfect society, or from bureaucrats thinking they can solve every problem given sufficient power.

How could the health insurance companies have come up with such an idea as forcing people by law to buy their product? Not even the most aggressive crony capitalist ever thought such a thing was possible. Officials in the Obama administration warned the insurance executives that they could either support the process of creating the health-care plan and obtain generous treatment (and new business), or suffer the consequences of a law that would punish them with excessive regulation. It was a deal they could not afford to refuse but did not initiate themselves.[18]

At a forum in Washington, D.C., in 2018, covered by C-SPAN, the dean of Johns Hopkins University's School of Medicine, the head of the Memorial Sloan Kettering Cancer Center, and the CEO of Children's National Health System were provoked by a very progressive moderator who charged that hospitals were bowing to business interests to secure funds. In response, the forum participants noted that their main source of income was actually the government, through Medicaid, Medicare, and research grants, and that most of their effort was spent in responding to government officials and meeting their demands.[19]

Most people believe that politicians are so in need of campaign funding that they must give favors to businesses in exchange for contributions. My experience is otherwise. More often than not, politicians receive campaign funds from different companies vying for favor, and from both anti- and pro-business groups on all sides of issues. That places the politician in the commanding position.

Bureaucrats cannot legally take bribes, and are mostly not foolish enough to try, but they can do favors in the hope of securing a job or a client later on. And they have more opportunities for mischief than politicians, since bureaucrats both write rules and enforce them. Congress writes laws so broadly—to be "fair" or "equitable"—that regulators have room to do pretty much what they wish in interpreting them. Congress requires agencies to publish their official rules, but most of the directives issued by agencies are not designated as official rules, and the top officials can make exceptions to them in "emergencies." Congress has little idea what the bureaucracy is really doing.

As Clyde Wayne Crews explained, the great majority of what the bureaucracy does can be characterized as "regulatory dark matter," which the rest of the world knows little or nothing about. Much of it is unwritten, and thus escapes judicial review, since courts deal with written material. The Consumer Financial Protection Bureau, created by the Dodd-Frank Financial Control Act, had no formal regulations at all, and the director, Richard Cordray, testified before Congress that his bureau would not even issue regulations that explain what practices might violate the law. So, as Crews asks, "how will a bank, credit union or other financial services business know if it has violated the law?"[20]

The simple answer is: the businessman probably won't know until it's too late. The capitalist is the weaker party when the rules cannot even be known, and when a mere telephone call from a low-level bureaucrat making a "suggestion" can carry the weight of law, but is really just a personal or political desire. An unwary executive can be sent to jail for violating an unknown rule, and whole businesses and industries can be destroyed.

So while big businesses can indeed sometimes get political favors, the all-powerful capitalist who controls politicians is a myth. Joseph Schumpeter, as usual, hit it on the head. A businessman, with his practical orientation and his focus on the stock market as the highest ideal, is no match for an intellectual or a public official or politician in the ability to manipulate public emotions, with inspirational holy-grail ideals and morality tales of greedy villains who should be punished by jail or at least public humiliation.[21]

Markets, Morality, and Complexity

What, then, is a moral capitalism? We have identified three possible types: the pragmatic mercantilism of Jean-Baptiste Colbert; the rationalized type inspired by David Ricardo and associated with Voltaire, centered on the idea of rationalizing markets; and the "natural liberty" pluralism of Adam Smith, which would eliminate coercive "preference or restraint" and establish itself "of its own accord." The first is excluded here as actually pre-capitalist, and the second as too narrowly constructivist. That leaves the third, which is closest to the popular understanding of capitalism, and also most open to moral considerations beyond crude scientific rationalization.[22]

Contrary to Pope Francis's charge that capitalism lacks a moral foundation, Adam Smith argued in favor of free markets not primarily on the grounds of pragmatic utility or scientific efficiency, but on the basis of moral goodness and justice.[23] In fact, most of the theorists in the pantheon of pluralist capitalism justified the market as the primary means for lifting people out of poverty and increasing prosperity around the world. They certainly did not believe that the essence of capitalism was "unbridled consumerism," as Pope Francis claimed. Their ideal of freedom was not limitless, but allowed individuals to make choices in markets operating under a limited government that established basic rules of justice.

Neither would they agree that social problems such as global hunger can be solved "simply by market growth," as the pope charged. Smith himself emphasized the necessity of free institutions in addition to the market, including voluntary charity and local welfare, which he called beneficence.[24] Defenders of capitalism today do not reject the

personal need to care about the less fortunate. Data show that supporters of small-government, free-market policies in the United States
contribute more to charity than do those on the left side of the political
spectrum.[25]

Rather than opposing all market regulation, even a libertarian icon like F. A. Hayek argued that a functioning market cannot
be purely laissez faire, but necessarily rests on moral custom and
traditional law.[26] At a minimum, market capitalism requires rules
forbidding theft and fraud, government administration of justice,
and protection of property. Most theorists identified the medieval
moral code as the source of the capitalist moral order, Schumpeter
even making the case that capitalist society could not survive without
those moral "protecting walls."[27]

Harold Berman listed the essential moral supports for capitalism:
respect for the human individual; severability of property; prohibitions
against coercion, theft, and fraud; a positive attitude toward work; and
a morality of thrift, encouraging capital savings and lending. As noted
earlier, all of these were developed during the Middle Ages or before:
the biblical moral code, property rights from Roman law, a Christian-
inspired individualism, the monastic valorization of work, and early
banking practice.[28] In *The Bourgeois Virtues*, Deirdre McCloskey summarized the essential moral code for capitalism as: the feminine virtues
of love, hope, and faith; the masculine virtues of courage and temper-
ance; and the androgynous virtues of prudence and justice.[29]

So capitalism rests on a moral code and on institutions that promote and enforce it. But the real test is in the results. Pope Francis
claimed that wealth from an expanding market economy never "trickles
down" to the poor. "Trickle-down economics" may be widely derided,
but free markets are in fact the primary way that wealth has reached
the less fortunate. Responding to the pope's criticism, Rev. Robert
Sirico of the Acton Institute noted that the International Labor Office
had estimated that the number of people on earth earning only $1.25 a
day or less fell from 811 million in 1991 down to 375 million in 2013. No
other explanation for that success is more plausible than the spread of
market capitalism during that time.[30]

Laurence Chandy and Geoffrey Gertz of the Brookings Institution
estimated that the total number of poor people in the world fell by half

a billion between 2005 and 2010, after India, China, and other countries had adopted market-oriented reforms.[31] This progress cannot be credited to a governmental redistribution of global wealth, or a colossal outpouring of charity, as those things did not happen. It could only have come about through market productivity and more open trade.

Pluralist capitalism, while not a panacea, has lifted millions up from abject poverty or bare subsistence, and has led to a remarkable increase in life expectancy worldwide. Would things have turned out better under mercantilism, or the philosophy of Thomas Malthus (who opposed both free trade and poor relief)? Have the rationalized command economies done better in sustaining and bettering the lives of ordinary people? Given the clear moral benefits of free markets—and the failures of government regulators to improve upon them very much—would it not be immoral to deny a capitalist order to the poorer nations of the world?

A Ricardian model of a rationalized welfare-state capitalism, controlled by bureaucratic expertise, promises to reduce poverty but has achieved little, especially in comparison with markets. It is also subject to moral abuse, through what John Paul II criticized as "bureaucratic ways of thinking" in public agencies rather than "concern for serving their clients."[32] One example is the "culture of complacency" that the inspector general found at the U.S. Environmental Protection Agency, which led to delays in taking action against serious misconduct that might have threatened the agency's control of chemical risks.[33] The question should be: which is more moral and effective in most cases, Smithian natural-liberty pluralist capitalism or Ricardian rationalized expertise?

Underlying the contrast of pluralist capitalism and rational centralization are two distinct visions of society. Two great philosophers of the eighteenth century set out the two alternatives, and Yuval Levin nicely clarified the crucial differences between them in *The Great Debate: Edmund Burke, Thomas Paine, and the Birth of Left and Right*. Burke and Paine differed on reason vs. prescription, justice vs. order, nature vs. history, choice vs. obligation, revolution vs. reform—Paine emphasizing the first in each pair and Burke the second. These contrasts are a useful way to summarize the two visions of society and to understand the moral basis for each.[34]

Running through all these particular contrasts, writes Levin, is "a disagreement about the authority of the given past in political life." More than "a staid and simple dispute between tradition and progress," it is fundamentally a disagreement over the means of arriving at truth—a question of epistemology, if you will. Paine, an admirer of Isaac Newton, devoted himself to scientific pursuits and inventions, and believed that a constructivist science "seemed to open endless possibilities for the conquest of nature and the empowerment of man." Burke was more critically rationalist, emphasizing an "experimental science" rooted in the concrete experiences of generations that develop into lasting traditions.

This, again, was at base a difference on the question of complexity. Paine argued in *The Age of Reason* that if individuals had more freedom to exercise their reason and apply scientific understanding to policy questions, it would "free liberal societies of countless ancient prejudices and open the way to a new politics of liberty," as Levin explained. Burke, on the other hand, "thought the governing of human communities was much too complex a task to be simplified into a series of pseudoscientific questions and resolved by logical exercises." In his view, governance depended on "a degree of knowledge and wisdom about human affairs that could only be gathered from the experience of society itself."[35]

Burke followed his friend Adam Smith in maintaining that the way to solve complex problems is to work them out experimentally. As we have seen, Hayek took the same view, drawing a distinction between the abstract, constructivist rationalism of Paine and Descartes and the more critical rationalism of Burke and Locke, and favoring the latter as promoting a balance between reason and tradition.[36]

Levin explains that ideal of balance powerfully in his introduction, answering the charge that Burke was inconsistent in supporting liberty for the American colonies while opposing the French Revolution. But as Levin observes, Burke argued in favor of the American Revolution "both as a reformer and as a conserver of Britain's political tradition," and against the French Revolution because it repudiated tradition. Burke consistently aimed for "a balance between stability and change."

Levin quotes the concluding words of *Reflections on the Revolution in France*, where Burke described himself as:

"one who wishes to preserve consistency, but who would preserve consistency by varying his means to secure the unity of his end, and, when the equipoise of the vessel in which he sails may be endangered by overloading it upon one side, is desirous of carrying the small weight of his reasons to that which may preserve its equipoise."[37]

The questions concerning balance, equipoise, tension, synthesis, fusion, and pluralism are at the core of the Left/Right political distinction. On one side is constructivist rational utopianism. On the other is a tense balance of free reason and evolved tradition. This debate began long before Burke and Paine, and even before Locke and Descartes, as both Hayek and Schumpeter pointed out.

Reactionary Simplification

To the contemporary Left, the whole attempt at morally justifying a capitalist civilization based on freedom is simply a matter of reactionary self-interest on the part of the Right. Mark Lilla offered a most insightful philosophical discussion of this proposition in *The Shipwrecked Mind: On Political Reaction* (2016), even though it takes place within his framework that casts conservatives as tragicomically Quixotic figures, hopelessly shipwrecked in the past and unable to comprehend the modern world.[38]

In some ways, Lilla was more perceptive about the larger Right than most conservative commentators were. Discussing the Paris terrorism attacks of 2015, he warned about the rise of a new nationalist right, as exemplified by the journalist Eric Zemmour and his book *Le Suicide français*. In Lilla's account, Zemmour first appeared on the scene as a fresh, affable, cool performer on television, then transformed into "an omnipresent Jeremiah who telegraphed the same message, day in and day out, on all available media: France awake! You have been betrayed and your country has been stolen from you." He was a "self-declared patriot nostalgic for national grandeur," who assembled a litany of betrayals by a treacherous establishment class of academics, businessmen, bureaucrats, journalists, and elite radicals. The list of elite betrayals, "too eclectic to be labeled," included outsourcing jobs and "killing commerce in small towns and villages," supporting the internationalist Common Market, abandoning the gold standard, surrendering to German domination, encouraging Muslim immigration, imposing speech codes, promoting birth control and abortion, favoring no-fault divorce, promoting gender studies and divisive identity politics, and more.

There would be echoes of this litany in Donald Trump's inaugural address: The politicians were flourishing but the people suffered. Factories were shuttered. Businesses migrated abroad with no consideration for the millions of American workers left behind. Middle-class wealth was being redistributed around the world. But "from this moment on it's going to be America First. Every decision on trade, on taxes, on immigration, on foreign affairs, will be made to benefit American workers and American working families."[39] The new president promised to take the country back for the working class. He would restore productive high-end manufacturing, which had been losing ground to a fictional service economy of administrators and clerks, and would stand up to the countries such as China that were taking advantage of America.[40] He was as good as his word in his proposals, although they were mostly frustrated by Congress and somewhat by the counteractions of other nations.[41] Most of his successes were more conservatively capitalist, such as cutting taxes and trimming regulations. Even his tariff policies were arguably just a correction of past overregulation.[42]

While Lilla predicted this kind of nationalist reaction to a sense of betrayal by the establishment, his main focus is the more mainstream thinkers on the right. He begins with the two great modern conservative philosophers, Eric Voegelin and Leo Strauss, whom he treats in an admirably evenhanded way. A leftist himself, Lilla sees Voegelin and Strauss as true philosophers, working in the abstract world of the mind. He distinguishes them radically from their more simplistic disciples, whom he labels "theoconservatives" in the case of Voegelin and "neoconservatives" in the case of Strauss. These secondary intellectuals, keeping one eye on politics, create "stories" to simplify ideas and rally a popular following.

Lilla identifies Voegelin as the first philosopher to inspire modern conservative intellectuals to defend Western tradition. Raised and educated in Austria, he won a fellowship that enabled him to study at various American universities, including Columbia, where he attended courses taught by John Dewey. But through his lifetime he was most influenced by German philosophers and historians. Returning to Vienna "with an abiding hatred of racism," he opposed Nazi racial theories, and escaped back to the United States on the very day the Gestapo were searching his home. His breakout work was *The Political Religions*, published in 1938, focusing particularly on Nazism and communism. This book was the core of a larger history that Voegelin "would elaborate and refine over the next three decades." Voegelin's work especially captivated the intellectuals who launched the conservative *National Review* in 1955.

As Lilla explains Voegelin's history, the "cosmological" civilizations that were characteristic of the agricultural age, uniting belief and power, were mostly unchallenged, except for short intervals in Athens and Jerusalem, until the rise of Christianity, "the first world religion to offer theological principles for distinguishing divine and political orders." While this distinction, elaborated in St. Augustine's *City of God*, was often "honored mainly in the breach," the principle worked its way into the substance of European society. It set the roots for an ethic of individual freedom under God, which flowered in the philosophy of John Locke. But the fragile Lockean balance succumbed to the "radical Enlightenment," which succeeded in "decapitating" its God and launching the modern age of skepticism and materialism.

Voegelin was not explicitly religious, notes Lilla, but did believe "in his own way the evidence of a divine transcendent order" and "valued the power of religion itself as a vitalistic force" directing society to the "good ends" of controlling chaos and restraining oppression. Voegelin saw clearly that secular gnosticism by the twentieth century had radicalized into Marxism, fascism, alt-right nationalism, and progressive positivism, as "political religions" worshiping party or *Führer* or state or science, or all four. In William F. Buckley Jr.'s favorite Voegelinism, they had "immanentized the eschaton," forcing next-world perfectionism into this world by means of violence and state oppression. This is what Voegelin's unauthorized disciples were determined to confront.

The other great intellectual appropriated by the modern Right, Leo Strauss, was born in Germany, served in World War I, wrote for Zionist publications, and earned his doctorate at the University of Hamburg.

He was greatly influenced by Martin Heidegger, who taught him the importance of classical philosophy. Strauss then spent the rest of his life in the United States defending Socratic philosophy, or at least "the possibility of philosophy," as Lilla puts it. Strauss agreed with Voegelin that the source of Western civilization's vitality was the "tension between two incompatible ways of addressing the human condition," divine revelation and philosophy, but he insisted that true knowledge was attained "exclusively through human reason."

Intellectually, Strauss argued, one must choose between the two because "all societies require an authoritative account of ultimate matters—morality and mortality especially—if they are to legitimatize their political institutions and educate citizens." Legitimacy can come either from rational-Athens or from divine-Jerusalem, but not both, Strauss maintained. The two cannot be synthesized intellectually, as St. Augustine and St. Thomas attempted to do. Nor can both be rejected, denying truth altogether, as skeptics and relativists following Machiavelli would do. Strauss himself was "a proud Jew who respected his people's belief," and he "appreciated what religion at its highest development could offer." Revelation could offer comfort to ordinary people, but a reflective person must rationally choose Athens alone.

Lilla discusses other philosophers too, such as the intriguing Franz Rosenzweig, who tried to synthesize philosophy and religion. While he respects the philosophers, Lilla is critical of intellectuals of all stripes who rationalize history in "stories" about discrete periods defined by a single project or concept. These intellectuals are of two types: the right-reactionary, driven by "nostalgia," a longing for an idealized "past in all its splendor," and the left-revolutionary, motivated by a vision of a "radiant future." The reactionary believes himself to hold the "stronger position" because "he is the guardian of what really happened, not the prophet of what might be." But in Lilla's view, the reactionary nurtures "myths" about the past that "do nothing but inspire a more insidious dream that political action might help us find our way back." The great philosophers have something to teach us, but their Right popularizers are attempting to immanentize reactionary bedtime stories.

The archetypal popularizers of Voegelin were the "reactionary" editors of *National Review*, whose founding mission was to "stand athwart history, yelling Stop." Linking Voegelin to *National Review* is uncontroversial, since Buckley himself and the literary editor Frank Meyer had acknowledged his influence. But Lilla does not explain why "stop" is the same as *going back* to an idealized past. He correctly highlights Voegelin's concern that politically oriented intellectuals were using his ideas to force history into the service of conservative ideological ends, and points to Meyer in particular.[43] Lilla also criticizes Voegelin's "conservative readers" for disregarding his belief that Christianity's vision of heaven had inspired notions of earthly paradise that degenerated into destructive ideologies.

Lilla describes the Buckley-Meyer "story" as one of America emerging from World War II still "strong and virtuous, only to become a licentious society governed by a menacing secular state." Yet Buckley and Meyer were not simply looking back to 1950s traditionalism, as Lilla believes. Both gave credit to Hayek for awakening them to the need for a more direct defense of capitalism and freedom than was offered by earlier traditionalists or by the pragmatic capitalists who made peace with New Deal market regulation and the welfare state.[44] Certainly they were concerned that Western tradition was being undermined by state power, in the name of public order and human welfare.[45]

Lilla is likewise critical of Strauss's disciples, who admired his rationality but were "traumatized by the changes in American universities and society" that happened in the 1960s, so they gravitated to the circles of "mugged by reality" neoconservatives forming in New York and Washington, around Irving Kristol. These neoconservatives developed a new catechism that "begins with the assumption that the modern liberal West is in crisis, unable to defend itself intellectually against internal and external enemies who are abetted by historical relativism." The remedy was to return to classical philosophy as a means to "shore up the American polity" at home and to promote "liberal democracy everywhere." This notion of a "redemptive historical mission," writes Lilla, is "an idea nowhere articulated by Strauss himself."

The neoconservative Straussians might fit into Lilla's "reactionary" designation if they had really expected to revive ancient Athens in modern America, but they did not. If making the whole world more democratic was the top neoconservative goal, as Lilla says, that might be seen as revolutionary more than reactionary. Kristol himself, however, unlike many of his acolytes, was very skeptical of the possibility for democratizing the world.[46]

Where Lilla's definition of "reactionary" would seem to fit best is in the Muslim world. This is where we most clearly see a belief in a lost golden age, and the most "potent and consequential" reactionary force trying to go back, although Lilla insists that this is not a "uniquely Muslim" phenomenon.

Lilla's reactionary/revolutionary contrast is a useful intellectual tool, but it does not suffice to clarify the Right/Left division. In fact, Lilla seems to tire of the classification himself when he writes of White Russian survivors of the Bolshevik revolution. They are said to have looked back to the old regime while wanting to "move forward to a new age inspired by the golden one." Lilla sums up: "Their nostalgia is revolutionary." But wasn't nostalgia the defining characteristic of right-reactionaries, in contrast with left-revolutionaries?

Lilla also mentions a distinction among the Buckley-inspired conservatives, some wanting a return to an idealized traditional past, while others dream of a libertarian-capitalist future "where frontier virtues will be reborn and Internet speeds will be awesome." The latter would appear to be more revolutionary than reactionary. And that would make Buckley, Meyer, and indeed most of the early movement conservatives not really reactionary either. In his otherwise insightful examination of the Right, Lilla seems to have missed the crucial fact that the Buckley movement was *both* libertarian and traditionalist.

The Enduring Tension

A consistent explanation of the modern Right must indeed start with Voegelin and *National Review*, and must include the idea of reaction. But the story actually begins with the belief of Buckley and the early editors that an effective defense of Western freedom and tradition required an explicit rejection of the status quo and its crony capitalism. Russell Kirk's incredibly successful book *The Conservative Mind*, published in 1953, became the source of the descriptive term "conservative" to link the enterprise to the Western tradition. Kirk himself did this primarily through Edmund Burke rather than an economist like Adam Smith, while not being hostile to him.[47] Other early *National*

Review editors emphasized the central importance of freedom and free-market capitalism.

By 1962, Brent Bozell Jr., a *National Review* editor, had characterized Frank Meyer's efforts to synthesize freedom with tradition as a "fusionist conservatism." Bozell's characterization was meant as a criticism of what he regarded as the dualism of the new conservative movement (and of *National Review*), a dualism that he thought overemphasized freedom and capitalist markets at the expense of tradition and morality. In a way, "fusionist" did describe the endeavor to harmonize freedom and tradition, though it took Buckley about two decades to embrace the term.[48] Meyer was most identified with fusionism, but he preferred the word "tension," or the more technically correct "synthesis."[49]

Still, Buckley had regularly insisted that the new conservatism was not ideologically rationalistic, but was a synthesis of different principles and values. He had emphasized the concept as early as 1959, in *Up from Liberalism* (whose title signals not an absolute rejection of liberalism but a step up from it). He summarized those principles as: "freedom, individuality, the sense of community, the sanctity of the family, the supremacy of conscience, the spiritual view of life," free markets, decentralized political power, and free people living together "peaceably, in order, justice and harmony, guided by prescriptive and traditional norms."[50] With Buckley's constant repetition, and reinforced by Meyer among others, a synthesis of these principles became more or less accepted as the heart of modern capitalist conservativism.

The critics were correct that there was an emphasis, perhaps an overemphasis, on freedom and capitalism in relation to all the other values of the tradition. But the synthesis included a means/ends distinction, which could be summarized as using libertarian means to achieve traditional ends, although freedom was not merely a means to other ends.[51] Whatever the limitations of this synthesis, the most important intellectual institutions that supported both free markets and traditional values could not avoid the fusionist term. These included the Intercollegiate Studies Institute, the Heritage Foundation, the Philadelphia Society, the Fund for American Studies, the Leadership Institute, Young America's Foundation, and the American Conservative Union (which organizes the Conservative Political Action Conference), among other new organizations.[52]

This fusionist justification for a moral synthesis of tradition and free-market capitalism invigorated a movement and a party, and even elected a president. In his first address to his fellow conservatives after his inauguration, President Reagan precisely explained his philosophical heritage. After naming "intellectual leaders" like Russell Kirk, Friedrich Hayek, Henry Hazlitt, Milton Friedman, James Burnham, and Ludwig von Mises as influential in his thinking, he spoke of a favorite:

> It was Frank Meyer who reminded us that the robust individualism of the American experience was part of the deeper current of Western learning and culture. He pointed out that a respect for law, an appreciation for tradition, and regard for the social consensus that gives stability to our public and private institutions, these civilized ideas must still motivate us even as we seek a new economic prosperity based on reducing government interference in the marketplace.[53]

Reagan said that Meyer had "fashioned a vigorous new synthesis of traditional and libertarian thought—a synthesis that is recognized by many as modern conservatism."[54]

In Meyer's story of Western civilization, the tension between freedom and tradition generated a dynamism that resulted in capitalist prosperity resting on a moral substructure, which in turn led to dominance of the world order. In the Western world it began intellectually with Augustine's theology of the City of God and the City of Man, which held that "peace is not full" in this world, even with good leaders, but always in tension with "anxiety and effort."[55] Because humanity is fallen, freedom must always be kept under some limits. This ideal of individual freedom in tension with moral restraints developed through the Middle Ages, in Magna Carta and Aquinas, on through Locke and Burke, and to the American Founders.[56]

Meyer argued that the seeming dilemma of having two first principles can be resolved "only by the classical logical device of grasping both horns." Both premises are true, he said: "on the one hand freedom is essential to the nature of man and neutral to virtue and vice; on the other hand good ends are good ends, and it is the duty of man to pursue them." What's more, these two premises are not contradictory in the real circumstances of life. "Rather they are axioms true of different though interconnected realms of existence."[57] In any practical situation, no one principle overrides all others.

> Freedom remains the criterion principle, the guide; but the applica-
> tion of principle to circumstances demands a prudential art. The
> intricate fibers of tradition and civilization, carried in the minds of
> men from generation to generation, always affect the realization of
> any general principle. Furthermore, no practical situation can be the
> realization of a single principle, however important. The compelling,
> if secondary, claims of other principles, though not decisive to judg-
> ment in the political sphere in the way that freedom is, do nerverthe-
> less bear upon every concrete political problem.[58]

So freedom and capitalist markets are not sufficient for a moral society; they must be prudentially balanced with tradition.

Meyer and Buckley both acknowledged Hayek as their inspiration in reviving that synthesis.[59] In 1945, Hayek introduced the paradox that free societies must be based on tradition.[60] He expanded upon the idea in *The Constitution of Liberty* (1960).[61] Then he added further refinement in the conclusion of his final work, *Law, Legislation and Liberty* (1973), which was fusionist to its core, without actually using the terminology. Here, Hayek rejected both the notion that freedom can be attained by casting off conventional morality and the idea that a constructivist rationalism is superior to the market:

> I believe people will discover that the most widely held ideas which
> dominated the twentieth century, those of a planned economy
> with a just distribution, a freeing ourselves from repressive and
> conventional morals, of permissive education as a way to freedom,
> and the replacement of the market by a rational arrangement of a
> body with coercive powers, were all based on superstitions in the
> strict sense of the word.

At the root of those superstitions, he continued, was a disregard for complexity and "an overestimation of what science has achieved." Ironically, these superstitions had germinated in the supposed Age of

Reason, when the so-called Enlightenment had pushed aside the deeper rationality to be found in tradition:

> What the age of rationalism—and modern positivism—has taught us to regard as senseless and meaningless formations due to accident or human caprice, turn out in many instances to be the foundations on which our capacity for rational thought rests.[62]

Hayek's list of wrongheaded ideas may seem strange to many, from one standpoint or another. But as the first modern fusionist, Hayek viewed a tension between freedom and traditional moral values as the essential driving force of Western civilization.

Hayek pointed to Magna Carta as a key example of joining liberty to a traditional moral order.[63] When Daniel Hannan, then a member of the European Parliament for England, visited the United States on the 800th anniversary of the signing of Magna Carta, he underscored the supreme importance of that document in the development of a pluralist Western order. He also shocked his audience by revealing that there had not been one monument to that great document of liberty erected by the British people in their country. The Magna Carta Memorial that stands at Runnymede was actually placed there by the American Bar Association in 1957. Yet in his American travels, Hannan had found one or more such monuments in every U.S. city he visited.[64]

The irony is that the birthplace of the great charter of divided power came to be dominated by divine-right kings who subordinated the church and other institutions to the monarchy, so that it became difficult to invoke Magna Carta in defense of traditional liberties, politically or intellectually. But at the same time, its ideals were being carried across the ocean to America by British colonists. Hannan emphasized the Anglo-Saxon, Protestant roots of America's founding principles, yet along with Puritan, Presbyterian, and Anglican colonists there were also Baptists, Quakers, Catholics, and others, resulting in a Christian influence that was not sectarian but "broad and pluralistic."[65]

Roger Williams, the Baptist founder of the Rhode Island colony, actually worked with English parliamentarians in an unsuccessful effort to put aspects of Magna Carta in the colony's charter. Lord Baltimore, a Catholic, tried to have the entire Magna Carta printed in his Maryland charter, but was likewise denied. William Penn, a Quaker, drew upon Magna Carta in drafting the constitution for Pennsylvania (1682), and printed the entire document with commentary a few years later.[66]

When Americans wrote their Declaration of Independence, and eventually the constitution for the new nation, their main source for incorporating Magna Carta values was John Locke. Although Voegelin, among others, criticized Lockeans for separating religion too far from the state, the Founders believed that religion could moderate the state without merging into it, and stressed the practicality as well as the merit of religious tolerance under an effective government.[67] The visible success of the new nation seemed to bestow moral legitimacy.

The Founders' vision of a pluralist republic, with a market economy, mostly held sway in the United States until the progressive intellectual revolution led by Woodrow Wilson and John Dewey questioned its legitimacy. The pluralist consensus was then challenged by an ideology aiming to create a more perfect society through expert administration and scientific education. This constructivist doctrine was not seriously challenged intellectually until Hayek in the mid-1940s.[68] Over the next several decades, the constructivist and fusionist paradigms were in competition, alternating in power, with no other significant alternatives until the explicit pragmatism of George W. Bush, social identity politics,[69] and the nationalism of Donald Trump.

"Compassionate Conservatism"

The attraction of a synthesis is broad appeal to a variety of groups and individuals who put a priority on one or another of its different principles. But that ability to draw in divergent interests is also a weakness. Some libertarians have seen fusionism as simply libertarianism with a traditionalist false front.[70] Others have criticized it as overly traditionalist.[71] Many traditionalists, on the other hand, have found it too libertine or too antigovernment, and too short on compassion.

George H. W. Bush pointedly distinguished himself from his predecessor, Ronald Reagan, by promising a "kinder and gentler" presidency. His even more critical son George W. labeled himself a "compassionate conservative" to separate himself from supposedly uncompassionate Reaganites. Michael Gerson, as an adviser to the second President Bush, made this theme the center of his book *Heroic Conservatism* (2007), where he set out the doctrine of a pragmatic yet moralistic conservatism, with which both Bushes were identified.[72]

In Gerson's view, the fusionist conservatism of Reagan represented "a distain" for government and a "brutal indifference" to the consequences of eroding government's capacity to promote "idealistic" policies that could support "progress in other lands" and "social justice at home."[73] The justice of a society, he wrote, must be measured by its treatment of the "helpless and poor," while ideals such as limited and decentralized government, free markets, and individual freedom are secondary at best.[74] A legitimate politics requires prioritizing justice and tolerance under one absolute truth: a belief in human dignity, derived either from religion or from secular philosophy.[75]

During the Obama administration, Henry Olsen and Peter Wehner attempted to add the still popular Reagan to the cause of compassionate conservatism by distancing him from an overly libertarian fusionist capitalism and insisting that he "was not in fact antiestablishment." Unlike current activists on the right, he sought to win over the establishment. Reagan was "unwavering" in cutting marginal tax rates, promoting economic growth, firing the striking air traffic controllers, and winning the Cold War, but he did not actually "roll back government to the extent he promised." By the end of his presidency, "federal spending averaged 22 percent of GDP, higher than it was under Carter and the highest it had ever been." Only with the Obama administration and the Covid-19 pandemic would it climb higher still.[76]

But this does not quite get it right. According to a major academic study, Reagan's tax cut actually drove total spending down in proportion to GDP, from a projected 23.8 percent to 22 percent.[77] More important, total federal spending includes defense outlays, which Reagan had promised to increase, and did. Looking only at nondefense discretionary spending, which is what finances the normal operations of domestic government—and is what Reagan promised to cut and which a president can influence—I used government historical data to demonstrate that Reagan had actually reduced domestic spending in absolute dollar terms by 9.6 percent over his eight years in office. Even including entitlements, Reagan reduced total domestic spending in relative terms, from 17.4 to 15.6 percent of GDP.[78]

After I pointed out these figures, Wehner presented data showing that Reagan increased domestic spending from $149.9 billion in 1981 to $173 billion in 1988.[79] To explain the different numbers, one must get into the weeds of the U.S. budget. Wehner measured spending with "budget outlays," which include residual spending from decisions made by past presidents and Congresses. The more appropriate category to measure Reagan's spending is called "budget authority," which only includes new spending. Also, Wehner ignored Reagan's success in reducing the spending projected in the last Carter budget, for fiscal year 1981.

Looking at the appropriate data in the 2015 Budget Historical Table 5.6, one finds that nondefense discretionary budget-authority spending went from $166.7 billion in 1980 down to $160.7 billion in 1988, a 3.6 percent decline.[80] More important, a budget table from 1991, much closer to Reagan's time and the actual events, showed a 9.6 percent decrease, the number I used in my book and article.[81] Government budget figures change over time with revised data estimates, inflation adjustments, etc., which may serve current needs but can distort the historical reality after a quarter century.[82] In any case, Wehner was factually incorrect by either year's measurement. Reagan absolutely decreased nondefense budget-authority discretionary spending, which is the relevant measure. He was the only president in modern times to do so; everyone else posted increases—the two Republican Bushes more than the Democrats Jimmy Carter and Bill Clinton.

Besides pointing to Reagan's spending record, Wehner and Olsen asserted that he "was not a man for all conservative causes" because he sometimes compromised capitalist ideals on trade and regulation, emphasized "human dignity" and compassion, sympathized with the average man, and believed that we have an "obligation to help the aged, disabled and those unfortunates who, through no fault of their own, must depend on their fellow man," as Reagan himself put it. They simply assumed that compassion, dignity, and welfare are incompatible with fusionist, capitalist conservatism. But for Reagan, any conflict between compassion and spending reduction was usually about means rather than ends. Policies and principles were often based on his commitment to federalism, not a disregard for social welfare. In his first inaugural address he explained, "It is my intention to curb the size and influence of the Federal establishment and to demand recognition of the distinction between the powers granted to the Federal Government and those reserved to the States or to the people."[83]

In a later address, Reagan spelled out how federalism as a principle was his reason for reducing federal spending: "We're not cutting the budget simply for the sake of sounder financial management. This is only a first step toward returning power to the states and communities, only a first step toward reordering the relationship between citizen and government."[84] He regarded America's federalism as "the secret of our success."[85] He believed that compassion is best practiced locally, by government, private organizations, and individuals. So he proposed a "grand bargain" with Congress: Medicaid would become a wholly national program in exchange for welfare decentralization, sending forty programs to the states.[86]

Olsen and Wehner asserted that "human dignity—not human freedom—came first" for Reagan. But in support of this claim, they cited in their article a statement by Reagan including the word "choose," pointing to freedom as an aspect of human dignity, and they acknowledged that Reagan "was indeed a great champion of human freedom." They got closer to the core of Reagan's beliefs when they cited his rejection of "ideology" as "a rigid, irrational clinging to abstract theory in the face of reality," which they considered to be "the complete opposite to principled conservatism." Reagan's fusionist philosophy, however, elevated both freedom and dignity as crucial principles, not in a constructivist ideology where one must take precedence, but in a synthesis to be worked out in concrete applications.

The first principle for Reagan, in fact, was neither human dignity nor tradition nor freedom nor markets. Addressing members of the Royal Institute of International Affairs at the Guildhall in London near the end of his presidency in 1988, he said that "the strength of our civilization and our belief in the rights of humanity" derive from faith in "a higher law" and trust in the power of prayer. He described his foreign policy with respect to the Soviet Communist world as "a crusade for freedom" that was "not so much a struggle of armed might, not so much a test of bombs and rockets, as a test of faith and will."[87] As with Locke and even Burke, there was no single master principle except for that "higher law."[88]

That master principle, governing a creative tension between human freedom and traditional restraints, was the key fusionist moral justification for a capitalist order. Beyond that, there was no single first principle, but rather a synthesis of ideals allowing for empirical discovery of the best solutions in any given situation. While Leo Strauss questioned the rationality of this synthesis, he acknowledged that it brought a unique energy and vitality to Western civilization.[89]

Doubling Down on Ideology

Reagan's fusionism was succeeded politically by the "kinder and gentler" presidency of George H. W. Bush, the pragmatic centrism of Bill Clinton, and the "compassionate conservatism" of George W. Bush. Whatever the abstract merits of the latter, it delivered rather less than promised, and a more progressive president followed.

The Bush-Kennedy No Child Left Behind Act of 2002 had set the goal that all K–12 students would be proficient in math and reading by 2014, and it increased federal spending on education by more than 90 percent. President Obama added $7 billion more for school improvement grants and $4 billion in Race to the Top grants for schools that were judged to have promising plans for better instruction. In 2017, three years after the magic date, the Department of Education reported that test scores, graduation rates, and college enrollment rates were no better in schools that received the compassionate funding than in schools that did not.[90]

President Bush also added a whole new entitlement program for prescription drugs, even with entitlements already exploding the national debt. While Medicare Part D turned out to be less expensive than was projected, Bush's proposals still added an unfunded liability estimated at $18.7 trillion, higher than the entire official debt as estimated in 2016.[91] Overall entitlement spending increased from 5 percent of the economy in 1965 to 15 percent by the end of Bush's second term, while the personal savings rate dropped from 25 to 18 percent. President Bush compassionately added a Federal Reserve debt of multiple trillions, well above any previous levels, which had not reached a trillion dollars.[92] He ended his second term in the Great Recession, with many economists concluding that capitalism was in permanent decline.[93]

After Barack Obama's election to the presidency, the dominant response on the right to what was seen as the failure of "compassionate conservatism" was to try resurrecting the Reaganite golden age. Reporting on the 2014 Conservative Political Action Conference, an event that Reagan had often addressed, a former longtime chairman of the American Conservative Union who spoke at the conference himself described how the disparate groups in attendance had worked at finding common ground: "There were all kinds of conservatives in attendance: libertarians, free-marketers, social conservatives, national defense conservatives, paleoconservatives and neoconservatives." Libertarians favored legalization of marijuana and "homosexual rights." All factions aside from neoconservatives supposedly favored "a more restrained foreign policy," and "virtually everyone" wanted "reduced federal spending and budget discipline." Could they all unite in "applying their basic values and principles to our fast-changing world"?

Reagan's image of a three-legged stool was invoked to describe a coalition of "economic, social and strong national defense conservatives," each with a different emphasis:

> Reagan's "three-legged stool" was a graphic way of describing the successful effort of Frank Meyer and William F. Buckley Jr. to turn a movement from a sort of philosophical debating club into a political force. They thought that people who differed on some important policies or emphasized very different issues could be persuaded to work together.
>
> Meyer called it "fusionism" and argued successfully that these factions shared basic values—freedom, free markets and traditional values—and the same enemies. He believed correctly that each faction's vision of a free society was equally threatened domestically by a growing and intrusive government... and internationally by the world communist movement then centered in the Soviet Union.[94]

Notice that these different ideologies were called "factions" that needed to be convinced to work together. Yet they also supposedly shared basic values. The ones specified were "freedom, free markets and traditional values." But did all really share them? And what were the "traditional values" that all the factions supposedly shared?

Traditionalism was mentioned in the conference speech as one leg of the stool, but it was not given content and otherwise was ignored in discussions of policy preferences. There was talk of "values and principles" but no specific ones were offered. National defense was named as the second leg of the stool, though in the original fusionism it essentially meant defending against the threat from expansionist communism. Now the Soviet empire was long gone, and conservatives were disagreeing on national defense issues. The third leg was budgetary restraint under a limited government, but now social conservatives were questioning that principle. The only value mentioned that actually appeared to be generally agreed upon was: "It worked—fractious as ever, conservatives began to come together and actually elect people to public office." In the end, conservatives were to unite around the goal of electing people to positions of power, not around shared principles.

Uniting for the sake of power is not the same as uniting around a fusion of principles. Indeed, the speaker pointed to the difference between a "philosophical debating club" and a "political force" that could achieve electoral success, and praised moving beyond the debating club that had formed the movement. Reviving fusionism now would be about pragmatically building a political coalition, which does not require having shared principles but only "the same enemies."

The "fusionism" of the early conservative movement actually had two senses, philosophical and political. A coalition grew from the intellectual readership of Buckley's *National Review* and bloomed into political organizations of the more pragmatic for the campaigns of Barry Goldwater and Ronald Reagan. The idea of a natural political coalition between libertarian-individualists and traditionalist-conservatives, however, is not ideological at all, but originated with a political scientist, Aaron Wildavsky, himself a Democrat. He had noted that a linking of traditionalists and libertarians would create a powerful alliance, and he saw it as the most likely democratic governing coalition among his four basic political types.[95]

The new conservative coalition did become enormously successful politically, first getting Goldwater nominated and then getting Reagan elected as president. But over time, success attracted those more interested in power than principle. Reagan himself used the three-legged stool only as a political metaphor. But in philosophical terms, Reagan as well as Buckley and Meyer in fact limited the fusionist concept to two poles of tension: freedom and tradition. The philosophical problem with the third element is that it took opposition to Communist adversaries, and later to other foreign foes, and made it morally equal to a whole cultural tradition of which it was merely a transitory and pragmatic part. The Sharon Statement of 1960, which some consider the founding document of fusionism, declared that "the forces of international Communism are, at present, the greatest single threat" to American liberties.[96] In other words, the third leg of the stool might be only temporary.

In fact, it did first take a "philosophical debating club" at Buckley's *National Review* to draw out the implications of a moral defense of an American capitalist society centered on both tradition and freedom permanently in tension. Inspired by the Reagan presidency, this philosophical ideal became influential enough to have a brief but enormously consequential political life both domestically and internationally.[97] Once the charismatic figure with a clear conception of the fusionist synthesis had left, however, political leaders more interested in the coalition as a step to power began to break the sense of relatedness between the principles, which eventually turned into mere slogans.[98]

Over time, the political coalition would encompass so many different interests that its intellectual integrity eroded into simple pragmatism. This is what Max Weber called the "routinization of charisma," which befalls movements as they go from the ideals of the charismatic founder, to winning over followers, to gaining power, to bureaucratic routine, finally ending in replacement by another set of leaders with new visions.[99]

By 2015, with presidential election campaigns gearing up, it was obvious that something had gone seriously wrong with conservative thinking. A rather young journalist worried that in eight years of power under President George W. Bush the Republican Party had "steadily ruined its reputation, damaging the public conception of conservatism in the process." In *The Conservatarian Manifesto*, Charles C. W. Cooke wrote that "Republicans had spent too much, subsidized too much, spied too much, and controlled too much." The GOP had "abandoned its core principle of federalism, undermined free trade, favored the interests of big business over genuinely free markets, used government power to push social issues too aggressively."[100] Cooke's proposal for leading the Right back to principle was a fusion of conservative and libertarian ideals into a "conservatarian" synthesis.

Cooke identified something that distinguishes both conservatives and libertarians from the progressive Left: they do not insist on telling people hundreds or thousands of miles away how to go about their lives. Progressive philosophy "is built upon the core belief that an educated and well-staffed central authority can determine how citizens should live their lives." But Utah is very different from New York. Indeed, Buffalo is not much like New York City. Different kinds of places need different policies.

Cooke emphasized that it is the federalism of the Constitution that allows for differing policies to suit local needs and interests. Respect for federalism and the general limiting authority of the Constitution was supposed to be the unifying principle of conservatism. Before the progressive revolution, courts had not found guarantees of black or female suffrage hidden somewhere in the Constitution. These reforms came about by amendments to the Constitution, through the process provided in Article V, requiring a broad regional agreement. Today, courts simply discover new rights in the Constitution, so activists don't need to rally a consensus for an amendment.

Overall, Cooke presented a lively discussion of issues, but his argument for a fusion of traditional conservatism and libertarianism looks rather like Athena emerging full-grown from the skull of Zeus. Here was a writer for *National Review* who seemed to have missed that it was his magazine that had developed the concept of a pro-freedom, pro-capitalist, pluralist, decentralized conservative fusionism a half century earlier. Perhaps the earlier synthesis was so tattered that no one there remembered it.

Resetting the Tension

By the time of George W. Bush's presidency, the open-ended Hayek-Buckley-Meyer-Reagan synthesis had been dominated intellectually and politically by a more pragmatic neoconservatism. Free markets, low taxes, moderated government spending, compassion, a strong national defense, and advancing democracy worldwide were still part of the mix, but became little more than loosely connected political slogans. Donald Trump's success in overwhelming the ideology's political remnants in 2016, followed two years later by the closing of the *Weekly Standard*, the neoconservative flagship magazine, confirmed its demise. But, as Cooke recognized, even the fusionist conservative remnant had lost its way, and freedom and market capitalism had lost a convincing argument for their moral legitimacy.

Kenneth Minogue, a professor at the London School of Economics who would become president of Hayek's Mont Pelerin Society, had already identified the problem inherent in forming a conservative identity largely around the idea of freedom. By the early twentieth century, beginning in Europe, the terms *liberal* and *freedom* had been co-opted by rationalist progressives, and so the heirs of Lockean market capitalism became "conservatives" by default. The "liberals" now supported a "centrally fostered community" with a socially "positive," progressive, ends-defined ideal of government-rationalized freedom. The "conservatives" reacted by adopting Isaiah Berlin's ideal of a "negative freedom" that would limit a nation-state government to policing coercion rather than actively promoting end-values. This had the great advantage of distancing freedom-supporters from Nazi and Communist totalitarianism, and even from the failures of moderate progressive welfare-statism. But the problem, as Minogue argued, was that negative freedom could also be called the doctrine of the 1960s left-radicals who aimed for "the total dissolution of all authority and order."[101]

Criticism from the right focused on the leading progressive idea of "positive freedom," which demanded more state power over markets and other pluralist entities, and less individual liberty. Freedom from government became the leading conservative message, at the expense of the tradition and order needed even to secure private property. Minogue referenced his teacher and colleague Michael Oakeshott on the need for tradition, for "manners and dispositions," for various "arts and habits" including "taste, discrimination, mental courage and mental soberness," not just abstract rationalization, in any project to promote true liberty.[102]

History shows that freedom appeared "first and preeminently as the condition enjoyed by warriors who were free for not being subdued," for not becoming slaves, Minogue argued. This idea of freedom as a "legacy of courage" was carried through the cultures of ancient Greece and Rome well into the European feudal era. It was a belief that freedom survives only "when it is given up" in service to community and traditions, in a courageous leadership with a "morality of integrity." But in modern society, the "condition of freedom" devolved into "oppositionality," a new ideal of a rational man as one who exercises "independent judgment" on matters formerly inculcated by the community and religion. The result has been "a steady, if irregular, drift in a permissive direction." What would traditionally have been considered socially disruptive became the modern Left's ideal of "total oppositionality," or resistance to all social, moral, and political restraints on the individual.[103]

Religion once acted as a powerful guard against social disruption and even a spur to courageous action, but its influence was seriously eroded by the later twentieth century, and social disorder grew. Government responded by providing more material benefits as amelioration for the disorder, and the masses came to support this positive welfare state. But as Minogue observed, "most people are disenchanted with the way it works." One reason for the dissatisfaction is that "our rulers now manage so much of our lives that they cannot help but do it badly." They aim to protect us from "our own smoking, eating the wrong food, not reading to our children, borrowing too much, and having unsound views of other races and cultures."[104]

Material benefits from the welfare state have not brought freedom but have made us "more and more accountable" to those in power. Indeed, "the rulers we elect are losing patience with us," demanding ever more servility and allowing less freedom. People have accepted this servility because actual freedom requires "the capacity not only to choose but also to face the consequences of one's choice." Today, the majority appear to have insufficient courage for this, viewing themselves as "vulnerable people whose needs must be met by others."[105]

Thus Locke's heirs were correct to focus their opposition upon the "superstition" (to use Hayek's term) that experts in a centralized government could engineer heaven on earth without eroding human freedom and dignity. But this emphasis did tend to devolve into a moderate oppositionality, with no theory for when and how government should take responsibility and no replacement for it in the areas where it should not. As Voegelin predicted, Locke's heirs naively assumed that people would more or less naturally develop a tolerant consensus on social order, so that a free-market society could operate without much political direction or moral guidance. Yet, as Minogue argued, "what most people seem to want is to know exactly where they stand and be secure in their understanding" of where they fit in. They need some positive view of government, or else they will support someone who offers one. Freedom is thus "constantly vulnerable to those who try to seduce us with dreams of perfection."[106]

Minogue's great insight was that "human beings are irredeemably moral creatures"—not so much social as moral—who, if they are to enjoy freedom, need to understand the limits within which their natural liberty must operate.[107] To fill this need, he identified the three institutions that have historically supported freedom and market capitalism: the Magna Carta–inspired, limited-function nation-state; the institutionalized moral constraints derived from Judeo-Christianity; and a Burkean subsidiarity with diverse associations limiting the power of both state and religious institutions.[108] Berlin's insights about "negative freedom" are useful, but Minogue is correct that a positive concept of governance is essential.

Minogue was not optimistic. By the twenty-first century, the Western state so influenced by Rousseau had been much expanded in order to reduce inequality and heal societal ills. But in the process, the state treated its beneficiaries as "incompetents" and created "vulnerable classes" of people dependent on the state, rather than allowing them to act as capable moral citizens. This runs contrary to "the human condition," Minogue argued, so people are increasingly disenchanted with the results. Yet the likely future is still more expansion of state power and a more servile populace.[109]

At this point, one might concur with Pope Francis in his view that we moderns have too easily indulged in a "mockery of ethics, goodness, faith and honesty," and that this "light-hearted superficiality has done us no good."[110] Is it possible to rise above both superficiality and servility, through a renewed fusion of traditional Western insights and institutions that can remoralize capitalist freedom?

This is the question that will consume the remainder of the book.

CHAPTER 7

THE PLURALIST
ADMINISTRATION SOLUTION

P
rofessor Minogue's first institution that has historically supported capitalism, but is now problematic and needs reconsideration, is the nation-state. Some forms, such as outright tyranny and pure socialism, are obviously incompatible. But a constructivist rationalization that overly controls subsidiary institutions, both public and private, also hobbles pluralist capitalism. As the United States has moved further toward the expert welfare-state model, the national government has become more widely viewed as dysfunctional, and not only by the conservative or capitalist Right.

Dysfunctional Centralization

In the mainstream progressive view, the explanation for the ills of American governance today lies directly in the form of government, in the U.S. Constitution itself. A representative example of this perspective is *Broken Trust: Dysfunctional Government and Constitutional Reform*, by Stephen M. Griffin, a law professor at Tulane University. Griffin starts with four major problems or crisis arenas to assess why the U.S. government has become so dysfunctional, and concludes that the problem in each case begins with the Constitution's irrational institutional structure.

In his first example, the terror attacks of September 11, 2001, Griffin finds fault in a parochial-minded Congress that allowed national intelligence agencies to be organized in a way that discouraged sharing of information. Second is the flooding of New Orleans and loss of life during Hurricane Katrina in 2005, where Griffin blames the botched response on federalism, because national, state, and local agencies had overlapping responsibilities, which resulted in confusion and delay. The financial crisis of 2008, he maintains, was caused by the Constitution's "general spirit" of overreliance on capitalist market mechanisms, with insufficient government regulation. His fourth issue is growing economic inequality, which he attributes to a constitutionally supported "oligarchy run by and for the benefit of wealthy citizens."[1]

The most astonishing aspect of Griffin's criticism is the absence of blame on one of the Constitution's major institutions, the executive branch of government. Griffin blames free-market capitalism for the financial crisis even though the key precipitating decisions—such as supporting mortgage derivatives and subsidizing "affordable" housing loans—originated in the executive branch. All the crucial decisions thereafter were made by the Treasury Department and the Federal Reserve. When Congress passed the law creating the Troubled Asset Relief Program (TARP), Treasury ignored the law's actual provisions for regulating assets and followed a very different plan to regulate stock purchases instead, and the banks were forced to follow.[2]

In assessing blame for the Katrina response, Griffin reaches for a structural angle by citing the journalist David Broder on the fact that the Constitution provides no means to coordinate local and national governments. What he ignores is that the Constitution leaves most local emergency response functions to the states. Adherence to that principle would have prevented the confusing overlap of responsibilities and removed a duplicative federal bureaucracy from the decisions. Griffin blames Congress for the 9/11 intelligence failures where intelligence agencies did not coordinate information that might have predicted the attack, but mentions offhand that the Department of Defense (last known to be in the executive branch) had vigorously lobbied Congress against reforming how those agencies work. He even conceded that blaming the legislative branch for "an apparent massive failure in the executive branch might seem surprising."[3] Well, yes, it does seem surprising.

Griffin asserts that "the main source of dysfunctional government" overall is Congress, supposedly the people's branch. He maintains that its dysfunction is an "uncontested fact," but in support of this "fact" he merely refers to a speech on immigration by President Obama that criticizes congressional Republicans for a politics of obstruction.[4]

Throughout *Broken Trust*, the author promotes the idea of direct democracy as the more appropriate way to represent the people. Without popular involvement, he argues, government simply cannot implement the progressive policies needed for social and economic reform. Griffin is confident that the people do in fact support the "reform agenda" recommended by the experts but rejected by legislators. If democratic government is to regain trust, the people must be given a greater say in how they are governed, and a way to end the partisan deadlock that thwarts the policies they really want. Griffin devotes his whole penultimate chapter to extolling the populist reforms of California and other western states: referendum, initiative, and recall. Yet when he finally presents his own proposals, they turn out to be not very populist.[5]

The leading constitutional reform he offers is the creation of boards of experts to oversee Congress, with the goal of broadening parochial legislative viewpoints, overcoming congressional sensitivity to interest-group demands, augmenting expertise, and breaking through the political deadlock. A new super-political regulatory body would be "staffed by appointed experts rather than elected officials," while the latter would not even be "permitted to influence the new regulators." Griffin proposes "permanent investigative commissions" to manage the whole electoral process.[6] It is not clear who would appoint the expert regulators, but presumably it would not be "the people."

To give his proposals what he calls "a populist cast," Griffin says that "ordinary citizens" would "serve as a check and balance" on the whole constitutional process. Yet after all the build-up to empowering the people, out of the blue comes Griffin's admission that he "instinctively shares" experts' worries about "public opinion getting out of hand."[7] And then, remarkably, he rejects the same populist measures he earlier praised, especially popular ballot initiatives, which he calls "far too nondeliberative."[8] As proof, Griffin mentions initiative outcomes he considers misguided, such as inflexible tax limitations and civil unions rather than gay marriage.

What he proposes instead of popular referenda is a system of "deliberative polls" that pose questions to citizens nationwide in order to glean their policy preferences, which would then be translated into public policy. Proposals based on such canvassing would be more rational than congressional actions, he argues, since the poll questions would "be developed under the friendly nonpartisan auspices of nonprofit organizations, philanthropic foundations and universities." While the poll results should have great force, Congress and the Supreme Court would at least have to give formal approval to the final recommendations.[9]

Griffin does not mention John Dewey, but his proposals are basically an update of Dewey's *Liberalism and Social Action*, with its concept of democracy driven by nonpartisan experts who gather public input but then make the decisions pretty much by themselves.[10] In Griffin's version, experts design the questions in the deliberative polls, and then write the reports that these "transpolitical" institutions will use to guide Congress. As I noted in the professional journal *Polity* many years ago, the phrasing of poll questions goes a long way toward determining the answers.[11] Griffin assumes that he and his "friendly" nonpartisan experts know what "uncontroversial" policies are needed, and it should be easy to reverse-engineer them and come up with poll questions to find that the public really wants those policies.

Griffin also does not cite Woodrow Wilson, although his argument for giving executive experts a larger role in leading legislators and the public, in managing markets and guiding mores, reprises Wilson's article from 1886, "The Study of Administration."[12] He does mention today's top expert on public administration, Paul Light, but only to quote his statement to the *New Orleans Times-Picayune* that a federal system complicates decision making. The statement is indisputable, but it would have been helpful if Griffin had read Light's own writings, where, as noted above, he concludes that the national government is already so tangled in red tape that it can no longer meet the Constitution's charge to faithfully execute the laws.[13]

Griffin works in the comfortable cocoon of academia, where opinion falls safely within a narrow range of reasonable, pragmatic progressivism. His book expresses fraternity with thinkers like Ronald Brownstein, Tom Friedman, Norman Ornstein, Laurence Lessing,

Tom Allen, Robert Dahl (selectively), and many other denizens of the political science establishment, carefully ignoring radical leftists as well as anyone on the right. Griffin says he "wouldn't have written the book unless I agreed with the authors just reviewed." As a law professor, he faults those wise political men only for not taking their argument further to decry a failure that is "governmental *and* constitutional." He dismisses as "simplistic" the idea that the constitutional framers created a separation of powers because of a "distrust of government."[14] Would they have created a government at all if they so distrusted it? No, they must have wanted government to work, but more rationally.[15]

The Founders would hardly recognize their government today. In 1989, the National Commission on the Public Service, for which Paul Light was a major adviser, reported that the national bureaucracy had "too many decision-makers, too much central clearance, too many bases to touch, and too many regulators with conflicting agendas," which made accountability "hard to discern and harder still to enforce."[16] Griffin hardly mentions the bureaucracy, although he would need to rely upon it to carry out his ambitious plans. The good law professor should have been sentenced to read the three reports of the National Commission on the Public Service. He and his friendlies could then ponder the fact that, as important as the nation-state is, we now have a century of evidence that Dewey's and Wilson's expectations for centralized bureaucratic administration based on expertise were, to put it mildly, overly optimistic, perhaps even simplistic, and that their lofty promises might even be the major source of today's dysfunction.

Permanent Factionalism

From Woodrow Wilson's *Congressional Government* (1885), to James MacGregor Burns's *Deadlock of Democracy* (1963), to Thomas E. Mann and Norman Ornstein's *It's Even Worse Than It Looks* (2012), to Griffin's *Broken Trust* (2015), the leading progressive scholarship has sought a way to overcome America's "dysfunctional" separation of powers. With so many opportunities for public input, the constitutional system frustrates the progressives' rational proposals to tame capitalism and disorderly associational freedom. The rationalist goal since Wilson's time has been to end government "deadlock" by creating a national consensus policy that is (supposedly) acceptable to all and would be implemented by a powerful national executive, relying on expert bureaucrats, with oversight by impartial courts.[17]

Griffin and other progressives assume that the public is largely in agreement on policy issues, and so the only real question is how the neutral experts get through the dysfunctional machinery to make things work. To the extent there is public disagreement, individual objectors will be persuaded to conform, in the interests of civic peace, or will be compelled to go along by force of law. Either way, the progressive rationalists do not consider social divisions deep or wide enough to pose a serious threat to consensus. But how has the theory of a homogeneous or cowed citizenry worked out in practice?

Obviously, consensus is absent on many policy issues, as public opinion polls have amply demonstrated. The so-called father of the Constitution, James Madison, recognized the reality that a population will naturally divide into numerous factions, so that even the "most frivolous and fanciful distinctions" can lead to conflict. And the "most common and durable source of faction" has been the "unequal distribution of property."[18]

This divisive element in American society is reflected today in residential segregation by income. Kendra Bischoff and Sean F. Reardon found that it had increased from 1970 to 2009, despite very substantial government efforts to promote economic integration. In 1970, only 15 percent of American families lived in areas of either concentrated poverty or concentrated wealth. By 2009, a full 33 percent of families lived in one or the other. In 1970, 65 percent of families lived in middle-income neighborhoods; by 2009, it was only 42 percent. Overall income segregation increased by about 29 percent.[19]

Income is often connected with racial segregation in residential choice. Edward Glaeser and Jacob Vigdor of the Manhattan Institute claimed in 2012 that neighborhood segregation had "ended," pointing to a "dramatic" decline since 1970. "As of 2010, dissimilarity had declined to its lowest level in a century and isolation to its lowest level in 90 years." But they added,

This shift does not mean that segregation has disappeared: the typical urban African-American lives in a housing market where more than half the black population would need to move in order to achieve complete integration. The average African-American lives in a neighborhood where the share of population that is black exceeds the metropolitan average by roughly 30 percentage points.[20]

Residential patterns are echoed in schools, which are also divided along lines of income and race. Despite all the governmental efforts toward integration, low-income black schoolchildren have become more isolated through both Democratic and Republican administrations. The share of black students attending schools that were more than 90 percent minority grew from 34 percent in 1989 to 39 percent in 2007. In 1989, the average black student typically attended schools in which 43 percent of their fellow students were low-income; by 2007, this figure had risen to 59 percent.[21] Between 2000 and 2010, the number of predominantly black and Hispanic schools identified as "high poverty" doubled, from 7,009 to 15,089.[22] Between 1996 and 2016, the percentage of children of color in segregated schools increased from 59 to 71 percent.[23]

For decades, government and the private sector have made efforts to integrate the workplace but the results have usually been disappointing, even within progressive organizations. For example, a survey by Dorceta Taylor of the University of Michigan in 2014, paid for by the environmental group Green 2.0, found that racial disparities persisted in environmental nonprofits as well as environmental agencies in government. Of the 3,140 paid staff of leading environmental nonprofits such as the Sierra Club, the Natural Resources Defense Council, and the Nature Conservancy, 88 percent were white. On governing boards, 95 percent were white. At governmental agencies, only 15 percent of staff were nonwhite, even after an Obama executive order to increase diversity across the government. The Environmental Protection Agency launched an internship program as a way to increase the minority presence, but few hires were made as a result.[24]

Policies bearing on race and immigration can be highly divisive, and public opinion on these issues may be shaped by social experiences as much as by initial beliefs. That is one conclusion of a study published in 2013 by Ryan Enos of Harvard. Enos did an experiment to see how public attitudes about immigration policy might be influenced by more exposure to minorities in one's home territory. Over a two-week period he had young Mexican nationals in pairs wait on the platforms at commuter train stations in Boston, conversing in Spanish, and then board trains bound for mostly white suburbs. Before the experiment was done, regular commuters were incentivized to complete online questionnaires asking their opinions on amnesty programs for illegal immigrants, increased legal immigration from Mexico, and making English the official language of the United States. The same questions were later given again to passengers who had been on the trains with the Spanish-speaking Mexicans, and their answers were compared with a control group of commuters who had taken different trains.[25]

Enos found that the commuters who had ridden the train along with the Mexicans were "far more likely" to oppose both amnesty and additional legal immigration, and "somewhat more likely" to demand official English, and their views on average had shifted after the experiment. He concluded that "exclusionary attitudes can be stimulated by even very minor, noninvasive demographic changes." Contradicting hopes that diversity increases community sociability, Enos predicted that demographic changes in the future will increase "intergroup exclusion" and ethnic conflict, although he hoped this effect would diminish over time.

Enos demonstrated that "exclusionary attitudes" can be found even in a generally liberal place like Boston. There, it was not primarily conservatives who turned more anti-immigrant after viewing Spanish speakers entering the white suburbs. It was liberals and moderates who came to think more like the Right, a finding that Robert Sapolsky, a science writer for the *Wall Street Journal*, found "pretty depressing."[26] Three years later, Donald Trump was winning votes from people who worried about immigration and changing ethnic demographics.

The progressive belief is that more integration, even if it is coerced, is better for minority groups and for the larger society. But people often respond to minority status in a diverse neighborhood by self-isolating and becoming less sociable. In his classic *E Pluribus Unum: Diversity and Community in the Twenty-first Century*, Robert Putnam observed:

> Inhabitants of diverse communities tend to withdraw from collective life, to distrust their neighbors, regardless of the color of their skin, to

withdraw even from close friends, to expect the worst of their community and its leaders, to give less to charity and work on community
projects less often, to register to vote less, to agitate for social reform
more, but have less faith they can make a difference, and to huddle
unhappily in front of the television.[27]

In reviewing Putnam's earlier book *Bowling Alone,* James Q. Wilson
echoed these observations about a general loss of social trust in diverse
communities.[28]

Whether integration is forced or occurs naturally, minority individuals become discouraged from participating in the community, or
even prefer to remain apart, perhaps growing alienated and distrustful.
A half century of national policy to promote racial and ethnic harmony
has not been as successful as rationalizers hoped, but instead has confirmed Madison's expectations about faction.

Political and Religious Factionalism

The majority of Americans are not highly engaged in politics, yet polarization on political issues has been increasing too. The Pew Research
Center studied the question of political polarization in 2014 and found
that the majority of Americans were neither strongly liberal nor strongly conservative, but somewhere in between. "Yet many of those in the
center remain on the edges of the political playing field, relatively distant and disengaged," the report noted, "while the most ideologically
oriented and politically rancorous Americans make their voices heard
through greater participation in every stage of the political process."
To that minority on either side, political beliefs are very important: 30
percent of consistent conservatives and 23 percent of consistent liberals
said they would be unhappy if a family member married someone who
belonged to the other party.[29]

Political views shape many other important life choices, even
among those who are not strongly ideological. Liberals told the Pew
pollsters that in choosing where to live, they preferred a place with
ethnic and racial diversity: 76 percent of consistent liberals and 58
percent of those who were "mostly" liberal. Similar majorities were
found to prefer a neighborhood where houses are smaller and stores

are closer, and for valuing proximity to art museums and theaters. Majorities of conservatives preferred to live where houses are larger and farther apart, even if stores and schools are more distant. The majority of conservatives said that living near people who share their religious faith was an important factor in deciding where to purchase a home. Liberals and conservatives alike said it was important to live where people share their political views.

The Pew study found that the two major political parties have become more distinct along ideological lines. Twenty years earlier (in 1994), close to one-quarter (23 percent) of Republicans were actually more liberal in their views than the median Democrat, while 17 percent of Democrats were more conservative than the average Republican. In 2014, the corresponding figures were only 4 percent for Republicans and 5 percent for Democrats. Each party was significantly more uniform ideologically.[30] Pew found similar results a few years later under President Trump, in 2017.[31]

The source of faction that Madison named first in *The Federalist Papers* was "a zeal for different opinions concerning religion."[32] Even in our more secular age, religious beliefs and values are still a major source of division among Americans. Pew found that nearly half of all Americans (49 percent) would be unhappy if a family member married someone who did not believe in God, with the number higher among consistent conservatives (73 percent).[33]

The Gallup Presidential Job Approval Center illustrated the reality of religious faction in 2014 with a report provocatively titled: "U.S. Muslims Most Approving of Obama, Mormons Least." Based upon an extraordinarily large sample of 88,000, the poll showed that majorities of Muslims (72 percent), "other" non-Christian religions (59 percent), Jews (55 percent), and atheists (54 percent) supported President Obama. On the other side, majorities of Catholics (51 percent), Protestants (58 percent), and Mormons (78 percent) opposed the Democratic president.[34]

Religion in combination with other factors creates even starker divisions. For example, only 34 percent of weekly church attendees who were married supported President Obama, while majorities of unchurched single women approved of the job he was doing. The division was likewise dramatic when religion was combined with race:

religious African Americans supported the president by 82 percent, Hispanics by 53 percent, and whites by only 30 percent.[35] Similar divisions along religious lines showed up in the following presidential election.[36]

Religious issues are a major factor in politics because 80 percent or more of the U.S. population identify themselves as religious, and even those who are not religious themselves tend to have strong views about religion. Historically, a broad Protestantism was a unifying force in the national culture, and especially in its leadership. Protestants are still a majority, yet many feel like a persecuted remnant. A poll by the Anti-Defamation League as far back as 2005 found that 64 percent of Americans believed that religion was under attack in the United States. Among Christians who attended church regularly, it was 75 percent.[37]

A few years later, in a Heritage Foundation legal memorandum, Jay Alan Sekulow warned:

> All across America, religious institutions and individuals are being subjected to increasing restrictions on their free exercise of religion and freedom of speech—a crackdown that can be seen in a variety of different contexts ranging from employers or health care professionals being required to provide or facilitate abortions against the dictates of their faith to street evangelists and public school students seeking to share their religious viewpoints with others. This rising disregard for religious liberty represents a marked break from the long-standing American tradition of accommodating religious practice and expression that predates the ratification of the Constitution.[38]

The Affordable Care Act's mandate that all employers provide health insurance covering contraception, abortion-inducing drugs, and sterilization was vigorously opposed by many on religious or moral grounds. Houses of worship were exempted from the requirement, but not religious charities, schools, hospitals, or other institutions at the center of religious life. This mandate prompted lawsuits demanding exemption from the law on grounds of the constitutional guarantee of religious freedom, but the results were mixed. Religious believers were discouraged, feeling that the entire U.S. establishment was against them, and they voted heavily for Donald Trump in 2016.

Many people, of course, see demands for religious exceptions as unreasonable. Chai Feldblum, the lawyer and activist who would become an EEOC commissioner, had abandoned formal religion in her teens but was still sympathetic to religious-liberty concerns in principle. She was willing to grant narrow exemptions, but not to give wide deference to religious objections.[39] Others are more hostile to any religious belief. There is a sizeable American audience for Richard Dawkins, a militant British atheist and author of multiple best-selling books who has been a guest on NPR and PBS. Dawkins believes that the influence of religion in society is overwhelmingly negative, causing intolerance, violence, and destruction.[40]

Religious commitment on one side and antireligious sentiment on the other will continue to divide Americans into factions that will not willingly submit to each other's policy wishes. It is unlikely that this division will diminish any time soon.

Family and Gender Factionalism

Divisions along religious lines very often include family or gender issues, resulting in a type of factionalism that Madison would not have recognized, although the place of the family in society has been shifting over the centuries. In the ancient world, the patriarchal family was the foundation of society and state. Pope Gregory I took the moral responsibility for marriage away from the extended family and gave it to the marrying couples themselves, with the church as witness. This did not immediately end the patriarchal system, but eventually the nuclear family would be the center of an emerging bourgeois, capitalist society.[41]

In the early stages of capitalism, as Schumpeter explained, "the family and the family home" were "the mainspring of the typically bourgeois kind of profit motive," which was to "work and save primarily for wife and children." Over time, these motives "fade out from the moral vision of the businessman," in favor of a more calculating ethos. As a consequence, "he loses the only sort of romance and heroism that is left in the unromantic and unheroic civilization of capitalism—the heroism...of working for the future irrespective of whether or not one is going to harvest the crop oneself." Men and women in a capitalist society become open to a "utilitarian lesson" and may adopt "a sort of inarticulate system of cost accounting" in their private lives. They become more "aware of the heavy personal sacrifices that family ties and especially parenthood entail." Whereas children in an earlier agricultural era had been economic assets, now they were only an expense and an obligation.[42]

Allan C. Carlson of the Independent Institute takes the argument further, arguing that "capitalism—at the most basic level—has a vested interest in family weakness." In a bygone era, self-sufficient families living in small communities were considered the essential foundation of civilization—culturally, educationally, and economically. Most work was done at home, on farms or in cottage industries. As a capitalist economy grows, "it takes over tasks and functions once performed by families or within closely knit communities, and reorganizes them on the industrial model," starting with textile and clothing manufacture, then food processing, and finally almost every economic activity.[43]

The capitalist need for efficient, low-paid labor was frustrated by traditional marriage, which kept women and their children at the hearth rather than the factory, Carlson argued. For this reason, capitalists promoted feminist ideology, as well as immigration, and now LGBTQ rights too. On the consumption side, capitalists favor single-person households, which require more furnishings and supplies per individual. Family businesses and privately held companies have tended to be more supportive of traditional families, but they too have begun emulating their larger competitors on social and political matters.

Carlson criticized the *Wall Street Journal* for regularly advocating lower business taxation but opposing child tax credits because the latter do not promote economic growth. What is needed, Carlson argued, is a "family wage" for fathers, allowing mothers to stay at home and care for children at least part-time when they are young. He offered the New Deal years as proof that similar policies were successful although in a more culturally conservative period. Democrats had historically been the party of the family wage, social protection of mothers, and prohibition of abortion and birth control. While the Republican Party today claims to be the pro-family party, it has actually returned to its earlier pro-feminist role. Carlson urged social conservatives to come down from the "attic" of the Republican big tent, assert their numbers, and "politely but firmly show the Wall Streeters the back door." But do they really have the numbers in the Republican Party? Or would the new New Deal party be more accommodating?

Let us turn once again to the fair-minded Chai Feldblum. In 2006, before becoming an EEOC commissioner, she attended a Becket Fund conference on religious liberty issues to express respect for traditional religious views, but in the context of advocating for gay rights. Feldblum admitted that advocates often frame gay rights too simplistically, failing to understand the moral implications of their demands. "When we pass a law that says you may not discriminate on the basis of sexual orientation, we are burdening those who have an alternative moral assessment of gay men and lesbians," she acknowledged. Advocates should not pretend there are no such burdens, nor that the concerns of religious people don't matter. Yet she thought the imperative to protect the dignity of gays and lesbians usually justifies that burden when religious liberty comes into conflict with sexual liberty, saying: "I'm having a hard time coming up with any case in which religious liberty should win."[44]

Well before homosexual marriage became a public policy issue, much of heterosexual America had basically already turned marriage from a lifelong commitment centered on children into a free alliance of two individuals maximizing their own interests—an alliance that is dissoluble anytime either party desires. Once marriage has been redefined this way, what logic is there to forbid same-sex marriage, or a marriage of more than two people? In 2006, the Pew Research Center found a large majority of Americans saying that an affair outside of marriage is morally wrong.[45] Jay Michaelson wondered if this view might become more contested in the future. He cited a study conducted in San Francisco in 2014 finding that roughly half of gay marriages were not monogamous, and he estimated that the reality was closer to three-fourths. He thought it likely that this non-monogamous approach to marriage would transform how the institution is understood in the heterosexual world, leading to more acceptance of "open marriage" and other kinds of non-monogamous "arrangements."[46]

Justin Raimondo, a libertarian homosexual, had earlier posed the question why other gay people were so insistent on marriage and whether it was really what they wanted:

If and when gay marriage comes to pass, its advocates will have a much harder time convincing their fellow homosexuals to exercise their "right" than they did in persuading the rest of the country to grant it. That's because they have never explained—and never could explain—why it would make sense for gays to entangle themselves in a regulatory web and risk getting into legal disputes over divorce, alimony, and the division of property. Marriage evolved because of the existence of children: without them, the institution loses its biological, economic, and historical basis, its very reason for being.[47]

Children introduce complications, and once they enter the equation, the state enters too. Marriage is about obligations, not simply rights guaranteed by the state. As Raimondo pointed out, the whole reason those obligations developed in the first place was the public interest in protecting vulnerable offspring. So what ground is there to insist on gay marriage, endorsed and supported by the state?

Richard Reeves, at the Brookings Institution, is a supporter of marriage equality and believes there will be no return to the old norm of marriage based on traditional social rules and gender roles, and restricted to a man and a woman. But he sees a compromise between old and new emerging, a more "egalitarian" model of marriage but with a strong focus on children. He recommends encouraging a view of marriage centered on parenting, if otherwise not wholly traditional.[48] Yet this will not satisfy those who believe in open marriages or those who believe a child needs both a mother and a father in a traditional marriage. Whatever the rights and wrongs of all these positions, issues surrounding marriage and gender will certainly continue to divide people into opposing political factions.

Back to the Catacombs

What solutions are possible within such a divided nation-state? Progressives seem determined to double-down on social-justice government policies.[49] Many influential traditionalist conservatives, meanwhile, seem to have given up the fight. James Davison Hunter, a sociologist at the University of Virginia, recommended as early as 2010 that the only solution to the political weakening of Christian beliefs, the failure of the so-called Moral Majority, and the divisions over moral values was for believers to stop seeking political victories, especially at the national level, and instead focus on practicing their faith in the local community.[50]

Similarly, Rod Dreher considered withdrawal and conceded even before the *Obergefell* decision that "we have had a quiet revolution in this country on matters of sex and sexuality," and that "the Left has definitively won the culture war." But this success had only made the Left more intent on getting more. Thus, "traditional Christians and Jews and other religious believers" need to ask themselves how they might "live in a world that increasingly sees us as enemies of the people."[51]

Religious traditionalists have become like a persecuted minority, Dreher claimed. He quoted the theologian Peter Leithart from an article written shortly after the Supreme Court decision in *United States v. Windsor* (2013) basically overturning the 1996 Defense of Marriage Act (DOMA), which had defined marriage as a legal union between one man and one woman for purposes of federal law. "We've fooled ourselves for decades into believing that Christian America was derailed recently and by a small elite," wrote Leithart. "It's tough medicine to realize that principles inimical to traditional Christian morals are now deeply embedded in our laws, institutions and culture." As a result of *Windsor*, he predicted, "Tax exemption will be challenged, and so will accreditation for Christian colleges and schools that hold to traditional views of marriage. Once opposition to same-sex marriage is judged discriminatory, no institution that opposes it will be unaffected."[52]

Traditional Christians need to acknowledge that their opponents are not merely an activist minority abetted by a handful of judges, Leithart observed. "All this means that *Windsor* presents American Christians with a call to martyrdom," which in Greek means "witness." Christians should continue teaching their ethics "without compromise or apology," but also be aware that "There will be a cost for speaking the truth, a cost in reputation, opportunity, and funds if not in freedoms."

Leithart referred to Justice Antonin Scalia's dissent in *Windsor*, which supported his pessimistic view that a new age of oppression was beginning. Dreher quotes from Scalia's dissent: "In the majority's judgment, any resistance to its holding is beyond the pale of reasoned disagreement." Scalia observed that DOMA "did no more than codify an aspect of marriage that had been unquestioned in our society for most of its existence—indeed, had been unquestioned in virtually all societies for virtually all of human history." Yet the Court majority held that those who supported the statute were doing so to "disparage," "injure," "degrade," "demean," and "humiliate" other people. The Court was portraying defenders of a traditional definition of marriage as "enemies of the human race."[53]

Dreher quotes an even darker view of the future for the new "enemies of the people," from Cardinal Francis George:

> I am (correctly) quoted as saying that I expected to die in bed, my successor will die in prison and his successor will die a martyr in the public square. What is omitted from the reports is a final phrase I added about the bishop who follows a possibly martyred bishop: "His successor will pick up the shards of a ruined society and slowly help rebuild civilization, as the church has done so often in human history."

Dreher concludes with a "bleak thought," saying that "American Christians faithful to Scripture and tradition had better start thinking outside of the usual categories, and preparing themselves for what's to come, especially given the technological capabilities of the national security state. Now is the time for realism."[54]

Realism compels acknowledging that religious individuals will face long, expensive court battles and will often lose. Catholic dioceses in Boston and Washington, D.C., abandoned their longtime provision of adoption services after being forced to refer children to gay couples against church policy. The city of Coeur d'Alene, Idaho demanded that Protestant ministers running a wedding chapel marry gay couples. The New England Association of Schools and Colleges threatened to revoke the accreditation of Gordon College, a private evangelical school in Massachusetts, if it did not change its policy opposing the homosexual lifestyle on its campus.[55]

But conservatives are not alone in believing that politics and government today have turned against them and that their own values are now in danger. During the 2016 election campaign, the progressive *Washington Post* columnist Richard Cohen described what he saw as a threat to "his" America:

> Donald Trump has taught me to fear my fellow Americans. I don't mean the occasional yahoo who turns a Trump rally into a hate fest. I mean the ones who do nothing. Who are silent. Who look the other way. If you had told me a year ago that a hateful brat would be the presidential nominee of a major political party, I would have scoffed. Someone who denigrated women? Not possible. Someone who insulted Mexicans? No way. Someone who mocked the physically disabled? Not in America. Not in *my* America.[56]

After years of disproportionately shaping what constituted "*my* America," of having a major voice in deciding what is proper exercise of free speech, Cohen found that the progressive consensus suddenly seemed to be broken.

Many on the left, especially in the media bubble, registered shock that their beliefs were not as dominant as they had thought. In response to polls showing large percentages of Republicans and independents saying that men and women should play different societal roles, William Galston wrote of himself and fellow progressives:

> We had assumed that some beliefs had moved so far beyond the pale that those who continued to hold them would not dare to say so publicly. Mr. Trump has proved us wrong. His critique of political correctness has destroyed many taboos and has given his followers license to say what they really think.[57]

Imagine, saying what one really thinks!

Some progressives have recognized the value of hearing alternative viewpoints. Nicholas Kristof, a progressive *New York Times* columnist, suggested that a lack of conservative voices on today's college campus, to challenge the dominant viewpoint, was leading to intellectual stagnation on the left. But the reaction surprised him.

I wondered aloud whether universities stigmatize conservatives and undermine intellectual diversity. The scornful reaction from my fellow liberals proved the point. "Much of the 'conservative' worldview consists of ideas that are known empirically to be false," said Carmi. "The truth has a liberal slant," wrote Michelle. "Why stop there?" asked Steven. "How about we make faculties more diverse by hiring idiots?"[58]

Whether they openly express their opinions or not, Americans are divided in many ways: by class, race, sex, religion, politics, cultural values; coastal versus middle America, blue versus red states, sophisticates versus rednecks, activists and the unengaged, conservatives against progressives. Americans disagree about moral values and about governance. The rationalistic progressive assumption that all will basically agree in the end is simply wrong, and the Framers were correct about the permanence of faction. Without broad consensus on policy, there is little legitimacy in comprehensive planning of social life and little hope of resolving many public issues centrally without some degree of coercion. Or is government really all about treating dissenters as idiots and brats, forcing them to conform or retreat to the catacombs?

Fragile Centralization

Why not just unify a divided nation by gaining control of the levers of power and enforcing a central plan? The first problem is that the population largely does not trust the central government. A study by the Pew Research Center in 2015 titled "Beyond Distrust: How Americans View Their Government" found that only 19 percent of respondents said they "can trust the government always or most of the time." This was among the lowest levels of trust measured in the past half century. Just 20 percent would describe the federal government's programs as well run. Elected officials were held in such low regard that 55 percent of the public said that ordinary Americans would do a better job of solving national problems.[59]

Yet even though Americans distrust government, Pew reports they still have "a lengthy to-do list" for it. "Majorities want the federal government to have a major role in addressing issues ranging from terrorism and disaster response to education and the environment," the Pew pollsters found. And while "most Americans *like* the way the federal government handles many of these same issues," they are "broadly critical of its handling of others—especially poverty and immigration." On the other hand, there is disagreement on precisely how to handle these matters, and a wide partisan divide on the size and scope of government: 80 percent of Republicans and Republican-leaning independents expressed a preference for "a smaller government with fewer services," while only 31 percent of Democrats said likewise.[60]

In general, people will say they like many of the current policy ideas but express dissatisfaction with how they are carried out. This kind of dissatisfaction with government was relatively low, however, before the 1970s and the expansion of welfare-state centralization, the War on Poverty, and increasing market regulation. Jimmy Carter won the presidency in 1976 by promising to make the programs work with a reformed bureaucracy, but then Ronald Reagan campaigned against centralized big government per se, winning two presidential terms, followed by another Republican president.

Except for a short period under President Reagan, popular dissatisfaction with national government has mostly been the norm since the 1970s. Even some progressives have come to think the U.S. government tries to do more than it should. President Bill Clinton, in his 1996 State of the Union address, felt it necessary to declare that "the era of big government is over." Professor Light had conceded that the national government promised much but often could not deliver. This continued failure is one explanation for the election of the outsider Donald Trump in 2016: perhaps he would have new ways of approaching problems. But dissatisfaction only increased.[61]

Why has public dissatisfaction with the federal government become so pervasive, regardless of the party holding the presidency or Congress, while local governments are more often trusted? One reason is a problem inherent to centralized decisions for any large population: they are necessarily based on reducing complex and varied phenomena to simple generalities and averages that conceal many particular facts. Solutions based on such generalizations are probabilistic. Centralized decisions can never fit every situation, and they might be wrong in very big ways.

Experts make calculations to formulate rational plans, but all their averages hide a myriad of specific problems off the mean. Lower officials in a bureaucracy may not always report accurate information to higher-ups, but government centralization itself presents an information problem. Regulatory dark matter accumulates in the bowels of bureaucracies, unseen to outsiders. Ludwig von Mises explained how the market calculates needs through the simple mechanism of pricing, which allows the top executive in a private business to decentralize decisions, needing only to know whether a unit is making a profit or not. Even the best government bureaucracies have no similar mechanism linking decision to result.[62]

A lack of specific knowledge about facts on the ground by centralized institutions can be dangerous. This author would be the last to deny Ronald Reagan, Margaret Thatcher, and Pope John Paul II credit for undermining the Soviet Union morally. Yet Robert Skidelsky, in *The Road From Serfdom*, makes a convincing case that the proximate cause of the Soviet collapse was looming bankruptcy, which lower-level bureaucrats hid from those at the top, or the latter disregarded what they knew.[63] Niall Ferguson of Harvard upended conventional academic thinking on civilizational decline by demonstrating that bureaucratic and leadership failure, especially regarding debt, has brought down as many civilizations as war—including seemingly powerful states of all types and times.[64]

Edward Gibbon had described a protracted "decline and fall" of the Roman Empire, but Ferguson argued that the causes he adduced—barbarian invasions, epidemics, economic crises, rival empires—were basically the norm for Rome. Ferguson saw the real fall starting only in 406 AD, when officials failed to halt the advance of Germanic tribes across the Rhine and down into Italy. The Goths sacked the city of Rome in 410; the North African breadbasket was conquered between 429 and 439; and finally Britain was lost, along with most of Spain and Gaul. The world's leading centralized bureaucracy quickly lost revenue too, and accumulated debt. By 476, the Western Roman Empire was gone.[65]

In the modern era, the British and French empires ended World War I as victors holding a large slice of the world as colonies, the upkeep of which was managed by the best civil service bureaucracies of the time. Thirty-eight years later, President Dwight Eisenhower could blithely dismiss Britain and France as inconsequential and wave them out of empire at Suez. America's CIA thought the Soviet system so efficient that the USSR was believed to be more prosperous than the United States when Mikhail Gorbachev became general secretary of the Politburo in March 1985. But a mere six years later there was no Soviet Union.

Centralization hides mistakes both from the top and from the outside. It can conceal problems in spending and the management of debt, which is especially insidious because debt turns the future over to others—to creditors. The central government myth is that the budget process can alert leadership to potential problems, though in reality the budget process provides no effective signaling or quality control.[66] At the national level, deep problems are simply kicked down the road.

Public attention to budgetary issues almost always focuses on "discretionary" spending: the 16 percent or so of national government spending (in 2017) that goes into discretionary domestic programs and the 16 percent that goes into national defense. But that is merely one-third of what the government spends, and while it has been increasing by 5 percent or so annually over recent years, spending on entitlements and mandatory programs, which account for the remaining two-thirds of the federal budget, has been growing by 20 percent. Presidents and Congresses of both parties have pledged not to reduce entitlements or to increase taxes sufficiently to pay for them going forward.[67] The upshot is a real unfunded liability of $120 trillion, dwarfing the official debt of $20 trillion or so. Experts at the Federal Reserve, Treasury, and other financial institutions warn about the danger, but keep suggesting they can handle it if the worst happens.[68]

Ferguson looked into widely ignored Congressional Budget Office data and estimated that U.S. national government debt could reach 90 percent of gross domestic product by 2021, then 150 percent by 2031, and 300 percent in 2047. At the time, Greece had a debt of 312 percent, and it was bankrupt. The official reporting of debt, however, does not include the politically sensitive unfunded liabilities of Social Security, Medicare, other federal pensions, state debt and pensions, and potential liabilities from government-insured private pensions. With these included, Morgan Stanley estimated that the U.S. debt was already 358 percent of GDP. It is impossible to reduce that large a debt if it is politically impossible even to admit its existence.[69]

As Ferguson remarked, "one day a seemingly random piece of bad news, perhaps a negative report by a rating agency," could reach beyond the experts to panic the public at large or spook investors abroad, where half of U.S. public debt resides. One-fifth was owned by a fearful China, which had already called the Federal Reserve's "quantitative easing" a form of "financial protectionism." The Fed's power to control national and global economies is now acknowledged to be limited, so what could backstop a worldwide debt panic?[70]

The squeeze between entitlement demands, an enormous debt, and a limited ability to increase taxes will necessarily force a reassessment of the nation-state. Gary North, who calls himself a contrarian libertarian, was impolitic enough to conclude that entitlements will soon start crowding out the rest of the central government, which does "not have organized voting blocs comparable to AARP and the Gray Panthers." This means that "the nondiscretionary budgets of all national governments in the West must be re-allocated to meet the growing demands of the retired oldsters." In other words, "Granny is going to get an increasing share of the federal pie."

The federal government will first decimate the Pentagon and the administrative state, North predicts, and eventually it "will not be able to support granny." Before that happens, "Americans are going to have to re-think the relationship between Washington and local governments" and consider how to accomplish a reset to a less centralized and less bureaucratized kind of government.[71]

Balanced Centralization

Beyond the principles of progressive ideology, the justification for centralization of U.S. governance has rested on the clause in Article VI of the Constitution that reads:

This Constitution, and the Laws of the United States which shall be made in Pursuance thereof; and all Treaties made, or which shall be

made, under the Authority of the United States, shall be the supreme
Law of the Land; and the Judges in every State shall be bound thereby,
any Thing in the Constitution or Laws of any State to the Contrary
notwithstanding.

So, federal laws are "the supreme law of the land," the argument goes.
Yet in reality, it is not that simple, as politicians of both parties have
had reason to discover at one time or another throughout American
history.

Senator Mitch McConnell provoked a media firestorm in 2015
with his suggestion that states need not adhere to a new environmental
regulation. A *New York Times* headline blared: "McConnell Urges States
to Defy U.S. Plan to Cut Greenhouse Gas." A barrage of similar stories
followed, to the effect that the Republican majority leader was telling
the states to violate national law and therefore the Constitution! Senator
Barbara Boxer, the ranking Democrat on the environmental committee,
was appalled, saying she could not recall another top politician actually
"calling on states to disobey the law."[72]

Senator McConnell's statement was actually rather mild, though,
recommending merely that states "think twice before submitting"
a plan for curbing greenhouse gases, as directed in the regulation.
Presenting such a plan could lock states into a federal enforcement
regime and expose them to lawsuits. McConnell opined that the Obama
administration was "standing on shaky legal ground" with the regulation. "Refusing to go along at this time with such an extreme proposed
regulation would give the courts time to figure out if it is even legal."[73]
But as many saw it, the Senate majority leader was openly calling for
states to defy national law.

States defying federal law is a bipartisan worry, but it seems to
depend on the issue. Hugh Hewitt, a law professor and conservative
media personality, writing in the *Washington Post,* criticized state and
local governments for legalizing marijuana and for enacting "sanctuary
city" policies to protect illegal immigrants from federal enforcement.
Hewitt claimed that such actions were in violation of Article VI, clause
2 of the U.S. Constitution. "It is not called the 'supremacy clause' of the
Constitution for nothing," he observed, adding, "State and local governments should obey federal law on every subject."[74]

When the Constitutional Convention voted on that very question, however, the delegates rejected the idea of straightforward federal supremacy over state law. James Madison at first was devastated at the vote, fearing it would make the Constitution inoperable. But he came to realize that this outcome, providing for "two distinct governments," was essential to the new federalism—and necessary for getting the Constitution ratified, especially by the smaller states.[75] In *The Federalist Papers*, Madison emphasized that the Constitution set up a "dual" system of national and state institutions, in which the latter were to do most of the lawmaking, while the central government had only limited national functions specified mainly in Article I, Section 8.[76]

Madison later went much further in challenging the federal government's power over states. In his Virginia Resolutions of 1798, he declared the Alien and Sedition Acts unconstitutional and appealed to the other states to take part in "the necessary and proper measures" to defeat the laws.[77] The Kentucky Resolutions drafted by Thomas Jefferson echoed the Virginia version.[78] In the minority, these states persuaded others to join the effort, thwarting enforcement attempts and building an opposition party that won the next election and repealed the laws.

But were not Boxer and Hewitt correct on the supremacy of national law? The U.S. Constitution does set itself up as "the supreme law of the land," after all. And the Supreme Court is the Constitution's final interpreter, right? Indeed, the justices of the Court often seem like the ultimate power in the government, acting as umpire between the other branches and making the decisions as to what the Constitution permits. The Court appears to sit at the apex of a centralized government.

But earlier American leaders did not see it that way. President Andrew Jackson, for example, said the three branches of the federal government should equally be interpreters of the Constitution:

The Congress, the Executive, and the Court must each for itself be guided by its own opinion of the Constitution. Each public officer who takes an oath to support the Constitution swears that he will support it as he understands it, and not as it is understood by others. It is as much the duty of the House of Representatives, of the Senate, and of the President to decide upon the constitutionality of any bill or resolution which may be presented to them for passage or approval as it is of the supreme judges when it may be brought before them for judicial decision. The opinion of the judges has no more authority over Congress than the opinion of Congress has over the judges, and on that point the President is independent of both. The authority of the Supreme Court must not, therefore, be permitted to control the Congress or the Executive when acting in their legislative capacities, but to have only such influence as the force of their reasoning may deserve.[79]

Following the Supreme Court's pro-slavery decision in *Dred Scott v. Sanford* of 1857, Abraham Lincoln gave a speech asserting that "vital questions affecting the whole people" should not be "irrevocably fixed by decisions of the Supreme Court" made in "ordinary litigation between parties." When an opinion in one case can override the views of all elected officials, "the people will have ceased to be their own rulers, having to that extent practically resigned their Government into the hands of that eminent tribunal."[80] As president, Lincoln would effectively overrule *Dred Scott* through the Emancipation Proclamation, which was followed by a constitutional amendment.

In fact, the different branches of the government—both federal and state—have constantly been checking each other's view of constitutional law from the beginning. A classic study by Robert Dahl, a leading political scientist at Yale, demonstrated that Congress throughout its history had often "overruled" Supreme Court decisions when the issue was considered important enough by passing laws that overturned them de facto if not de jure.[81]

Nor did that end after Dahl wrote in 1957. Thirty years later, for example, Congress passed the Civil Rights Restoration Act specifically to de facto overturn the 1984 Supreme Court decision of *Grove City v. Bell*, which had narrowed the scope of antidiscrimination law. Then Congress passed the Civil Rights Act of 1991, specifying by name five Supreme Court decisions that had likewise put limits on the application of antidiscrimination statutes, which were de facto overruled. Challenges have come from the states, too. Each Christmas, states were finding new ways to get around the Supreme Court's restrictions on public religious displays. Finally the Court, with a swing vote from the progressive Stephen Breyer, relented on the issue in *Van Orden v. Perry*

(2005), which allowed a monument displaying the Ten Commandments to remain on the Texas state capitol grounds.[82]

Contrary to ideologues on all sides, it is by no means clear what constitutional law really is. The question is hashed out in practical situations, in untidy and unpredictable tensions, involving challenges and negotiations between the different branches and levels of government. Federalism is supposed to be complicated. All law emerges from the divided structure established by the Constitution, under which the fickle human individuals who populate its institutions make decisions. The justices of the Supreme Court who interpret the law can set precedent but cannot make it permanent. Presidents can refuse to enforce decisions and Congress can pass laws to frustrate them, and have done so throughout U.S. history.

Presidents with all their constitutional power and prestige certainly do not always get their way either. President Obama started his tenure with bold "transformational" plans to be like his hero Franklin Roosevelt. With Democratic supermajorities in Congress, he was able to achieve his major health-care reform, affecting one-sixth of the U.S. economy. But soon the opposition gained a majority in the House, and held the Senate too. So President Obama resolved to rule with "a pen and a phone," by executive order alone on matters he deemed important. But he ended his second term with limited success in his goal of "fundamentally transforming the United States of America," with other executive rules significantly modified by the following Congress and president.[83]

Blame all this jostling between institutions and contesting of constitutional law on the Constitution itself, the only legal center. Under its terms, no one institution is always in charge. The branches of the federal government check each other, as do the different levels of government. The states created the Constitution and could end it with three-fourths support through the amendment process. Over the years, national control has increased by Supreme Court decisions citing the commerce clause and the Fourteenth Amendment's incorporation doctrine. But the former has recently been somewhat limited by the Court,[84] and the latter's reach is still shifting.[85] The Supreme Court may issue precedent-shattering decisions, but the results may keep evolving indefinitely.

Alexander Hamilton argued that without a separation and division of powers there would be "anarchy, civil war, a perpetual alienation of the States from each other, and perhaps the military despotism of a victorious demagogue."[86] But with three centralized branches to check each other, and the "two distinct governments" of nation and states, the Constitution diffuses power rather than concentrating it in one center.

This pluralist federalism, with all its seeming disorder, is the longest-lasting constitution in the modern world, and it has nurtured the world's most dynamic capitalism.

Federalism

Divided and decentralized pluralism is in fact the political setting that gave rise to Western capitalism, and it has provided a clear alternative to what had otherwise been the global norm: governance through coercion and centralized bureaucracy.[87] The feudal system was fundamentally decentralized, and Magna Carta affirmed a pluralist understanding of governance in early England. But Europe offered another very substantial alternative to centralization, in the Holy Roman Empire. Its long history is instructive.

Voltaire's jibe that the Holy Roman Empire was neither holy, nor Roman, nor an empire is both clever and true as far as it goes. It was not holy since it separated church and state more than most other governments at the time. It was not Roman but German, which is why Frenchmen liked to repeat Voltaire's quip. Most important, it was not actually an empire but a pluralist federation, the most successful one before the United States. After the strong hand of Charlemagne, it became based on compact and decentralized power.[88] But the otherwise perceptive Max Weber considered it simply an empire, and did not think of federalism as a proper form of government.

In any case, the Holy Roman Empire was the longest-lasting European government and stretched across a large portion of the continent, though its internal and external boundaries constantly shifted in befuddling ways. Several dynasties held power during the course of its long life, beginning in the year 800 with the Carolingians (for 119 years), Ottonians (105), Salians (101), Lothar (12), Staufers (116), "Little Kings" (93), Luxembourgs (90), and the incredible Habsburgs (368), who finally ended it all in 1806, to deny the imperial title to Napoleon.[89]

In his comprehensive history, Peter H. Wilson emphasizes that the central structure of the federal empire was based on the only known ideal, the Roman Empire, which in some respects it was intended to revive. But three major new forces were present by this time: a fierce Germanic tribalism, a Christianity that was now deeply rooted in much of the realm, and finally an extensive river trade, protected by Europe's craggy terrain. The sinew of the federation was a network of vassalage that formed a human link from the very top out to the remotest outpost: from the emperor, kings, dukes, princes, and lords, down to the lowest knights. The commoners were mostly left to supervision by lesser nobles and clergy: bishops, priests, and abbots.

Dynasties notwithstanding, the emperor was selected by prince-electors, who constituted the first electoral college. The emperor had two main responsibilities. The first was to keep the borders secure, especially from invading Vikings, Slavs, Magyars, and Muslims. The second task, equally important, was to travel throughout the realm carrying the flag of the office and settling disputes between principalities with his traveling court, from which local courts emerged with more or less common legal principles. The princes in their scores of states could pretty much run their fiefs as they saw fit, so long as the emperor's interests were respected, and their own vassals operated on the same principle. By the high Middle Ages, principalities, manors, parishes. and villages had some degree of autonomy.

Certainly there were differences and disagreements among these many semi-independent powers, which called for negotiation, with warfare only as a last resort. Similar religious beliefs and institutions encouraged peaceful bargaining and compromise, including major "public peace" resolutions that often were binding on the emperor himself. There were agreements settling common law, rights, and obligations between emperor, lords, and church. The General Privileges of Aragon in 1283 were an echo of Magna Carta, with nobles securing recognition of their liberties from a king who needed their allegiance. The empire survived the Reformation and its violence, ending up with both Protestant and Catholic principalities in the federation.[90]

The theory of federalism basically originated with this pluralistic empire. Johannes Althusius is often identified as the first philosopher to promote the federalist creative tension between center and periphery, by contrast with the strongly centralizing theory of his contemporary Thomas Hobbes. Althusius used the existing tension between his Catholic emperor, Lutheran prince, and Calvinist city to explain his theory.[91] Alain de Benoist summarized the federalism of Althusius this way:

> The state is defined as a real organic community, formed by many symbiotic "consociations," public and private, with two sets of agencies on each level, one representing the lower levels, which must retain as much power as possible, the other representing the higher levels, whose jurisdiction is limited by the lower levels. Thus, freedom within society does not emanate from the sovereignty of the higher levels, but from the autonomy of the lower levels. The articulation of and equilibrium between different levels is guaranteed by the principle of subsidiarity.[92]

Whether or not Montesquieu actually read Althusius, he surely employed similar pluralist arguments, which greatly influenced the American Founders.

Most importantly for the thesis of this book, "the most advanced pole of up-and-coming capitalism in Europe" between the 1470s and the 1520s was located in that pluralist federation, especially the principalities of Flanders, Tyrol, Saxony, and West Bohemia, as Henry Heller points out. In fact, "decentralization facilitated the initial emergence of markets and manufacturing based on hundreds of economically interconnected free or imperial cities operating within the web of late medieval feudal relations."[93] It is worth noting that early capitalist England was also more decentralized and pluralist than its formal structure would suggest up until the reign of Henry VII.[94]

Decentralization likewise encouraged economic development in the American colonies. During the period of benign neglect, the colonies adopted the pluralist Magna Carta doctrines as they were being abandoned in the mother country, and also turned against hereditary aristocracy in favor of elections and popular representation. Lord Acton regarded its popularly based federalism as America's unique contribution to free government, since other elements—such as individualism, property rights, rule of law, contract, markets, moral equality, democracy, parliaments, separation of powers—had all come before.[95] The new federal nation soon left more centralized Europe far behind economically, as the Maddison data presented earlier illustrate.

States still retain an important voice today even in national matters. Amending the U.S. Constitution requires state action. States have proposed amendments and federal laws, and have brought lawsuits challenging national laws and policies on marriage, abortion, racial preferences, gun restrictions, the Real ID Act, immigration, and many other issues. More than half the states raised challenges to provisions of the Affordable Care Act. Even President Obama recognized state authority by announcing he would not enforce federal drug laws against states that had legalized marijuana. President Trump basically turned over the handling of the Covid-19 pandemic to the states. Even on purely national programs, states administer many federal laws, and state courts dispose of many federal court decisions. One can be assured that states like Alabama and California go about it quite differently.[96]

Decentralizing Pluralism

Federalism is pluralist in that it allows regional entities to tailor decisions to a narrower range of preferences, with a better chance of satisfying the wishes of individual inhabitants. As Rousseau argued, though, one simply cannot have the same social solidarity in a nation or state of millions as in a city of 20,000 or a town of 2,000.[97] Individuals may be able to vote in larger entities, but they will find their wishes not well represented. The Marquis de Condorcet, a great mathematician, demonstrated that the relationship between what people want and voting results becomes more distorted as the numbers of persons and issues increase.[98] Markets avoid this difficulty because individual customers and businessmen make their own choices to trade or not. Governments cannot let every individual select policies, but in smaller units there will be a better correlation between individual preferences and community decisions.[99]

Localizing decisions as much as possible is called subsidiarity, a principle recognized as a central value of social ethics, even by Pope Francis.[100] Some American states, such as California and Texas, have populations larger than many nation-states. Even many counties and cities have populations in the millions. Large cities have their particular internal problems given their diverse populations, especially the cities with poor ghettos rimmed by affluent neighborhoods and suburbs. In the 1960s, these conditions helped lead to a series of riots in Los Angeles, Detroit, and Newark—followed by more in the wake of Martin Luther King Jr.'s assassination—prompting various national urban-renewal initiatives, including the Model Cities Program, the Job Corps, the Community Action Program, and even a whole new federal agency, the Department of Housing and Urban Development. But many of these imposed changes that were not solicited by residents and did not help them.

When President Trump announced his choice for secretary of HUD soon after the 2016 election, he was criticized by Lawrence Vale, a professor of urban design and planning at MIT, for his use of the term "urban renewal":

> When I talk to people about contemporary redevelopment of public housing in certain cities, the words they use are, "We don't want another round of urban renewal in our neighborhoods." Even though it's a long time ago, I'm struck by how frequently people who are being impacted by changes they don't want to see in their neighborhoods will evoke that earlier era.[101]

"Renewal" of urban neighborhoods initiated by bureaucrats in Washington had become so unpopular that center-city residents came to prefer their very imperfect status quo.

Crime in large cities is still a significant problem, even while crime rates across the United States have declined since the violent 1970s. There were still over a million violent crimes and 15,000 murders nationwide in 2015, mostly concentrated in big central cities.[102] In fact, there had been an increase in murders that year in the nation's thirty largest cities, and about half of that increase came from Chicago, Baltimore, and Washington, D.C.[103] Apparently this was partially the perverse result of the so-called "Ferguson effect," when proactive policing of troubled areas was pulled back in response to the anti-police riots of 2014. The cost was borne mostly by young black men between ages 15 and 34, for whom murder was the leading cause of death.[104] But there was no new central plan for cities, and few locals trust that any such plan would be helpful.

Policing has become a major problem in today's cities, as demonstrated in the new wave of protests and riots that followed the police killing of George Floyd in Minneapolis in May 2020. The police are often seen as an alien presence in communities, with little understanding of the people who live there.[105] In a bygone era, the people who policed city neighborhoods were intimately familiar with them. The local cop walked a small beat and was lightly armed, mostly just keeping a watchful eye. By and large, local policing seemed to work. In old England, the parish constable was hardly armed at all, perhaps only with a nightstick. He was more part of the community than an agent of an external institution, and he had to rely upon the community for his own safety.

All of that changed with greater public mobility, especially the automobile, and the introduction of rationalized professional policing by the early progressive movement, which instituted citywide rather than local standard operating procedures for police.[106] Even the bravest officers learned they were safest in a car moving quickly in and out of a neighborhood, stopping only to roust those who looked dangerous. Politicians learned it was cheaper too, requiring fewer personnel. Police forces became more militarized, often resembling an alien army. Police districts grew larger, with more levels of expertise but less local knowledge and fewer eyes on the street. In 2016, it was a resident of Linden, New Jersey, who noticed that a man sleeping in the doorway of his bar resembled the New York City bombing suspect he had just seen on a television news report, and alerted the police.

In recent years, there have been many attempts to revive "community policing," with some additional police walking a beat here and there, but the efforts have not been very extensive, with the exception of New York City (although that effort may be waning).[107] Community policing has become a byword for modern practice, but often the concept gets lost in bureaucracy. As chief of police in the District of Columbia, Cathy L. Lanier started a community outreach program that garnered acclaim for defusing anger at police. She was a popular leader, and said that crime had fallen under her watch. But she had some complaints herself, and apparently felt free to be forthcoming in her exit interview as she stepped down in 2016 after a decade in the job. The highlight of her remarks was this surprisingly frank statement: "The criminal justice system in this city is broken. It is beyond broken."[108] The courts keep releasing violent offenders back into neighborhoods, free to do more harm. "Where the hell is the outrage?" Lanier asked. "People are being victimized who shouldn't be."

She insisted that residents "want more police. They want more arrests. But if we're arresting the same people over and over again, there's got to be some questions being asked." She mentioned the recent case of a man on judicially ordered home detention whose GPS tracking device malfunctioned, and he then went on a violent crime spree—robbery, shooting, car theft—which left an innocent bystander critically injured. "The agency that supervises that person didn't tell anybody or do anything with it. That shouldn't happen. And it's happening over and over and over again. You can't police the city if the rest of the justice system is not accountable."

Lanier's main complaint was that a large number of local and federal bureaucracies were all pulling in different directions. She objected to federal control of bail, detention, the filing of charges, and the monitoring of suspects under court supervision. Lanier hopefully predicted that police work in the future would become "much more service-oriented and much more collaborative than it is now," but also suggested that many current police duties should be shifted to other social agencies, such as caring for the mentally ill and dealing with minor rule violations. "A lot of the things we deal with right now, you don't need a police officer. And it is putting us in confrontational positions with people who are not criminals that are causing a lot of the turmoil we see right now." She urged the community to take up part of the burden.[109]

The problem is that policing has become highly rationalized in bureaucratic institutions that have little connection to real people. It needs less expert rationalization and more community contact. A new style of community policing, with a less "confrontational" and bureaucratic character, would involve a constable system, separate from the professional police but working with it. There are constabularies in nine states, with differing limited functions. A real community constable would primarily be a listener rather than an enforcer, seeking to understand the neighborhood and learn what is needed to keep it safe and secure. He would keep an eye on the bad guys, whom the locals know and will complain about if someone is listening and not a threat. A constable could also alert others to medical emergencies, domestic disputes, runaway children, public intoxication, mental problems—all the multiple difficulties of neighborhood life.

A constabulary could somewhat replace yesterday's network of mothers who stayed at home—whom Jane Jacobs recognized as the glue then holding city neighborhoods together.[110] Constables would be attached to the most local government possible, or even engaged by private residential organizations. They could be elected or appointed, and would live in the neighborhood. Compensation would be minimal, and a very small office space even in one's basement would suffice and keep things close to residents. Constables would have a list of agencies and potential helpers, and links to the police to provide warnings about possible violent activity, at the constable's discretion, to guard individuals' privacy and build trust in the community.

Constables could even be recruited from among welfare and disability recipients, to serve as good neighbors with no hard duties other than talking, walking (or even doing rounds in a wheelchair), making phone calls, texting, and emailing. They could be exempted from welfare work requirements with this public service—and, more importantly, could build a sense of self-worth and responsibility, perhaps opening the door to higher employment.

Suburban areas already have their own versions of constable policing. Private homeowner associations often have volunteers performing constable functions, which is one reason those neighborhoods are safe. Large cities are most needful of constables, perhaps one to every block in some areas. There would be abuses, but there are ways to minimize them. Constables would be limited by police professionals working around them, and by the legal system.[111] Ideally, they would be responsible to local justices of the peace with power to oversee and dismiss them for cause, subject to appeal to local common law courts. Multiple constables could check each other. Higher local officials could watch too. Professional community organizers should be welcome to join the constabulary to help solve problems they see.

Reviving a constabulary and community policing would be one way of returning to the more personal and local governance of an earlier time. Until the early twentieth century, almost all American government was local, remaining powerful even as markets nationalized. At that time, local governments accounted for the greater part of total government spending and states spent another 10 percent. Today the central government alone directly dispenses 60 percent of total U.S. government spending, the states 14 percent, and local governments 26 percent. And the result? A majority of Americans are still convinced that the national government is dysfunctional in most of what it does.[112]

Pluralist Administration

Max Weber, the rationalizer par excellence, recognized a pluralist type of governance although he considered it a much inferior type and not really an alternative to centralized bureaucratic administration.[113] Such a pluralist concept did not really get a sophisticated theoretical foundation until 1973, when Vincent Ostrom published *The Intellectual Crisis in American Public Administration*. Ostrom, a professor of political science at Indiana University, constructed a new pluralist theory of administration for the modern nation-state together with his spouse, Elinor, a winner of the Nobel Prize in Economics.[114]

Ostrom acknowledged that national and state governments have major responsibilities and that centralization is sometimes necessary. But he demonstrated that centralized rule by command is inferior to decentralized cooperation, which requires plural powers to adjust decisions for mutual benefit and can result in broader social satisfaction. Central governments, even down to multifunction counties that leave little independence to towns, can simply issue orders to subordinate levels of government. But where counties (or metropolitan areas with more than one county) contain multiple independent local governments, they must cooperate to achieve any common goals.

Being closer to the facts of the situation, local governments can represent their constituents' interests more accurately. At the same time, close interaction with constituents can gain leaders enough trust to make compromises. Local officials are more directly observed and they know their constituents have easy exit to nearby governmental alternatives, so they are less likely to betray constituents, or impose unpopular restrictions, or disadvantage minority groups.

Progressive rationalizers were as opposed to having many independent "parochial" local government institutions as they were to having a market of many uncontrolled capitalist corporations, and on the same principle. Economies of scale in the private sector permit larger businesses to operate more efficiently and allow higher-level oversight, and the same should work in government, the thinking went. Likewise, centralized regional governments should provide efficiencies in overseeing and regulating local governments. Woodrow Wilson's goal was for "town, city, county, state and federal government [to] live with a like strength making all interdependent and cooperative" under common rational principles.[115] Consolidated municipalities, combined school districts, larger legislative districts, and multifunctioned county governments were as important to progressive centralizing ideology as was national regulation of capitalist businesses. If not for the progressive laws passed over the last century favoring consolidation of smaller units, the number of municipalities and townships per capita would be almost three times what it is today.[116]

Ostrom observed that centralization is often preferred because of a tendency to confuse *similar* problems with *common* problems. All people have more or less common needs: food, shelter, clothing, health, education, child care, and so forth, but these common needs are not really the *same* across large numbers of people, and no one solution will cover all common problems effectively. For example, homelessness is a common problem and often thought to result from a shortage of housing. Perhaps the cause is market forces, or poor zoning or construction codes. But in other cases it may be caused by mental or physical illness or lack of employment. A local government can more easily see the specifics in particular situations and distinguish and adjust between various causes.[117]

The economist Charles Tiebout has been credited for recognizing that multiple, diverse local governments actually act in a type of market that allows individuals to choose among them. Also like a market, locals keep benefits closer to the individuals who bear the costs. Even county governments have become large and bureaucratic, but when locals are forced to compete with nearby alternatives they must be more market-responsible. Locals cannot create fiat money like the central government, so they face limits to profligacy. Market-like competition encourages setting priorities and prudent decision making.[118]

Stephen Goldsmith, a Harvard professor who previously served as mayor of Indianapolis and deputy mayor of New York City, was especially successful in introducing the Ostrom view into local administration. His major contribution was the concept of "municipal federalism," to break down counties and large or even small cities into still smaller communities where individual citizens have more ability to affect policy.[119] There is no incentive for local participation in community affairs when the real decisions come from faraway governments, especially from distant national agencies.

Unusually for a government agency, the U.S. Advisory Commission on Intergovernmental Relations acknowledged in 1989 that it had been wrong in saying that small governments could not attain efficiencies to compete with national, state, or county governments. The commission had not previously considered that locals could use private contracting or joint municipal operations to match efficiencies of scale.[120] Today, hundreds of thousands of free contracts exist between local governments, allowing them to do pretty much anything a larger unit can.[121] Moreover, efficiencies of scale at some point dissipate even in the private sector.[122]

In Ferguson, Missouri, African Americans represented 70 percent of the population and could easily have been in charge of an abusive local police force if they thought it was worth the effort. During the riots of 2014, the governor sent in state troopers and National Guard forces, who appeared to residents like a foreign occupying power. The U.S. attorney general came to the town representing national power, to little effect. Eventually the Ferguson community took charge themselves and elected a more responsive and representative local council.[123]

As George W. Liebmann demonstrated in *Neighborhood Futures*, even smaller local institutions within communities can resolve internal conflicts better, because individuals are in fact facing each other and are more able to address differences through dialogue and compromise. Whether governmental or private, face-to-face institutions can go deeper into the community to form very small communities of interest:

neighborhood councils, street governance committees, block associations, amenity cooperatives, and the popular residential community (or homeowners) associations.[124] More people in the United States have chosen to live under self-governing residential community associations than live in cities over 200,000 in population.[125]

Alexis de Tocqueville noted that private voluntary associations were the unique way that Americans organized their local communities.[126] In the early days of settlement, individuals and families typically created voluntary communities, some of which evolved into formal government institutions. In most places, local governance functioned before receiving formal recognition from state authorities. For this period, local governments were de facto voluntary associations without formal legal status or powers. When chartered, many municipalities were called "corporations" and held contract-like charters, and early local communities were mostly just loosely overseen by limited-function county and state governments.[127]

Edmund Burke's voluntary "little platoons" are still active in the United States today. About 60 percent of Americans say they keep in touch with their neighbors at least once a month.[128] A quarter of Americans volunteer with charitable organizations, although the numbers have declined somewhat in recent years.[129] The variety of voluntary associations is overwhelming: civic and fraternal groups, alumni associations, sports clubs, private schools, book clubs, cooking clubs, independent business districts, myriad charitable associations, and many religious organizations. Robert Putnam and David Campbell found that participation in local religious institutions is strongly correlated with participation in all other types of public and private associations, including the most important one, the functioning family.[130]

Really empowering local institutions would be radical—challenging the whole progressive assumption that the top experts know better than the locals. Professor Ostrom's work and the examples presented here suggest that very often they do not.[131] Central balancing of tensions is sometimes necessary, but forcing resolution of conflict from the top can cause more discord when a nation is as divided as the United States is today.

Indeed, the continuing vitality of the nation-state itself may rest on more decentralization. F. H. Buckley, a professor at George Mason

University, has put this concept into statistical form, relating the size of governments around the world to societal attributes and demonstrating that larger nations have more corruption, less wealth and freedom, and a more unhappy population.[132] Much internal factionalism can be minimized simply by taking advantage of the fact that differing viewpoints on contentious issues tend to fall into geographically localized patterns.[133]

Federalism, decentralization, and privatization can allow "a variety of outlooks" to be heard and perhaps to prevail locally. Yuval Levin suggests that "human sized" institutions of various forms can even "help the whole see the good" by testing things locally and then building up to some truly agreed-upon national approaches on a limited number of important matters, rather than trying to force a single moral vision on all.[134] Worldwide, as the editors of the *Economist* have reported, most successful reforms of government in recent times have resulted from decentralization, moving decisions down to local authorities or to markets.[135]

Decentralization is actually very much in harmony with most Americans' opinions about government organization. Polls by Gallup and the Pew Research Center for decades have found that substantial majorities of Americans think local government works and national government does not.[136] Is it not possible that free people in free communities can better solve their own problems and resolve disputes between themselves than any faraway, dominating experts could do?

We can now respond to Professor Minogue's challenge to a conservative fusionism that tends to give too much importance to freedom and too little to government. We have taken his third institution that has historically been compatible with capitalism, a Burkean subsidiarity, and applied it to his first requirement, a limited-function nation-state. The result is a positive, moral vision of government in the form of a federal state that is decentralized, both locally and privately, as far as is consonant with public order.

Very importantly for the major theme of this book, localized governance and subsidiarity also provide a means to control the "creative destruction" of capitalism in a dynamic tension that is congruent with popular cultural and moral values. Free individual choices in the marketplace can set the broad parameters for efficient economic life, while a less-overwhelmed national government can better perform its essential functions in serious interstate and international matters.[137] The central government can even provide a sense of national identity, but with a Tocquevillian understanding that the burden of governance will primarily be local.[138]

Pluralist local governments and associations can go far in allowing people with diverse values and preferences to find solutions to problems without undue coercion, which people of many different perspectives find morally preferable. Still, some would continue to fear that minimizing central supervision only results in more "unruly freedom" and a less moral public order. The concluding chapter takes on that concern.

CHAPTER 8

THE CIVILIZATIONAL CHOICE

Forty-eight years ago, I published *The Political Culture of the United States* in the prestigious Little, Brown Analytic Studies in Comparative Politics series. The editor warned that its thesis asserting the existence of a broad moral consensus on traditional Lockean values among Americans would antagonize fellow professors who assumed that the U.S. population was inclining leftward. But the book's use of virtually all public opinion polls taken in the United States up to then was so empirically persuasive that academic reviews in a profession dominated by progressive intellectuals accepted its factual if not its moral conclusions.[1]

What were the elements of that consensus? A proud but not arrogant emotional identification with the American nation. Support for limited government and a division of powers, but favoring Congress. Federalism and decentralization of power. Moral and legal equality, with a guarantee of equal opportunity for all but not equal results. Free markets and property rights. Faith in God and commitment to traditional moral values. Attachment to family and community. A premium on education and work achievement.[2]

Many elements of that old consensus are now almost fighting words for a large part of the American population, facing off against the other part across a deep cultural and political divide. Today there is no national consensus on the values necessary to sustain a free democratic society, much less on the morality of pluralist capitalism.

But if Minogue's and others' social critiques are correct, freedom and order both require some agreement on the need for institutionalized moral constraints.

The Endgame of Traditional Culture

One of America's most sophisticated social scientists, Charles Murray, is not hopeful that a broadly shared sense of national identity can be restored. In 2016, he saw the enthusiastic support for the socialist Bernie Sanders on one side and the nationalist Donald Trump on the other as expressions of "the legitimate anger that many Americans feel about the state of the country, the endgame of a process that has been going on for a half-century of America's divestment of its historic national identity."

Our "Anglo-Protestant heritage," Murray wrote, was once the core of a shared culture, but has "inevitably faded in an America that is now home to many cultural and religious traditions." Even the secular part of the national creed, "the very idea of America" as a unique blend of "egalitarianism, liberty and individualism," has seriously eroded. There is weakening commitment to "equality before the law, equality of opportunity, freedom of speech and association, self-reliance, limited government, free-market economics, decentralized and devolved political authority."[3]

Murray saw the erosion of America's secular culture as beginning after World War II with the emergence of a new class system based primarily on education rather than elite Anglo-Protestant affiliations, although there was some overlap at least initially. Brainpower and academic credentials became "radically more valuable in the marketplace," and the highly educated tended to marry each other, producing an elite class. Among them, there has been a "large-scale ideological defection from the principles of liberty and individualism, two of the pillars of the American creed." By the 1980s, cultural elites "overwhelmingly subscribed to an ideology in open conflict with liberty and individualism as the masses understood it."[4]

Strangely, this ideological defection came about "in large measure because of the civil rights and feminist movements, both of which began as classic invocations of the creed, rightly demanding that

America make good on its ideals for blacks and women." The success of these movements, however, soon led to policies in direct contradiction with liberty and individualism. "Affirmative action demanded that people be treated as groups. Equality of outcome trumped equality before the law."[5]

In the past, America has done much better than other countries at integrating people from many places into the national identity, Murray continued. But when a large part of the American public no longer believe in the secular creed, we "will have detached ourselves from the bedrock that has made us unique in the history of the world."[6] Even patriotism and the very idea of national identity have become divisive, especially for the young.[7] The More in Common poll mentioned in an earlier chapter found a bare majority in the United States saying they were proud to be American.

Shortly after World War II, the distinguished political commentator Walter Lippmann already saw a decline in cultural consensus and faith in liberal democracy. He described a traditional public philosophy of civility whose basic elements were: loyalty, community harmony, representative assemblies, popular elections, majority rule, free speech, private property and free trade. Underlying these values was belief in a natural law of civility, or restraint on impulses, which was almost universally thought necessary to sustain and legitimize them. By the twentieth century, these beliefs came under serious questioning by a large number of radical and progressive intellectuals.

Lippmann began his story by tracing a natural law of civility derived from the Greeks and Romans as it evolved through the Middle Ages until around 1500. A new version appeared under the divine-right monarchies, and it worked reasonably well at keeping social order until the end of the eighteenth century. Then the aristocracies which had sustained the law of civility were overwhelmed by the Industrial Revolution and the enfranchisement of the masses under the guidance of a radicalized intellectual elite.[8] Inspired by Rousseau, "the Jacobins and their successors made a political religion founded upon the reversal of civility," wrote Lippmann.

> Instead of ruling the elemental impulses, they stimulated and armed them. Instead of treating the pretension to being a god as the mortal

sin original, they proclaimed it to be the glory and destiny of man. Upon this gospel they founded a popular religion on the rise of the masses to power. Lenin, Hitler and Stalin, the hard totalitarian Jacobins of the twentieth century, carried this movement and the logical implications of its gospel further and further towards the very bitter end.[9]

Lippmann characterized Rousseau's Jacobinism as a "Christian heresy" that sought perfection in this material world rather than the next.[10]

Friedrich Nietzsche and Jean-Paul Sartre later destroyed God and all laws above the self. But Western capitalist society needed something higher than mere self-interest as a defense against the rule of raw coercion, Lippmann argued. English jurists invoked an "ancient constitution" of common law going back many centuries, as a way to restrain power and legitimize its use. But once the myth was exposed by progressive intellectuals, the legitimacy of originalist constitutionalism became suspect.

The natural law of civility survived in the United States under the protection of its WASP aristocracy. But the ideal of its natural law code ceased to be taught in the elite universities. The result after World War II was a new generation, and particularly its leaders, with "a low capacity to believe in precepts that restrict and restrain private interests and desires."[11]

Lippmann considered a common secular culture to be absolutely essential for the survival of the "good society under modern conditions," and he thought the need to provide such a culture to support Western civilization was "acute." In a pragmatic age with few illusions, his solution was to recover a commitment to natural law by rationally rejecting the impossible dream of perfection, and instead demonstrating the "practical relevance" of a public philosophy that promotes a decent and prosperous civil life. With religion weakened, the only way to recover natural law was to go back to its source, to the Stoics, to Alexander the Great and Zeno. The Stoic ideal had fortified the West in ancient times, and it was the only secular doctrine that might conceivably support a common moral culture in the modern world.

The Materialist Illusion

John Nicholas Gray, a British philosopher, likewise suggests some form of stoicism as a plausible moral code for modern culture, but questions the materialist assumptions typically behind it. Gray is probably one of the most broadly respected intellectuals in the world today. He won acclaim from the right for his book on Isaiah Berlin and for his influence on Margaret Thatcher, and then from the left for his books criticizing "the delusions of global capitalism" and supporting an agnostic liberalism. In *The Soul of the Marionette: A Short Inquiry into Human Freedom*, he confronts both sides with hard truths.

Gray begins by drawing out the implications of secular materialism. If all reality is material, he asks, how can we have any mental sense of it? Our mental image of the material must be an illusion. Thus there is only one rational stoic truth: the "fact of unknowing," the fact of a deep ignorance that can be escaped only through an "inner freedom" that rejects any effort to "impose sense on your life." To avoid "becoming an unfaltering puppet" of some mythic higher rationale, a true stoic must simply make one's way as best one can in this "stumbling human world."[12]

This stoical ethic is a difficult sell today, Gray acknowledges. It was commonplace in ancient Greece and Rome, but was outbid by Christianity, with its promise of happiness in another world. In turn, Christianity was outbid by the "curdled brew of Socratism and scraps of decayed Christianity" called modern scientific materialism, promising Nirvana in this world. Nietzsche tried escaping by criticizing both the "brew" and the "scraps," but only concocted a new materialist myth with his "absurd figure of the Übermensch," embodying "the fantasy that history can be given meaning by the force of human will."[13]

Gray takes on the modern epitome of atheist materialists, Steven Pinker, a professor of psychology at Harvard who is considered one of the world's leading intellectuals, with awards from the National Academy of Sciences, the American Psychological Association, the Royal Institution, and the Cognitive Neuroscience Society. Pinker has written a best-selling book, *The Better Angels of Our Nature: Why Violence Has Declined*, with a thousand pages and innumerable statistical tables aiming to demonstrate that things are fine and getting even better in this world, so we can all live a good and meaningful life without needing stoicism, a god or an afterlife.[14]

Gray observes that Pinker's optimism echoes that of Auguste Comte, the nineteenth-century philosopher who invented the term "altruism" and founded a "religion of humanity," a secular creed that drew its hope from science. Comte's optimism—before the world wars, the Holocaust, and the Gulag—is understandable to Gray, but Pinker's is not. Gray regards the optimism of Pinker and other positivists as a secular "faith," reinforced by reciting "amulets" of encouraging data to "exorcise any disturbing thoughts from their minds." In this way, positivists find the "meaning in life" that was "long shrouded in myth and superstition."[15]

Materialist intellectuals like Pinker believe in valuing humanity, even while believing that humans are mere flesh and that consciousness came about by chance through chemical processes resting on interactions of subatomic particles. But how could material processes alone create a self-aware human spirit that seeks meaning in life and has a capacity for such things as tolerance and empathy?

Pure materialism simply cannot generate those values and capacities, Gray argues. There must be a higher source for the distinctive qualities of human beings. "If you want to reject any idea of God, you must accept that 'humanity'—the universal subject that finds redemption in history—also does not exist." But he expected few materialists to follow his logic. Materialists retain a belief in humanity by resorting to a modernized version of historical Gnosticism, one that promises full knowledge to the few through reason and science. "Belief in the liberating power of knowledge and science has become the ruling illusion of modern mankind," writes Gray. Far from being "annihilated by Christianity as assumed by earlier historians, Gnosticism has actually conquered the world."[16]

The Gnostics of the early Christian era believed that humans had been created by an evil demiurge, and therefore human freedom necessarily results in evil. Only a very few could escape the world's grossness of deteriorated matter into primal spirit through a profound, "unnatural" knowledge, or *gnosis*. Today's gnosticism is more optimistic and successful among intellectuals but less logical. It assumes that randomly evolved humans can rise above all other humans and turn them all into well-behaved puppets.

In *The Soul of the Marionette*, Gray rejects the optimists' belief that science and rationality have subdued human violence and mayhem, and adduces enough facts to show that things may be even worse now than in the horribly violent twentieth century. Much that today goes under the rubric of popular movements for freedom and democracy is actually manipulated by gnostic intellectual shamans to produce the same violence, or more. They simply use the "sorcery of numbers" to hide the true extent of human brutality today.

Like Rousseau, Gray sees humans in ancient times as living in ignorance, with no concept of perfection. Christianity "undermined this tolerant acceptance of illusion" and pushed aside philosophy's paralyzing uncertainty, replacing it with its claimed universal truth.[17] As Christianity weakened intellectually over time, its claim to universal truth was taken up by science, and by political ideologies promising perfection. But the scientific revolution was, "in many ways, a by-product of mysticism and magic," Gray argued. Kepler was a mystic. Newton believed in alchemy and numerology. "In fact, once the tangled origins of modern science are unraveled, it is doubtful whether a 'scientific revolution' occurred." Even the Enlightenment developed within Christian assumptions.[18]

So there wasn't such a clean break from religious thinking as the rationalists have supposed. In particular, the idea that humans are uniquely self-aware is "a prejudice inherited from monotheism."[19] There is no scientific basis for the belief that humans alone have consciousness or self-awareness, Gray argues. In fact, dolphins and chimpanzees do seem to display self-awareness. If they do not fully comprehend their own consciousness, the same is true of humans. The illusion of a unique human consciousness is an extension of the Judeo-Christian concept of the soul, and of a Creator who specifically endowed humans with higher faculties and moral status.

As for the foundational scientific belief that the world is composed only of matter, it is purely a "metaphysical speculation." Yet the materialist ideology remains dominant among intellectuals. With unquestioning faith in their own assumptions, the materialists arrogantly suppress the "valuable insights" that come from other perspectives, particularly from religion. They reject religion's "idea of evil while being obsessed by it," and never ask "why humankind is so fond of the dark." Unlike religious believers, they do not realize that "they face an insoluble difficulty." Traditional believers, "aware of evil in themselves," know that human action cannot expel it from the world, while "secular believers dream of creating a higher species." The "fatal flaw in their scheme" is that humans as they now exist would have to create that higher, more perfect being.[20]

How can materialists rationally believe it possible? Logically, they should deny that they even have the ability to carry out such a project, but most do not. In fact, the "predominant religion" of today's "boldest secular thinkers" is that humans can escape their natural limitations by means of modern technology. This ambition, ironically, is held by people who view humankind as "puppets on genetic strings, which by an accident of evolution have become self-aware."[21] Gray quips that gnosticism is "the faith of people who believe themselves to be machines," specifically mentioning the futurist visions of Ray Kurzweil, with his aspiration that people will become perfectly wise human machines.[22] Since Kurzweil is a billionaire with control of Google's vast resources for genetic research, he might have the means to finance the transformation.

In comparison with this technological gnosticism and the materialistic view of humankind that underlies it, Gray finds that even Christianity presents a "more truthful rendering of the human situation." With its doctrine of salvation in another world, Christianity is anti-tragic, yet its account of life in this world is "closer to ancient understandings of tragedy than it is to modern ways of thinking."[23] In the end, it too is an illusion—indeed, the one that begot the dominant modern illusions, though it may be "the least harmful illusion."[24] The better reflection on humanity is a true sense of tragedy, without any illusion of perfection either in this life or another.

Gray sees the world today facing a new period of instability, as "faith in political solutions is fading and renascent religion contends with the ruling faith in science to replace it."[25] But science has no solution for the first requirement of civilization, which is to restrain human impulses. "Before it means anything else, civilization implies restraint on the use of force; but when it serves noble-sounding goals, violence has a glamour that is irresistible." A true freedom among humans comes down to "mutual non-interference," but this is not natural; it is a "rare skill that is slowly learnt and quickly forgotten." Practices that have allowed such freedom in the past, such as the rule of law, are now being compromised or "junked" altogether.[26]

For true stoics, Gray's answer is an inner freedom that is indifferent to the type of society in which one lives or the government that rules it—as long as it does not prevent individuals "from turning within themselves."[27] But even Marcus Aurelius, the emperor who was also a model Stoic philosopher, did not allow such freedom to Christians, among others. Should one take for granted that a modern stoic ruler would be different?

Is It All Power?

If a common humanity and religion are illusions, and there is no scientific basis for moral restraints on human impulses, then does it all come down to the ability and willingness to exert power? Machiavelli advised the prince to make good use of "beastly qualities," being fierce like a lion and cunning like a fox.[28] Nietzsche saw power as the highest value. Even Voltaire thought a powerful ruler was essential for rationalized capitalism. Indeed, most of the modern world is ruled more by force and manipulation than by social consensus.[29]

Even in relatively free nation-states, the population is subject in substantial measure to the moral dictates of those who control the levers of power at any given time. The ruling powers in the United States formerly held slavery to be justified and made it legal, but later governing powers de-moralized and outlawed it. Segregation was legalized by one Supreme Court but outlawed by another. Homosexuality was once illegal and medically disapproved, but then Supreme Court decisions mandated that it be accommodated even by people with religious objections.

Power in itself has no standard to legitimize any social or moral order. And its results are ephemeral. As Yuval Levin observed, recent national government efforts to establish moral order by legal dictate have gone against conservative cultural morality on marriage, sexuality, religion, and patriotism, but also against progressive moral beliefs in

Great Societies, large-scale social policies, a powerful central government, and political correctness.[30]

For example, President Obama was more than willing to respond
to calls for more national oversight of local police practices in
the wake of the tragic shooting of Trayvon Martin in 2014, and
he increased criminal investigations of local police by a factor of
four.[31] In the very next administration, the new attorney general, Jeff
Sessions, charged that the increase in federal oversight had intimidated local police into being "more reluctant to get out of their squad
cars and do the hard but necessary work of up-close policing that
builds trust and prevents violent crime."[32] Police officers were not
doing "the kind of community-based policing that we have found
so effective," as they feared being second-guessed by overbearing
federal officials. The federal government must stop "dictating" to
local authorities.

> States and cities and towns have always played the lead role in crimi-
> nal law enforcement in our country—and that must continue, for two
> reasons. First, it's the most effective way for us to work. Many of the
> law enforcement techniques that helped make our neighborhoods
> safer in recent decades—such as community-based policing—were
> developed and refined at the local level. So it's a good bet that the
> best new ideas for meeting this current crisis will also come from
> innovators in state and local law enforcement. The second reason is
> sheer numbers. About 85 percent of all law enforcement officers in
> our nation are not federal, but state and local.[33]

During his two years as attorney general, Sessions did not open any new
investigations into local police activities.[34] His successor did.[35]

At the same time, however, Sessions criticized the previous administration for not cracking down on local authorities that refused to
cooperate with federal immigration enforcement. Sessions took a
tougher approach, but one result was to invigorate the "sanctuary
city" movement and local policies of noncooperation.[36] Intriguingly,
progressives justified their sanctuary movement on federalist grounds.
The attorney general responded, correctly, that the Constitution gives
the national government the authority to regulate immigration. But

the Constitution does not seem to authorize the national government to compel state or local governments to help it do so.[37]

Pluralist regimes like the United States have built-in restraints on the use of power. The subsidiarity principle by which local communities control their own peacekeeping goes back to the nation's origins. The Insurrection Act of 1807 and subsequent amendments, along with the Posse Comitatus Act of 1878, set strict conditions for the use of federal military forces in local law enforcement.[38] Local problems, however, often become nationalized through ubiquitous media attention and ambitious elites forcing their pet issues into the national agenda. But if New York City's 50,000 police officers do not enforce national immigration laws, what can the federal government do using raw power alone? The FBI often pressures local law enforcement to cooperate, but it has only 17,000 agents. The national government has a powerful army, although its use in domestic law enforcement is narrowly restricted by law. Even the Soviet Army refused to fire upon protesters in Moscow, and the U.S. Army is probably not more ruthless.

When it comes to ordinary violations of law by average citizens, enforcement has limits at every level. The police cannot be everywhere at once. They can enforce speed limits to some degree on city streets or even rural roads, but anyone who ever drives on a freeway knows that speeding is commonplace and cannot really be prevented. Federal regulations are even more difficult to enforce. With 300,000 or so on the books, no one can possibly know or even guess what is forbidden, and the most honest citizen may unknowingly commit several crimes in a day.[39] It has been proposed that people simply ignore the least rational rules, and create an insurance-like fund to help those who are unreasonably harassed, the hope being that the government might be shamed into abandoning indefensible rules. But it would certainly make laws harder to enforce.[40]

As Lord Acton famously warned, power does seem to corrupt.[41] But corruption limits the reliability of enforcers and the possibility of gaining popular cooperation. The FBI inspector general found that the bureau's top administrators had improperly tried to affect the 2016 presidential election. It was Bob Kerrey, a Democratic former governor and former U.S. senator, who called for a full investigation, and then the Republican attorney general initiated one.[42] The FBI sent heavily armed agents to arrest nonthreatening white-collar suspects in the dark of night, terrifying innocent spouses.[43] As mentioned earlier, Secret Service officers have leaked confidential personnel files, and the IRS has also violated confidentiality laws. Top U.S. attorneys and SEC investigators apparently seek to indict people who will garner the biggest headlines rather than the most guilty.[44]

Abuses of power occur at all levels of government, but seldom of the magnitude or with the impunity found in centralized bureaucracies. Members of Congress and even executive-branch officials may hesitate to rein in bureaucrats who are pursuing their own interests, for fear of retaliation—as when Secret Service agents leaked confidential information on Rep. Jason Chaffetz.

With America divided almost equally between reds and blues, one half will try to force the other half to follow its own preferences. But any victory for raw power will tend to be fleeting. In eight of the ten national elections before 2020, the political party in control lost either the presidency or the Senate or the House of Representatives, so the other party had at least some opportunity to reverse its accomplishments or frustrate any long-term agenda.[45]

Moralizing Power

Victories for political power in the United States are, it must be acknowledged, largely a consequence of democratic elections, and surely have a kind of legitimacy on that account. Indeed, it was once widely believed that democracy itself, broadly understood, legitimizes power.[46] But confidence in democratic values is on the decline: the World Values Survey recently found that more that 70 percent of Americans born in 1930 said it was "essential" to live in a country governed democratically, but the numbers declined sharply in subsequent age cohorts, to slightly over 30 percent among those born in 1980.[47] Support for the democratic right to freedom of speech is declining even (or especially) in schools of higher learning.[48] When organizations headed by George W. Bush and Joe Biden commissioned a large-scale national poll on the ideals of democracy, 55 percent of respondents said that American democracy is "weak," and 68 percent believed it is "getting weaker." More concerningly, young adults place less value on democracy than do older people, while some groups, especially racial minorities, find less benefit from democracy than do others.[49]

Serious observers discussed earlier have focused on different values that appear to be weakening among the public. Eric Voegelin wrote of a declining concern for social order. Walter Lippmann saw an erosion of civility. Charles Murray was concerned about decreasing regard for liberty and individualism. Jonah Goldberg feared the loss of support for free markets, private property, and the work ethic. Patrick Deneen worried about diminishing commitment to family, community, and natural law.

Yet polling data reveal a surprising fact: it is actually religious values that have remained widely supported over time. According to Gallup, 96 percent of Americans in 1944 believed in God, and 89 percent did in 2014, still an extraordinarily high level of public agreement on anything. In a 1968 poll, 85 percent said they believed in heaven, and the figure was unchanged in 2011. In a 1992 poll, 87 percent said that religion was very or fairly important in their life, as did 78 percent in 2014. Support for organized religion declined modestly from 86 percent in 1973 to 72 percent in 2016. Those saying they had no religious affiliation did increase from only 2 percent in 1948 to 17 percent in 2015, but 30 percent of these so-called "nones" said that religion was important to them in some way. Pew found in 2014 that 55 percent of Americans said they prayed every day and an additional 21 percent did so at least monthly. Even among those with no religious affiliation, 37 percent said they prayed.[50]

Is the United States unique in this respect? In 2009, John Micklethwait and Adrian Wooldridge shocked the secular West by declaring that "God is back." These two epitomes of British reasonableness offered evidence that Europe was the only part of the world where God was thought to be dead. Everywhere else they looked, religion was booming, notably in China, of all places.

China's officially atheistic government itself reported that its Christian population had doubled over the preceding decade, reaching 21 million people worshiping in 55,000 officially approved Protestant churches and 4,600 Catholic churches. The underground church was much larger, perhaps 77 million by foreign estimates—if so, exceeding the membership of the Chinese Communist Party. A Pew Global

Attitudes survey had found 31 percent of Chinese people saying that religion was very or somewhat important in their lives and only 11 percent saying it was not important at all.[51]

Statistics from WIN-Gallup International in the early twenty-first century showed 59 percent of the world's population saying they were religious and only 13 percent atheist. The latter were heavily concentrated in four countries: China, Japan, the Czech Republic, and France. The study found majorities even in Europe saying they believed in God (Sweden being the only country where just a minority did), and majorities saying they prayed. The people of Africa, Latin America, India, Asia (other than India, China, and Japan), and the Muslim world overwhelmingly considered themselves religious.[52]

The very systematic and sophisticated Pew global studies projected that all the traditional religions except Buddhism will continue to grow at least until 2060, reaching 3.1 billion Christians, 3.0 billion Muslims, 1.4 billion Hindus, 0.9 billion of other religions (including Buddhism), and 1.2 billion unaffiliated. The unaffiliated "nones" are expected to decline proportionately from 16 percent to 13 percent of world population by 2060, increasing as a share of population in much of Europe and North America but declining in Asia, where 75 percent of reported "nones" now live.[53] They are mostly in China, where religious affiliation has come under scrutiny by leaders worried about the large increase in Christian belief and practice. Pew suggested that correcting for the Chinese government's suppression of Christianity would significantly lower the estimate of nonbelievers in the world.[54] In some repressive countries, on the other hand, respondents might be fearful of identifying themselves as unbelievers.

Christianity will grow substantially in South America, Africa, and Asia, mainly due to high birth rates. In Asia and Africa, Christians may come into more conflict with a Muslim population that is growing even faster worldwide. This is mainly because the Christian population will shrink dramatically in western Europe, and slightly in the United States, but even there in numbers only rather than population share.[55]

Even Europe is more religious than has generally been thought, according to a very large study by Pew in 2018. While only one-fifth of its population attend religious services regularly, 71 percent identify as Christian, 65 percent believe in God or a higher power, 51 percent pray, and 70 percent are raising their children as Christians. The study found that those who identify as Christian but do not attend church are generally more traditional in cultural attitudes than those who claim no religious identity. In Italy, 40 percent attend church regularly, 35 percent in Portugal, 34 percent in Ireland, 28 percent in Austria, and 27 percent in Switzerland, similar to U.S. attendance rates. The countries with the lowest attendance and affiliation are northern and Protestant: the Netherlands, Norway, Sweden, Belgium, and Denmark. Only the Netherlands has a plurality claiming no particular religion, although only 17 percent are specifically atheist or agnostic.[56]

Not all are pleased with the persistence of religion. Richard Dawkins, the militant British atheist, set up a Foundation for Reason and Science, with an American branch, devoted to undermining the influence of religion in society. In 2006 he released a documentary first aired on British television under the title *The Root of All Evil?* Soon he expanded it into a book, *The God Delusion*. Dawkins criticizes all three Abrahamic religions but his main target is Christianity, since he considers it the most powerful religion. He characterizes the Christian belief that Jesus had to be "hideously tortured and killed so that we might be redeemed" as a "nasty sadomasochistic doctrine." He says that private Christian schools promote a "poisonous system of morals." He compares the teaching of religion to a virus that infects young people and spreads from generation to generation. He asserts that parents who teach religion to their youngsters are committing "child abuse."[57]

Yet a great deal of evidence refutes this dark picture of how Christianity in particular and religion in general affect society. Throughout the world, religion is positively associated with neighborliness, marital happiness, childbearing, work habits, longevity, voting and political participation.[58] Many surveys in the United States particularly show similar correlations.[59]

A majority of Americans report that religion is "very" important in their lives, well above the levels in western Europe and even South America, though below Africa, the Middle East, and Asia (excluding China).[60] Only about 10 percent of Americans say that religion is not at all important in their lives.[61] It is difficult to know precisely the degree to which religious belief motivates the actions of individuals in the United States today, but in normal times a hundred million Americans go to religious services every week, and 61 percent attend church on the most popular holiday, Christmas. These figures dwarf the numbers of those who regularly attend college or professional football games or other sports events, or musical performances. Only 48 percent of the public watch the top secular event, the Super Bowl, on television.[62]

The Necessity of Mythos

David Martin, a professor of sociology at the London School of Economics, challenged the intellectual assumption that the world's future is secular in *Religion and Power: No Logos without Mythos*. What impressed him the most was the religious revivals in modern times: an almost moribund Islam had resurged in the Middle East and resurrected in once-secular Turkey; Orthodoxy had returned to Russia; Pentecostalism had erupted in Latin America; Catholicism and Protestantism had revived in Poland, Lithuania, Ukraine, and even in China. This religious resurgence helps explain why a state's rational logos must have a sacred mythos to grant it legitimacy, or it cannot survive.[63]

Alternatives to formal religion are possible. One in three western Europeans say that "science" has replaced religion in their personal decision making, while substantial numbers believe in a variety of supernatural practices such as fortune tellers or magic.[64] The secular creeds of Nazism and Soviet and Chinese communism have had considerable appeal, but romantic, nationalistic, and pseudoscientific myths were mixed in to gain more support, and even then an extraordinary degree of coercion was required.[65] Nationalism prospered much longer, but mostly with traditional religious support. Indeed, the global data we have reviewed support Martin's belief that most of the world's people find moral authority in traditional religion. Muslims, Christians, and Hindus are all increasing in numbers and relative to population.[66] Christians will outnumber "nones" even in Europe (except narrowly in France and the Netherlands) at least until 2050.[67]

Just as logos power needs a legitimizing mythos, Martin argues, mythos conversely cannot do without support from logos power. In the short run, power is dominant, so religions that absolutely reject it cannot survive. Zoroastrianism basically dissolved in the face of militant Islam, and Buddhism bent to warlike powers with more aggressive myths. Christianity first grew independently of state power, even in defiance of it. After the persecution under Diocletian, the church was rescued by Constantine, but then constrained by him. Christianity survived by making accommodations to militaristic Rome, fierce Germanic tribes, and the medieval warrior nobility.

This might seem to have presented a predicament to those who believed the foundational doctrines, as Martin pointed out:

> How could Christianity, which in its foundation documents has no honor code and categorically rejects reciprocal violence, become implicated in what it most vehemently rejects? How could the Sermon on the Mount become the sacred Scripture of the crusaders without the evident contradiction sparking a social convulsion or the direct and unequivocal repudiation of Christianity by the principles and the powers? How could aristocracies based on blood ties and feudal obligations of service owed by serf to lord tolerate a liturgy that in the *Magnificat* daily promised "to put down the mighty from their seats and exalt the humble and meek"? How could merchants bent upon accumulating wealth at all costs regularly recite a scripture which sends the rich "empty away" and preaches good news to the poor?

But in Martin's view, the remarkable fact is "not the extent to which the Gospel message is overridden but the extent to which it remains on the books to be read, recited and sometimes heeded," even by someone like Athelstan, a tenth-century English king who sought public penance after murdering a rival.[68]

One basis for the accommodation with logos was the Christian understanding that there is an evil power in the world that can never be fully subdued, though it can be challenged and perhaps diminished over time.[69] This required acknowledging the reality of power in this world, and working with it. Some strains of Christianity, such as Pentecostalism with its individualized gifts of the spirit, formally spurn any accommodation to secular power as immoral. That may have helped it become perhaps today's fastest-growing creed, though when such denominations become large enough to be politically relevant, they do tend to make accommodations to the state.[70]

In the Western synthesis of mythos and logos, medieval warriors did find justification for their fighting in the more martial books of the Old Testament. Christian merchants could justify seeking profit by emphasizing the peaceful, voluntary nature of market trade for mutual benefit. One should not be surprised at the compromises, Martin argues, as they are part of the universal pattern of mythos meeting logos. Those who do not understand this reality will criticize religion from opposite directions. Some have charged that Christianity promotes weakness and passivity, from the early Christian refusal of military service, to the pity that Nietzsche scorned. But pacifists claim that Christianity has been too warlike, from the Crusades to recent Western military ventures.[71] Rousseau, in *The Social Contract*, criticized Christianity for *both* passivity and aggression.[72]

From the perspective of a scholar of ancient history, Donald Kagan of Yale believes the United States to be weakened by an "unsubtle Christianity" with a strong strain of pacifism that "makes it hard for us to behave rationally when the rational thing is to be tough." Kagan asked rhetorically, "Who else has a religion filled with the notion 'turn the other cheek'? Who ever heard of such a thing? If you're going to turn the other cheek, go home. Give up the ball."[73] As an appointee of President Reagan, Kagan had not agreed with him that the Cold War should be mostly a peaceful engagement.[74]

Western Europe differs from the United States in the explicit secular orientation of its leadership, but even there, majorities identify as Christian and believe in God, and majorities say that traditional religion has significance in their lives.[75] Pew found that two-thirds of Europeans rejected the idea that science makes religion superfluous.[76] Worldwide, relatively few people do not identify with some traditional religion, and their share of the global population is shrinking. Samuel Huntington reminded us that differences in religion still mark the lines between the major civilizations.[77] The world's people will apparently remain mostly religious, which will continue to have social and political consequences.[78]

The Faith Instinct

Nicholas Wade had been an editor of the Science Times section of the *New York Times* and of the prestigious journals *Nature* and *Science* when he published *The Faith Instinct: How Religion Evolved and Why It Endures*. Wade argued that religion first originated from a genetic instinct that promoted behaviors useful for survival. These behaviors evolved into culture, or learned practices, and the cultural behavior fed back into material evolution. It was the development of culture, including religion, that enabled early humans to advance rapidly beyond their nearest genetic relatives and become masters of the world.[79]

Wade observed that chimpanzees, the presumably closest living cousins to humankind, have not evolved much since they diverged from hominids some 5 million years ago; but humans have changed dramatically in just the last 50,000 years. "Such substantial and fairly steady progress cannot have been directed by evolution, a blind and largely random process with not a flicker of interest in human welfare," Wade concluded. "Surely the only possible origin of progress is human choice." When early hominids began choosing certain kinds of behavior, they were creating culture.

Secular evolutionists had only recently introduced culture into their analyses. They had learned that culture can feed back into the genome, "accounting at least in some part for the vast differences between people and chimpanzees."[80] Other animals have instincts for restraining their behavior in the interest of group survival, and that is the genetic root of morality. But only in humans was religious behavior "grafted" onto morality. Early humans developed rules to restrain self-interest for the sake of group survival, which was necessary for individual survival. These moral rules were reinforced with taboos and invocations of higher powers that made people accountable. Such rules and taboos, along with the rituals that support them, provide the "primal glue" that unifies a group and promotes its survival. Humans then came to be "kinder, more prosperous, less warlike, less profligate of the environment and more knowledgeable" than chimpanzees. They had lower rates of combat deaths, and they still do, even in the horrible twentieth century.[81]

The genetic root of religion explains why it is similar from one society to another, Wade argues, while cultural choices explain the differences between religions. The major similarities that he found among religions were prohibitions of murder, theft, and incest, along with a general principle of "Do as you would be done by," which Darwin regarded as "the foundation of morality."

Religious rules and taboos, and the stories surrounding them, have been crucial for human advancement and for social cohesion. Mythos sustained the society and supported the state. Roman emperors thought it necessary to title themselves *pontifex maximus*, or chief priest of the state cult, and bureaucracies probably started out in temples. The unifying mythos did not need to be really true, but only to be accepted as the legitimate foundation of moral rules.

J. Budziszewski, a professor of political science at the University of Texas, had earlier argued that we *must* actually know that our moral sense has a real truth. In a provocatively titled book, *What We Can't Not Know*, Budziszewski wrote that all humans have a sense of a moral law written in their conscience, although some deny or ignore it through sloth, or apathy, or self-deception. Furthermore, just as Aristotle was certain there must be a first cause of all things, it must be obvious to any thinking person that our moral conscience has a real force behind it.[82]

It might be considered arrogant for Budziszewski to have told his fellow professors that they actually knew the truth of what most of them claimed to disbelieve. But like Wade after him, he began by observing there is a "universal common sense of the human race" concerning morality, which started in our genes and then was further developed by culture to promote group survival. Budziszewski called this a "natural law," a set of moral principles that are "not only right for all, but at some level known to all."[83] Without some common moral sense, mutually understood, no communication could ever get started. What is included in this natural law of morality? In addition to the items on Wade's short list of universal moral values, Budziszewski specified caring for one's children, and being opposed to maiming, slander, and most adultery.

If this natural law is written on all human hearts, enforced by conscience, how can so many deny both its existence and its dictates? Humans generally want to be thought moral, and to think themselves moral, Budziszewski maintained. They might resist the demands of conscience, but can escape it only through psychological defense mechanisms such as remorse or blurted confession, reflexive atonement, pursuing reconciliation or seeking justification. To ignore the promptings of conscience carries a psychic cost. The executioners at the Treblinka extermination camp had to undergo psychiatric conditioning to carry out their monstrous instructions. It turns out that "the flotsam of natural law—all those corks of truth...can't all be kept down at once."[84]

Budziszewski acknowledged that the monotheistic traditions and pre-Noahide covenant pretty much summarized what he saw as natural law. Adam Smith too found certain values like beneficence to be "natural," although he also anchored his moral concepts in a specifically Western sense of justice.[85] If there is in fact a cultural element in this "natural" moral law, then does it really hold sway over what René Girard described as perhaps more fundamental human attributes such as acquisitive desire? Budziszewski's critics have charged that his theory is based on faith more than reason, but he noted that honest rationalists like Thomas Nagel admit that their atheism is at least partly based on "hope."[86]

Surely, though, it is impertinent to insist that everyone must really believe in God, especially as the source of moral law, when so many intellectuals assume that God is safely dead. Even so, why would a sophisticated professed agnostic such as Charles Krauthammer have said, "I don't believe in God but I fear him greatly"?[87] In response, George Will said he thought his friend Krauthammer was actually an "amiable low-voltage atheist" like himself, but "flinches from saying it." Asked if he believed in God, Will replied:

> No. I'm an atheist. An agnostic is someone who is not sure; I'm pretty sure. I see no evidence of God. The basic question in life is not, "Is there a God," but "Why does anything exist?" St. Thomas Aquinas said that there must be a first cause for everything, and we call the first cause God. Fine, but it just has no hold on me.[88]

So the committed atheist seems to leave an opening for the possibility of a higher power that could explain our existence, a power that many people call God, and even had a kind word for Thomas Aquinas. And the agnostic still feared what he doubted. How does one explain these concessions to religious belief on the part of nonbelievers?

The Western Story

Among many social scientists, David Martin believes that we humans need a story to tell ourselves and others about who we are, or we simply have no identity, although the story need not be literally true to serve its purpose. As we have seen, William F. Buckley Jr. and Frank Meyer both believed that a *story* of Western civilization was essential to a fusionist conservatism that supports capitalism. Kenneth Minogue once thought any such story should not be taken seriously, saying, "The way to kill a tradition is to believe that any particular theory of what the story means is true." But later he came to believe that this level of skepticism would make all knowledge frivolous and would be rationally destructive. So he called for "an imaginative sympathy with the deeds and customs" of the past, and even a "piety" toward the stories, mindful of what we have learned as those old customs became our traditions. And he challenged those seeking a story legitimizing capitalism to include institutionalized moral constraints.[89]

Martin argues that what made the West unique was the idea of a common humanity, a universal moral equality for every individual.[90] The concept of human equality is not universal. Indeed, Aristotle's first political principle was inequality. Slaves obviously had no rights in ancient society, while most women and even the majority of men had few. The paterfamilias ruled over an extended family and was the exclusive owner of its property. He was the spiritual leader of the clan, the only one allowed to maintain the sacred flame connecting the living to the ancestral spirits resting on his property and to invoke their protection. Early cities were collections of powerful families, which brought their own gods to the mix. The dominant patriarchal families created the story for the hierarchical, cosmological state.[91]

We saw earlier how the break from that cosmological story began with Abraham and Moses, and to a lesser extent, Homer and Socrates, since Socrates himself was defeated by the enforcers of the traditional gods. Both the Jews and the Greeks were absorbed forcibly into great empires.[92] Alexander crushed Athens, and the Jews challenged his heirs but finally lost their temple to a more powerful caesar.

The more persistent challenge was a spiritual one, starting in ancient Jerusalem and elaborated under the Roman Empire by a lowly provincial carpenter preaching an ethic of individual moral duty, especially an obligation to the weak and powerless, but first to a heavenly Father he claimed as his own, an obligation transcending any allegiance even to an earthly father or mother (Matthew 10:37).[93] His challenge to the status quo was not a directly political one, for he instructed his followers: "Render to Caesar the things that are Caesar's, and to God the things that are God's." (Matthew 22:21) This became the foundation of the dual-powers tension that is at the heart of the fusionist synthesis. Caesar dispatched the Goad, but his followers proclaimed him God's Son, now raised from the dead, having been sent into the world with a salvific message that transformed hope (*elpis*) from an uncertain future into the promise of a joyful afterlife.

His apostle Paul too publicly emphasized that he was not challenging the secular powers: "Let every person be subject to the governing authorities; for there is no authority except from God, and those authorities that exist have been instituted by God." (Romans 13:1) Yet the first obligation of each person was to obey the laws of God, which were written in every individual heart, and to which conscience bears witness (Romans 2:15). This commitment to a higher power made imperial authorities wary and led to persecution.

This mythos was put into concrete form by a community of followers who built an institutional church, with a moral authority separate from the state. It was eventually accepted and promoted by the emperor Constantine, losing some independence but gaining legitimacy. Theodosius, the first emperor not to use the title *pontifex maximus*, vigorously promoted Christianity as the official creed, but after the massacre at Thessalonica in 390 even he was convinced to do penance by Ambrose, the bishop of Milan. The church survived the fall of the Western Roman Empire, sometimes co-opted or dominated by secular powers but sometimes challenging them.

Most crucially, the Christian idea of the soul with a "moral status" equal to all other souls was translated into "a social role," as Larry Siedentop put it, resulting in "the invention of the individual."[94] Moral individuality led to many changes in the family and society: the power of the paterfamilias waned, primogeniture eroded, and the rule of free contract protected property and promoted trade.[95]

Through the Middle Ages, a social church offering individual salvation needed defense by a secular, military power, but never fully bowed to it. The church had autonomous moral power, acknowledged by the community and even by emperors, such as Henry IV, who in 1078 knelt in the snow seeking forgiveness from the pope, and by kings like Henry II of England, who in 1174 walked three miles barefoot to Canterbury Cathedral and submitted to a scourging as penance for his role in precipitating the murder of the archbishop Thomas Becket. The church lost much of its moral authority as a consequence of its inability to mitigate the devastating fourteenth-century plague, and also as a result of the "Babylonian Captivity" of the papacy in Avignon and the following papal schism. The Reformation and Counter-Reformation brought a remoralization but also a permanent division, allowing divine-right monarchs to pull national churches firmly under their wings. Social order and economic growth resulted, but mythos became subordinated to logos.[96]

The pluralist story might have ended there. But the divine-right kings, in humbling the churches, had weakened their own supporting mythos, which was soon displaced by nationalism, parliaments, and mass political parties. In England, the Magna Carta ideal barely survived the Civil War and the Restoration, with the weak 1689 Bill of Rights making the monarchy conditional on the will of Parliament, which was dominated by the aristocracy.

But the ideal was reborn more successfully in America, in its Declaration of Independence and its Constitution, its pluralist capitalism and a degree of religious toleration. Later came a muted legal equality for race and sex. The evolving results were higher per capita income and more pleasant jobs, increased life expectancy and better health, and high rates of literacy. Religious participation remained high, apparently encouraging more capitalist risk taking.[97] The individuals in this story were generally proud to call themselves Americans.

That was the story for the majority, anyway, though of course there were ambiguities and injustices, and loss of memory about abuses along the way.[98] But today the idea of individual rights and responsibilities being granted by a creator and centering on a few fixed principles has lost considerable ground in Western nation-states, including the United States, being replaced to a large degree by a utilitarian, positivist legalism whose legitimacy rests solely on practical success rather than moral principle.[99] That ethic fueled the welfare state, which has now become crippled by bureaucratic sclerosis, overregulated markets, protracted wars, bankrupt treasuries, and populations questioning the legitimacy of the story.

The Alternative Individualist Story

A variety of alternative value systems have appeared, with their own logos and mythos: socialism, communism, utilitarianism, positivism, nationalism, syndicalism, Nazism. These have had their successes and failures, and some long-term consequences. Even Richard Dawkins, while committed to defeating religion, appeared to sense the social importance of mythos when he asked his fellow atheist Christopher Hitchens, "Did you ever worry that if we win and so to speak, destroy Christianity, that vacuum would be filled by Islam?"[100]

The most comprehensive and forceful repudiation of the Western story that originated in Athens and Judeo-Christianity came from Friedrich Nietzsche. He presented his alternative individualist story in *The Anti-Christ*, whose preface announces at the outset that it is a book for "the very few." Those who understand him are his "true readers," his "predestined readers," wrote Nietzsche, adding, "and who cares for the rest of them? They are just humanity. You need to be far above humanity."[101]

Nietzsche began his alternative individualist story with a new definition of good and evil. He found power to be supreme over all else, and therefore good. All "failures" are evil and "should perish." He asked, "What is more harmful than any vice?" His answer: "pity for all failures and weakness," a pity that was embodied in Christianity.

Christianity has taken the side of everything weak, base, failed; it has made an ideal out of whatever contradicts the preservation instincts of a strong life....

Christianity is called the religion of pity.... Pity is the opposite of the tonic affects that heighten the energy of vital feelings: pity has a depressive effect....

Pity preserves things that are ripe for decline, it defends things that have been disowned and condemned by life, and it gives a depressive and questionable character to life itself by keeping alive an abundance of failures of every type.[102]

This pity had seeped into the very pores of Western civilization, corrupting individual views of secular power and economic life. Exorcising it would require great effort. A superior minority, the "very few" who can overcome the weakness of pitying the weak, must be "honest to the point of hardness." Nietzsche exhorts his readers: "You need to become indifferent, you need never to ask whether truth does any good, whether it will be our undoing." Rejecting God, pity, and civilization takes "more courage than anyone possesses today; a courage for the *forbidden*; a predestination for the labyrinth.... New ears for new music."[103] Those with this courage can become the "perfect caste" and have "the privilege of the few: this includes happiness, beauty, and goodness on earth. Only the few most spiritual human beings are allowed to be beautiful: only among them is goodness not a weakness."[104]

Could Nietzsche's Dionysian individualist supermen actually create a new mythos, devise their own beauty, fashion a new music? Is this just arrogance? In fact, a disciple of Nietzsche did fashion a new music that rejected foundational elements of the art. Music plays a large role in culture generally, and very notably in the Western tradition. In ancient Greece, Plato warned that poets controlled the polis, and today the poets are mostly musicians, who might have similar power to reshape society.[105]

Arnold Schoenberg was the dominant individual force in twentieth-century musical composition. He originally tried to reconcile the music of Brahms and Wagner, and in a few early works he carried forward the tradition, receiving praise from the likes of Richard Strauss and Gustav Mahler. Then he broke away from the foundations of Western music with two works in 1908. Three years later he published his revolutionary *Theory of Harmony*, which remains one of the most influential music treatises ever written. Soon a large school of artists were spreading his ideas.[106]

Even an artistic illiterate like your author can hear that music took a radical turn in the early twentieth century. The change first appeared in classical music but greatly affected popular genres too. The story is told by Robert R. Reilly, a specialist in foreign policy whose true expertise is in music. He wrote a wonderful book called *Surprised by Beauty*, first published in 2002 and recently reissued in an updated version.[107]

The center of Schoenberg's revolt against the Western musical tradition was his rejection of tonality. He denied that tonality exists in nature as a property of sound itself, which had been accepted since the time of Pythagoras. Schoenberg's revolutionary idea did not result from any new knowledge in the science of acoustics, but simply a desire to escape all the ancient rules of sound. "I am conscious of having removed all traces of a past aesthetic," he declared. Reilly explained how Schoenberg went about this and what he intended to achieve:

> Schoenberg took the twelve equal semitones from the chromatic scale and commanded that music be written in such a way that each of these twelve semitones be used before any one of them is repeated. If one of the semitones is repeated before all eleven others are sounded it might create an anchor for the ear, which could then recognize what was going on in the music harmonically. The twelve-tone system guarantees the listener's disorientation. Schoenberg proposed to erase the distinction between tonality and atonality by immersing man in atonal music until, through habituation, it became the new convention. Then discords would be heard as concords. As he wrote, "The emancipation of dissonance is at present accomplished and twelve-tone music in the near future will no longer be rejected because of discords."[108]

Schoenberg was not overstating what he had done. His twelve-tone system overcame centuries-old conceptions of tonality, concord, and harmony, making discordance the new ideal for music, including popular music. But once beauty in the classical sense is upended, why stop there? Schoenberg's disciple Pierre Boulez went after all the other elements of music: pitch, duration, tone production, intensity, and timbre. Another French disciple, Edgar Varese, asked why it was even necessary to follow the twelve-tone rule, though he could not abandon order altogether. It was the American composer John Cage who went the whole way and simply "created noise through chance operations." Now you know why it sounds so ugly. That is the point.

This dreary story of the nihilistic revolution in music is somewhat well known, but Reilly lets us in on a little secret: it has already run its course, at least in classical music. He interviewed the leading counter-revolutionaries, particularly George Rochberg, who was a prominent twelve-toner when his son died in 1964. Rochberg became frustrated that he could not express his deep emotions in the language of the reigning musical orthodoxy. He recalled:

> I couldn't breathe any more. I needed air. I was tired of the same round of manipulating the pitches, vertically and horizontal-ly.... What I finally realized was that there were no cadences, that you can't come to a natural pause, that you can't write a musical comma, colon, semicolon, dash for dramatic, expressive purposes or to enclose a thought.

In short, he could not really *say* anything through the sterile, ratio-nalistic twelve-tone method. He rejected it and returned to tonality so he could express grief, and create beauty. Asked about Schoenberg's remark that artists must be "cured of the delusion that the aim of art is beauty," Rochberg replied: "I have re-embraced the art of beauty but with a madness, absolutely. That is the only reason to want to write music."[109]

Gian Carlo Menotti had ignored the new orthodoxy from the beginning when he wrote his operas. In his interview with Reilly, he recounted his difficulties with critics' hostility to his beautiful scores, but he had the last laugh because they were popular with audiences, and even won some critical acclaim. His story of meeting the saintly Padre Pio may be disorienting in these skeptical times, although Menotti refused to tell the full outcome, as it was too "private" to reveal.

Reilly also interviewed David Diamond, perhaps America's most honored composer, a few years before his death. Reilly recalled an ear-lier statement by Diamond saying that music should uplift and nourish the spirit, or else it is merely "aural sensation," not music. Diamond replied:

I still think that is true. It is one reason so much of the music that was written during the 1950s and part of the 1960s, music that was basically textural in the sense that the sonorities were the important thing, or patterns of sound, have not lived on. In fact, they've disappeared. Nobody ever listens to them. It's because there is a lack of real musical language which communicates. In other words, there is no melodic substance and there's no feeling in the sense of emotion, whether it's lighthearted emotion or whether it's dark and profound emotion.[110]

Diamond claimed that even Schoenberg at the end told him that twelve-tone was perhaps not for everyone.

Listening to the music critics as opposed to the public today, you would not know that twelve-tone is passé. But Robert Craft, an associate of Stravinsky, remarked that audiences will simply walk out at discord they find unpleasant, regardless of what their betters tell them. The beautiful classics still dominate almost every program for which people have to buy tickets. Most of the modern composers that Reilly discusses do incorporate some dissonance into their works. In fact, the composer John Borstlap has observed that even Beethoven used dissonance in his compositions, but would not allow it to dominate their tonal beauty.[111] But the atonal orthodoxy is dead, and that is good news indeed for those seeking beauty and appreciating tradition.

Why had composers abandoned beauty in the first place? Rochberg said that an older colleague of his was reported to have asked, "Why does George want to write beautiful music? We have done that already."[112] Was it only nihilism, about doing anything as long as it was different?

This author's own view, echoing hints from Reilly and Rochberg, is that composers of the stature of Bach and Mozart, Beethoven and Brahms, Schubert and Mahler were so steeped in the tradition and produced such beauty that it took the heart from those who followed. The masters had gone so far in understanding and attaining the ideal of beauty, who could equal them or possibly carry that tradition forward?

But the reasons go deeper. World War I had destroyed Western idealism and hope, said Rochberg, and a rationalistic nihilism followed in music, and in most arts as well as other intellectual endeavors. The avant-garde in visual arts too had embraced disorder, ugliness, and absurdity. Today, museum visitors often walk quickly past the contemporary works after lingering over those of Raphael and Rembrandt and Renoir.

Each of the counterrevolutionary composers interviewed by Reilly had deeply individual reasons for becoming dissatisfied with the nihilistic emptiness that flowed from Nietzsche's attack on aesthetic and moral order. Each began searching earlier traditions for something beyond arid abstractions. The pursuit was religious for Menotti, "almost" religious for Diamond, and vaguely spiritual for Rochberg.

Classical music is returning to traditional ideas of truth and beauty. Other arts lag behind, yet the revival of beauty in classical music gives hope. If the beautiful can truly be a surprise in music, perhaps elsewhere too?

Can Human Individuality Survive?

The core of Western civilization and its pluralist capitalism is the free but moral human individual shaped by a tradition that has a positive answer to both nihilism and coercion. Some have dismissed this ideal as merely an antisocial "possessive individualism."[113] Others regard the whole pluralist vision of society as irrational.[114] But the story of the free individual responsible to a moral order starts as a founding truth of the civilization.

As intellectuals came to question that truth in the twentieth century, the ideals of individual freedom and innate human worth faded too, and the primitive sacrificial mythos reemerged. Lenin sacrificed the capitalist bourgeoisie to the proletariat. Freud scapegoated fathers as the cause of children's mental disorders. Hitler scapegoated Jews for the ills of the fatherland. Rousseau and Nietzsche scapegoated Christianity and its slave morality for hindering the general will or the godly supermen. But Girard was right that the ancient sacrificial myth had been exposed to the light, so the proponents of modern myths claim to reject scapegoating even as they use it to gain power against unpopular minorities and weak majorities.

Dionysian-inspired elites cleverly appropriate to themselves the status of innocent sacrificial victim to advance their own power.[115] More progressive democratic elites adopt the guise of scientific experts to pied-piper the masses and scapegoat opposing elites and populations. Finally came an identity-politics elite mimicking George Orwell's Media Ministry of Truth, which scapegoated innocents like Emmanuel Goldstein for standing in the way of its deeper truth and subjected them to a daily "two minute hate," to unite the crowd against any enemy of the brotherly-love state.[116]

In contrast, the Goldstein-Emmanuel story teaches universal pity and respect for every individual's freedom. This command comes from a Creator who himself gave humans the liberty even to disobey him. As Girard asked, how could a loving God coerce individuals to be moral?[117]

Was John Gray correct then in saying, "If you want to reject any idea of God, you must accept that 'humanity'—the universal subject that finds redemption in history—also does not exist"?[118] Most Americans have no difficulty identifying their first principle, finding it in the Declaration of Independence: "We hold these truths to be self-evident, that all men are created equal; that they are endowed by their Creator with inherent and inalienable rights; that among these are life, liberty, and the pursuit of happiness."

In his *Tales of a New America*, Robert Reich acknowledges that a popularly supported "morality tale" is necessary for governing a society effectively, but suggests that it can consist of secular "parables" rather than truths.[119] Hayek regarded simple utilitarianism as "insufficient." Even if the society's traditional beliefs are only symbolically true, he argued, those like himself who were "not prepared to accept the anthropomorphic conception of a personal divinity ought to admit that the premature loss of what we regard as nonfactual beliefs would have deprived mankind of a powerful support in the long development of the extended order we now enjoy, and that even now loss of these beliefs, whether true or false, creates great difficulties."[120]

Hayek thought some conception of a higher responsibility was necessary to legitimize secular law in a free society.[121] As we have seen, Kenneth Minogue observed that the Christian moral order, despite its "problems," was in fact "the source" of the traditions that resulted in Western order, openness, and productivity. He regarded institutionalized moral constraints like those derived from Judeo-Christianity as essential for the survival of capitalist economies based on freedom.[122]

The first philosopher of capitalism, John Locke, emphasized that the great ancient philosophers had attempted to legitimize their logos regimes with abstract wisdom and idealist speculations alone, but their teachings had no bite. "The philosophers showed the beauty of virtue" but they "left her unendowed," so that "few were willing to espouse her" until "the immortal weight of glory" that was the Incarnation made it real to common people. This gave rise to a concrete civilization based on truths made plain.[123]

It is difficult to tell the story of Western capitalist civilization without its foundational religious elements. Jonah Goldberg starts his account of Western success by announcing in the first sentence: "There is no God in this book." He then tells the story of how the overwhelming majority of human beings, for thousands of years, lived in poverty and squalor, until a secular "Miracle" and a rationalist Enlightenment led to democracy, free-market capitalism, and unprecedented material abundance. But after about 330 pages, Goldberg admits: "I have tried to keep God out of this book but as a sociological entity God cannot be removed from it."[124]

In fact, the capitalist Miracle had a backstory, which began with a "Hebrew god [who] recognized the moral sanctity of the individual Jew—both male and female," and a "Christian god [who] universalized that moral sanctity," providing the essential support for a culture of individual freedom that produced Western capitalism and prosperity. In earlier times people had chosen gods to grant their desires. But the Hebrews were to serve God instead, and earn their reward. Their God was a "single omniscient being looking at us in all of our private moments," and this God became the "primary source" of the morals underlying the secular Miracle of capitalist plenty.[125]

The French philosopher Pierre Manent argues that it is not necessary to accept the Western story in all its particulars, but the secular nations of Europe today need to acknowledge their historic "Christian mark" and "find a place for the collaboration of human prudence and divine providence." He sees the basic principles for such a fusionist collaboration in the thought of Thomas Aquinas, although not the details for how to apply them in specific situations. The essentially religious values of hope, love, and individual freedom are indispensable to the perpetuation of the European culture.[126]

Without some concept of a higher responsibility, materialists cannot logically maintain that the human individual has inherent rights not possessed by other creatures. Richard Dawkins, for example, supports the Great Ape Project, an international movement to secure legal rights for primates. The project was launched in part by Peter Singer, a rationalist philosopher who advocates for universal animal liberation.[127] An Argentine court ruled that an orangutan is a "non-human person" with some rights.[128] The Nonhuman Rights Project filed lawsuits in New York to establish the "legal personhood" of chimpanzees named Tommy and Kiko, to permit them to relocate to sanctuaries. Courts have ruled against the organization, but it is pursuing other cases for several animals.[129]

Apes do seem human in many ways, with some demonstrable intelligence and perhaps even self-consciousness. If that is the criterion for granting rights, why should apes not have individual rights at least equal to humans, or perhaps to cognitively impaired humans or infants? Orcas too have attractive qualities, and some argue they have intelligence. People for the Ethical Treatment of Animals filed a lawsuit against the SeaWorld corporation in 2011, alleging that five wild-captured orcas were treated like "slaves" at its marine parks in Orlando and San Diego. A U.S. district court dismissed the case in early 2012.[130] But a documentary called *Blackfish* released a year later shifted public opinion and led to a ruling against orca breeding from the California Coastal Commission. SeaWorld agreed to end the breeding and public performances of orcas.[131]

Should we really discriminate against other creatures on the basis of intelligence, or our sense of what is attractive and appealing? Bees and maggots have life. Some say that viruses do too.[132] From the materialist perspective, why should not all life have equal rights? And what would be the harm anyway?

Keith Mano explored that question in *The Bridge: A Novel about the Last Man on Earth*. It starts with a civil war pitting a secular force that supports absolute equality for all life forms against a force of traditionalists and libertarians who place the free human individual above all other earthly creatures. The first group believes that even stepping on grass is an act of assault. It defeats the other side, and decrees that humans will subsist on a weak liquid chemical nutrient consumed without waste and laced with narcotics to induce quiescence.

Communication may be done only by hand signals, since noise is harmful to some living things. Finally, the side of equality for all life eliminates humans altogether so their offensive breath can no longer kill or injure the much more numerous germs and viruses.[133]

Why choose human life over other life? The idea of rights that are distinctively human and apply to all human persons equally comes exclusively from the Western story. Can a belief in unique human rights and freedoms—or love or beauty or individuality—survive without some foundational mythos that elevates humans above other creatures and sanctifies the human individual? From a strictly materialist perspective, Jeremy Waldron of Cambridge notes that secular progressives have been confused in seeing Locke's principles of individual rights, freedom, and toleration as purely rationalist and liberating. Those principles in fact depend on Locke's belief in a Creator who bestows rights, imposes moral rules, and promises rewards in the afterlife.[134]

The conception of humanity that sustains and justifies a pluralist, capitalist social order derives from Judeo-Christian beliefs, beginning with the doctrine of a caring Creator who made humans in his own image, endowing them uniquely with a moral worth that inheres in every person. This was the faith of the American Founders, not in the form of a sectarian Christianity but as a broad consensus on a few basic doctrines. Even Thomas Jefferson rested natural rights on the idea of a beneficent creator.[135]

Schumpeter warned that capitalism and even a viable socialism must have a tradition, a mythos that limits demands, or its logos will be overwhelmed.[136] As Lippmann put it earlier, the "radical error of the modern democratic gospel is that it promises not the good life of this world but the perfect life of heaven," realized here on earth. This promise raises desires that are impossible to satisfy and that only a totalitarian state would claim to fulfill.

The moral assumptions of the Western traditional mythos, in which individuals have been created free and equal, are indispensable to legitimizing a pluralist, federalist, traditionalist, capitalist society with free markets and localized powers under a limited central state—a society where liberty and order coexist in creative tension. If its legitimizing source is forgotten or denied, one can assume, with Schumpeter, that the civilization will likely fail.

NOTES

INTRODUCTION

1 Joseph Schumpeter, *Capitalism, Socialism and Democracy*, 3rd ed. (New York: Harper & Row, 1950).

2 Francis X. Rocca, "Pope Francis Says Ills of Global Economy, Not Islam, Inspire Terrorism," *Wall Street Journal*, August 1, 2016.

3 John Burtka, "A very different conservative agenda emerges," *Washington Post*, July 23, 2019.

4 Robert Samuelson, "Schumpeter: The Prophet," *Newsweek*, November 9, 1992.

CHAPTER 1 — WHICH CAPITALISM?

1 Science AMA Series, "Stephen Hawking AMA Answers!" *New Reddit Journal of Science*, November 2015, https://www.reddit.com/r/science/comments/3nyn5i/science_ama_series_stephen_hawking_ama_answers/cvsdmkv.

2 Karl Marx, *Capital*, vol. 1 (New York: Penguin, 1976), 27.

3 Karl Marx and Friedrich Engels, *The Communist Manifesto*, 1848, trans. Samuel Moore (Chicago: C. H. Kerr & Co., 1912).

4 Angus Maddison, *The World Economy: Historical Statistics* (Paris: Organization for Economic Cooperation and Development, 2006). See also Donald Devine, "The Real John Locke—and Why He Matters," *Law and Liberty*, May 21, 2014.

5 Tom G. Palmer, "Poverty, Morality, and Liberty," in *After the Welfare State: Politicians Stole Your Future, You Can Get It Back*, ed. Palmer (Ottawa, Ill.: Jameson Books, 2012).

6 F. A. Hayek, *The Constitution of Liberty*, ed. Ronald Hamowy, vol. 17 of *The Collected Works of F. A. Hayek* (Chicago: University of Chicago Press, 1960), chap. 11.

7 Palmer, "Poverty, Morality, and Liberty," 119.

8 Jonah Goldberg, *Suicide of the West: How the Rebirth of Tribalism, Populism, Nationalism, and Identity Politics Is Destroying America* (New York: Crown Forum, 2018), 7, 12, 107, 331. See also Donald Devine, "Goldberg's Intricate Story of Romantic Suicide," *Imaginative Conservative*, June 7, 2018.

9 Niall Ferguson, *Civilization: The West and the Rest* (New York: Penguin, 2011).

10 Nathan Rosenberg and L. E. Birdzell, *How the West Grew Rich: The Economic Transformation of the Industrial World* (New York: Basic Books, 1986).

11 Christopher Dyer, *An Age of Transition: Economy and Society in the Later Middle Ages* (New York: Oxford University Press, 2005).

12 Jean Gimpel, *The Medieval Machine: The Industrial Revolution of the Middle Ages* (1976; New York: Pimlico Books, 1992).

13 Harold J. Berman, *Law and Revolution: The Formation of the Western Legal Tradition* (Cambridge, Mass.: Harvard University Press, 1983).

14 Derek Thompson, "The economic history of the past 2,000 years: Part II," *Atlantic*, June 19, 2012, with data from Michael Cembalest.

15 Deirdre McCloskey, *The Bourgeois Virtues: Ethics for an Age of Commerce* (Chicago: University of Chicago Press, 2006).

16 Palmer, "Poverty, Morality, and Liberty."

17 Rodney Stark, *For the Glory of God: How Monotheism Led to Reformations, Science, Witch-Hunts, and the End of Slavery* (Princeton: Princeton University Press, 2002), chap. 2.

18 Terry Miller, Anthony B. Kim, and James M. Roberts, *2020 Index of Economic Freedom* (Washington, D.C.: Heritage Foundation, 2020).

19 Fareed Zakaria, "Capitalism, not culture, drives economies," *Washington Post*, August 1, 2012.

20 Svetozar Pejovich, *Law, Informal Rules and Economic Performance: The Case for Common Law* (Northampton, Mass.: Edward Elgar Publishing, 2008).

21 Marx and Engels, *The Communist Manifesto*, 66–101.

22 Joseph Schumpeter, *Capitalism, Socialism and Democracy*, 3rd ed. (New York: Harper & Row, 1950).

23 Pope Francis, *Evangelii Gaudium,* Apostolic Exhortation on the Proclamation of the Gospel in the Modern World, November 24, 2013.

24 José Ignacio García Hamilton, "Historical Reflections on the Splendor and Decline of Argentina," *Cato Journal*, 25:3 (Fall 2005), 521–40.

25 Jorge Mario Bergoglio and Abraham Skorka, *On Heaven and Earth: Pope Francis on Faith, Family, and the Church in the Twenty-first Century* (New York: Transaction, 2013), 64.

26 World Bank, GDP per capita (current US$), http://data.worldbank.org/indicator/NY.GDP.PCAP.CD.

27 Jacob T. Levy, *Rationalism, Pluralism, and Freedom* (Oxford: Oxford University Press, 2015).

28 Jerry Z. Muller, *The Mind and the Market: Capitalism in Western Thought* (New York: Borzon Books, 2002), chap. 2.

29 Levy, *Rationalism, Pluralism, and Freedom*, chap. 5.

30 Ibid., chap. 3.

31 Sheldon Richman, "Two Tyrannies," *American Conservative*, April 20, 2016.

32 Sheldon Richman, "A Note on the Limits of Pluralism," *American Conservative*, May 20, 2016.

33 Woodrow Wilson, "The Study of Administration," *Political Science Quarterly*, 2:2 (June 1887), 197–222. See also Donald J. Devine, *America's Way Back: Reclaiming Freedom, Tradition, and Constitution* (Wilmington, Del.: ISI Books, 2013), 9–10, 63–64, 175, 202.

34 Levy, *Rationalism, Pluralism, and Freedom*, 122ff.

35 Ibid., 89.

36 Ibid., 223.

37 John Stuart Mill, "Chapters on Socialism," in *Essays on Economics and Society Part II (1850)*, vol. 5 of *The Collected Works of John Stuart Mill*, ed. J. M. Robson (Toronto: University of Toronto Press, 2002); Linda Raeder, *John Stuart Mill and the Religion of Humanity* (Columbia: University of Missouri Press, 2002).

38 Levy, *Rationalism, Pluralism, and Freedom*, 295.

39 Donald Devine, "Pluralism vs. Bureaucracy," *American Conservative*, May 12, 2016.

40 Louis Hartz, *The Liberal Tradition in America* (New York: Harcourt, Brace & World, 1955), x; Donald J. Devine, *The Political Culture of the United States: The Influence of Member Values on Regime Maintenance* (Boston: Little, Brown, 1972), 60.

41 Goldberg, *Suicide of the West*, 130.

42 Yoram Hazony, *The Virtue of Nationalism* (New York: Basic Books, 2018), 30–32.

43 Patrick J. Deneen, *Why Liberalism Failed* (New Haven: Yale University Press, 2018), 34–47; Donald Devine, "Patrick Deneen on 'Why Liberalism Failed,'" *Imaginative Conservative*, March 26, 2018.

44 Donald Devine, "Critiquing Robert Kagan's Enlightenment Liberalism," *Imaginative Conservative*, May 6, 2019.

45 Hartz, *The Liberal Tradition in America*, 6.

46 Peter Laslett, Introduction to John Locke, *Two Treatises of Government*, 2nd ed. (Cambridge: Cambridge University Press, 1967), 99.

47 George H. Sabine, *A History of Political Theory*, 3rd ed. (New York: Holt, Rinehart & Winston, 1961).

48 Kenneth Minogue, "The Conditions of Freedom and the Condition of Freedom," in *On Liberty and Its Enemies*, ed. Timothy Fuller (New York: Encounter Books, 2017), 99.

49 Edward Feser, *Locke* (Oxford: Oneworld, 2007), 170.

50 C. B. Macpherson, *The Political Theory of Possessive Individualism* (Oxford: Clarendon Press, 1962), chap. 5.

51 Leo Strauss, *Natural Right and History* (Chicago: University of Chicago Press, 1953), 249–51.

52 Eric Voegelin, "John Locke," in *History of Political Ideas*, vol. 7, *The New Order and the Last Orientation*, ed. Jurgen Gebhardt and Thomas A. Hollweck (Columbia: University of Missouri Press, 1999), 137–52. See also Donald Devine, "Understanding Voegelin's Critique of Locke," *Imaginative Conservative*, November 30, 2018.

53 Voegelin, "John Locke," 140–41.

54 John Locke, *A Letter Concerning Toleration*, ed. James H. Tully (Indianapolis: Hackett, 1983), 26. By "indolency" he meant the absence of pain.

55 Voegelin, "John Locke," 142.

56 Ibid., 142–43.

57 Ibid., 143.

58 Ibid., 146.

59 Ibid., 152.

60 Ibid., 154.

61 Karen Vaughn, "John Locke's Theory of Property," *Literature of Liberty: A Review of Contemporary Liberal Thought*, 3:1 (Spring 1980).

62 John Locke, *An Essay Concerning the True Original, Extent and End of Civil Government*, in *Social Contract: Essays by Locke, Hume and Rousseau*, ed. Sir Ernest Baker (New York: Oxford University Press, 1962), sec. 123–24.

63 Sabine, *A History of Political Theory*, 519, 528, 530–31.

64 Feser, *Locke*, 99, 113, 168.

65 Brad S. Gregory, *The Unintended Reformation: How a Religious Revolution Secularized Society* (Cambridge, Mass.: Harvard University Press, 2012), chap. 4.

66 John Locke, *An Essay Concerning Human Understanding*, ed. A. D. Woozley (Cleveland: Meridian, 1969), chap. 18, sec. 9.

67 John Courtney Murray, *The Problem of God* (New Haven: Yale University Press, 1964), 89–90. See also Samuel Gregg, *Reason, Faith, and the Struggle for Western Civilization* (Washington, D.C.: Regnery Gateway, 2019), chap. 2; George Weigel, *The Irony of Modern Catholic History* (New York: Hachette, 2019), 125–28.

68 Gregory, *The Unintended Reformation*.

69 F. A. Hayek, "Individualism: True and False," *Individualism and Economic Order* (Chicago: University of Chicago Press, 1948), esp. 48–54.

70 F. A. Hayek, "Kinds of Rationalism," in *Studies in Philosophy, Politics and Economics* (Chicago: University of Chicago Press, 1967), 84.

71 Karl R. Popper, "Towards a Rational Theory of Tradition," 1948 lecture, in *Conjectures and Refutations: The Growth of Scientific Knowledge* (New York: Harper & Row, 1968), 129–31. For the dynamic relationship between Hayek and Popper, see Bruce Caldwell, Introduction to *Studies on the Abuse and Decline of Reason*, vol. 13 of *The Collected Works of F. A. Hayek* (Chicago: University of Chicago Press, 2010), 36–38.

72 Strauss, *Natural Right and History*, 8.

73 Feser, *Locke*, 64, 97; cf. Goldberg, *Suicide of the West*, 130.

74 Alasdair MacIntyre, *After Virtue: A Study in Moral Theory*, 3rd ed. (South Bend, Ind.: Notre Dame University Press, 2007), chap. 14.

75 Richard Rudner, "The Scientist Qua Scientist Makes Value Judgements," *Philosophy of Science*, 20:1 (January 1953), 1–6.

76 Donald J. Devine, "John Locke: His Harmony Between Liberty and Virtue," *Modern Age*, 22:3 (Summer 1978), 246–56. The Locke citations not referenced here may be found in this article. See also Locke, *An Essay Concerning Human Understanding*, chap. 4, sec. 18 & 29. On the probabilistic nature of understanding for Locke, see *Essay*, chap. 4, sec. 11 & 15.

77 See Devine, "Goldberg's Intricate Story of Romantic Suicide."

78 John Locke, *Some Thoughts Concerning Reading and Study for a Gentleman*, in *The Works of John Locke in Nine Volumes*, 12th ed. (London: Rivington, 1824), vol. 2: 407, https://oll.libertyfund.org/titles/762; *The Reasonableness of Christianity*, in *The Works of John Locke* (London: Tegg et al., 1823), vol. 7: 135.

79 Paul E. Sigmund, "Jeremy Waldron and the Religious Turn in Locke Scholarship," *Review of Politics*, 67:3 (Summer 2005), 407–18; see also David Wootton, *Power, Pleasure and Profit: Insatiable Appetites from Machiavelli to Madison* (Cambridge, Mass.: Belknap Press, 2018), esp. 227; Ben Shapiro, *The Right Side of History* (New York: Broadside Books, 2019), 108–12.

80 Jeremy Waldron, *God, Locke and Equality: Christian Foundations of John Locke's Political Thought* (Cambridge: Cambridge University Press, 2002), 13.

81 Steven Forde, *Locke, Science, and Politics* (Cambridge: Cambridge University Press, 2013), chap. 2.

82 Waldron, *God, Locke and Equality*, 13; and see Bertrand Russell, *Russell on Ethics*, ed. Charles Plaunt (New York: Routledge, 1999), 165.

83 See also Devine, "The Real John Locke—and Why He Matters."

84 Berman, *Law and Revolution*, chap. 5.

85 Ibid.

86 Ryan Cooper, "Bernie Sanders Is Right," *The Week*, June 8, 2015.

87 Frank Newport, "Americans Continue to Say U.S. Wealth Distribution Is Unfair," Gallup, May 4, 2015.

88 Pew Research Center, "Most See Inequality Growing," January 23, 2014.

89 Steve Chapman, "Why millennials are drawn to socialism," *Chicago Tribune*, May 18, 2018; Emily Ekins, "What Americans Think About Poverty, Wealth, and Work," Cato Institute, September 24, 2019.

90 Kevin C. Rhoades, "Bishop Reflects on Catholic Teaching on the Economy," *Today's Catholic*, March 9, 2016.

91 "Dharna outside the Deputy Commissioner's office," *The Hindu*, December 20, 2015.

92 Rogers Mugabo, "Women's land, inheritance and property rights in Uganda," Land Portal, December 16, 2015, https://landportal.org/news/2015/12/women%E2%80%99s-land-inheritance-and-property-rights-uganda.

93 Armen A. Alchian, "Private Property," *The Concise Encyclopedia of Economics*, December 2007.

94 Janet Beales Kaidantzis, "Property Rights for 'Sesame Street,'" Professional Land Systems, https://plsurvey.com/property-rights-for-sesame-street/.

95 Konstantin Sonin, "Why the Rich May Favor Poor Protection of Property Rights," *Journal of Comparative Economics*, 31:4 (December 1, 2002), 715–31.

96 Hernando de Soto, *The Mystery of Capital* (New York: Basic Books, 2000), esp. chap. 6.

97 Hayek, *The Constitution of Liberty*, 231–33.

98 Henry Thomas Buckle, *History of Civilization in England* (1857; BiblioLife, 2009), 39.

99 Devine, "Critiquing Robert Kagan's Enlightenment Liberalism."

100 Stark, *For the Glory of God*, 128–50.

101 Karen Armstrong, *Fields of Blood: Religion and the History of Violence* (New York: Random House, 2014).

102 The following is from ibid.; Jerry Z. Muller, *Capitalism and the Jews* (Princeton: Princeton University Press, 2010), esp. p. 8; William Cavanaugh, *The Myth of Religious Violence* (New York: Oxford University Press, 2009); Donald J. Devine, *In Defense of the West: American Values Under Siege* (Dallas: University Press of America, 2004).

103 Jacob Rader Marcus, *The Jew in the Medieval World* (Cincinnati: Sinai Press, 1938), 111–13.

104 Robert Chazan, *Reassessing Jewish Life in Medieval Europe* (New York: Cambridge University Press, 2010), 1–6, 193–99.

105 David B. Green, "This Day in Jewish History, 1348: Jews Aren't Behind the Black Death, Pope Clarifies," *Haaretz*, July 6, 2016.

106 John A. Crowe, *Spain: The Root and the Flower* (Berkeley: University of California Press, 1963), 153.

107 Michael Coren, "The Reformation at 500: Grappling with Martin Luther's anti-Semitic legacy," *Maclean's*, October 25, 2017.

108 Sara Lipton, *Dark Mirror: The Medieval Origins of Anti-Jewish Iconography* (New York: Henry Holt, 2014), 1–4.

109 Ibid., 1, 16–33.

110   Ibid., 57–58.
111   Ibid., 93–94.
112   Ibid., 173–74.
113   Ibid., 245.
114   Ibid., 263–67, 278.
115   Ibid., 279–81.
116   Joseph Jacobs, "Statistics," *Jewish Encyclopedia*, http://www.jewishencyclopedia.com/articles/13992-statistics.
117   Armstrong, *Fields of Blood*, 394.
118   Donald Devine, "Why Liberalism Failed," *Imaginative Conservative*, March 26, 2018.
119   Rod Dreher, "Sex after Christianity," *American Conservative*, April 11, 2013.
120   Ibid.; Angela Nagel, *Kill All Normies: Online Culture Wars from 4Chan and Tumblr to Trump and the Alt-Right* (UK: Zero Books, 2017), chap. 4.
121   See discussion and citations in Devine, *America's Way Back*, 123–27.
122   Jean-Jacques Rousseau, *The Social Contract*, in *The Social Contract and Discourses*, trans. G. D. H. Cole (New York: E. P. Dutton, 1950), Bk. IV, chap. 8.
123   Rousseau and Nietzsche will be considered extensively below. For Machiavelli, see Niccolò Machiavelli, *The Art of War*, trans. Ellis Farnsworth, ed. Neal Wood (Cambridge, Mass.: Da Capo Press, 1965), Bk. II, 79.
124   Larry Siedentop, *Inventing the Individual: The Origins of Western Liberalism* (Cambridge Mass.: Harvard University Press, 2014), chap. 4; Rodney Stark, *How the West Won: The Neglected Story of the Triumph of Modernity* (Wilmington, Del.: ISI Press, 2014), esp. chap. 6.
125   Eric Voegelin, *Science, Politics, and Gnosticism*, in *Modernity Without Restraint*, ed. Manfred Henningsen, vol. 5 of *The Collected Works of Eric Voegelin* (Columbia: University of Missouri Press, 2000), 260–61.
126   Leo Strauss, *The Rebirth of Classical Political Rationalism*, ed. Thomas L. Pangle (Chicago: University of Chicago Press, 1989), 270.
127   Frank S. Meyer, "Western Civilization: The Problem of Political Freedom," in *In Defense of Freedom and Related Essays* (Indianapolis: Liberty Fund, 1996), 209–24.
128   Ibid.; Eric Voegelin, *The Political Religions*, in *Modernity Without Restraint*; Voegelin, Mary Algozin, and Keith Algozin, "Liberalism and Its History," *Review of Politics*, 36:4 (October 1974); Kenneth Minogue, "The Irresponsibility of Rights," in *On Liberty and Its Enemies*, 251–53.
129   Eric Voegelin, *The New Science of Politics*, in *Modernity Without Restraint*, 150–52.

CHAPTER 2—BEGINNING AT THE BEGINNING

1   Eric Voegelin, *Order and History*, vol. 1, *Israel and Revelation*, ed. Maurice P. Hogan, vol. 14 of *The Collected Works of Eric Voegelin* (Columbia: University of Missouri Press, 2001); Voegelin, *Order and History*, vol. 2, *The World of the Polis*, ed. Athanasios Moulakis, vol. 15 of *The Collected Works of Eric Voegelin* (2000).
2   Meyer names Voegelin as the source for his conception of history, while emphasizing that he presents it "according to his own lights," without Voegelin "being responsible for the result." Frank S. Meyer, "Western Civilization: The Problem of Political Freedom," in *In Defense of Freedom and Related Essays* (Indianapolis: Liberty Fund, 1996), 210.

3   Ibid., 209–24.
4   However, see Eric Voegelin, *Order and History*, vol. 4, *The Ecumenic Age*, ed. Michael Franz, vol. 17 of *The Collected Works of Eric Voegelin* (Columbia: University of Missouri Press, 2000), chap. 1.
5   Meyer, "Western Civilization: The Problem of Political Freedom," 210ff.
6   Ian Tattersall, *Masters of the Planet: The Search for Our Human Origins* (New York: St. Martin's Press, 2012), esp. 204–5.
7   T. Douglas Price and Ofer Bar-Yosef, "The Origins of Agriculture: New Data, New Ideas," Introduction to Supplement 4, *Current Anthropology*, 52:S4 (October 2011), S171.
8   Robert L. Kelly, *Lifeways of Hunter-Gatherers* (Cambridge: Cambridge University Press, 2013).
9   Jean-Jacques Rousseau, *A Discourse on the Origin of Inequality*, in *The Social Contract and Discourses*, trans. G. D. H. Cole (New York: E. P. Dutton, 1950), 84.
10 Ibid., 100.
11 Ibid., 123.
12 René Girard, *Things Hidden Since the Foundations of the World*, trans. and with commentary by Stephen Bann and Michael Metteer (Stanford: Stanford University Press, 1978), Bk. I, chap. 3; Donald Devine, "René Girard's Challenge to Fusionism," *Modern Age*, 60:1 (Winter 2018), 23–32.
13 Yuval Noah Harari, *Sapiens: A Brief History of Humankind* (New York: HarperCollins, 2015), 20–25, comparing *Homo sapiens* with other humanoids, and 52ff, seeing a cognitive revolution occurring by "pure chance" around 30,000 years ago.
14 Girard, *Things Hidden Since the Foundations of the World*, Bk. I, chap. 2–3.
15 Ibid., Bk. III, chap. 1.
16 Ibid., Bk. I, chap. 1.
17 René Girard, *I See Satan Fall Like Lightning*, trans. James D. Williams (Maryknoll, N.Y.: Orbus Books, 2001), 83.
18 There are some modern holdouts, although they seem mostly to ignore or seriously downplay violence. See Graeme Barker, *The Agricultural Revolution in Pre-History* (Oxford: Oxford University Press, 2006), esp. chap. 10.
19 René Girard, *The One by Whom Scandal Comes*, trans. Malcolm B. DeBevoise (East Lansing: Michigan State University Press, 2014), 32.
20 Sigmund Freud, *Totem and Taboo: Resemblances Between the Psychic Lives of Savages and Neurotics*, trans. A. A. Brill (London: Routledge & Son, 1919).
21 Girard, *The One by Whom Scandal Comes*, 17.
22 Mircea Eliade, *The Myth of the Eternal Return* (Princeton: Princeton University Press, 1954).
23 Freud, *Totem and Taboo*, 6, 69–73.
24 René Girard, *The Scapegoat*, trans. Yvonne Freccero (Baltimore: Johns Hopkins Press, 1986), 112–14.
25 Girard, *I See Satan Fall Like Lightning*, chap. 1.
26 Girard, *Things Hidden Since the Foundations of the World*, Bk. I, chap. 4–5.
27 Ibid., Bk. II.
28 Ibid., Bk. II, chap. 3.
29 Friedrich Nietzsche, *The Will to Power*, trans. Walter Kaufman and R. J. Hollingdale (New York: Vintage, 1968), No. 1052 (542–43). See also René Girard, "Dionysus versus the Crucified," *MLN*, 99:4 (September 1984), 816–35.

30 Friedrich Nietzsche, *On the Genealogy of Morals*, trans. Douglas Smith (Oxford: Oxford University Press, 1996), chap. 1.

31 Girard, *Things Hidden Since the Foundations of the World*, 249–52.

32 Joshua Mitchell, "Why Conservatives Struggle with Identity Politics," *National Affairs*, 40 (Summer 2019).

33 Girard, *I See Satan Fall Like Lightning*, 75; for Hitler, see Herman Rauschning, Preface to *The Ten Commandments: Ten Short Novels of Hitler's War Against the Moral Code*, ed. Armin Robinson (New York: Simon & Schuster, 1944).

34 Jean-Jacques Rousseau, *The Social Contract*, in *The Social Contract and Discourses*, trans. G. D. H. Cole (New York: E. P. Dutton, 1950), Bk. IV, chap. 8.

35 Harold J. Berman, *Law and Revolution: The Formation of the Western Legal Tradition* (Cambridge, Mass.: Harvard University Press, 1983).

36 See Rodney Stark, *The Rise of Christianity* (Princeton: Princeton University Press, 1997), chap. 1; Rodney Stark, *How the West Won: The Neglected Story of the Triumph of Modernity* (Wilmington: ISI Books, 2014), chap. 7. Cf. chap. 13.

37 A. R. Bridbury, "The Black Death," *Economic History Review*, 26:4 (1973), 577–92.

38 Alec Ryrie, *Protestants: The Faith That Made the Modern World* (New York: Viking Penguin, 2017), chap. 1.

39 See Ryrie, *Protestants*, esp. 33, where he notes that it was Philip Melanchthon who systematized Lutheran doctrine. Also see, Donald Devine, "The Unbridled Faith," *Law and Liberty*, August 7, 2017, for this and the following sections.

40 Karen Armstrong, *The Lost Art of Scripture: Rescuing the Sacred Texts* (New York: Knopf, 2019), 333–48.

41 Ryrie, *Protestants*, 4. Rodney Stark, in *How the West Won,* chap. 13, shows that Protestantism paradoxically was most successful where the Catholic Church had been powerful, as the rulers had a strong interest in gaining its wealth to build their treasuries—Henry VIII of England being the most obvious example. Where the church's wealth was already under royal control, as in Spain or France, there was no such incentive.

42 Johannes Althusius, *Politica* (1614), abridged trans. by Frederick S. Carney (Indianapolis: Liberty Fund, 1995).

43 John Neville Figgis, *The Divine Right of Kings* (Cambridge: Cambridge University Press, 1914), chap. 5–7.

44 Anne Stott, "Europe 1700–1914: A Continent Transformed," http://europetransformed.blogspot.com/2006/10/war-and-empire-2-dynastic-wars.html.

45 Rousseau, *The Social Contract*, Bk. IV, chap. 8; Donald J. Devine, *In Defense of the West: American Values Under Siege* (Dallas: University Press of America, 2004), 144–47.

46 Michael A. Heilperin, *Studies in Economic Nationalism* (Auburn, Ala.: Ludwig von Mises Institute, 2011), chap. 3.

47 Isaiah Berlin, "Nationalism: Past Neglect and Present Power," in *Against the Current: Essays in the History of Ideas*, ed. Henry Hardy (London: Pimlico, 1979), 333–55.

48 John Lukacs, *The Passing of the Modern Age* (New York: Harper & Row, 1970), 17.

49 Karen Armstrong, *Fields of Blood: Religion and the History of Violence* (New York: Random House, 2014), 301.

50 Alexis de Tocqueville, *Democracy in America*, trans. Henry Reeve, vol. 1 (London: Saunders & Otley, 1835).

51 Murray M. Rothbard, *An Austrian Perspective on the History of Economic Thought* (Northampton, Mass.: Edward Elgar Publishing, 1995), chap. 7–8.

52 Adam Smith, *An Inquiry into the Nature and Causes of the Wealth of Nations* (New York: Modern Library, 1937), Part II, Bk. IV, chap. 8.

53 Ibid., Part II, Bk. IV, chap. 9.

54 Jerry Z. Muller, *The Mind and the Market* (New York: Borzon Books, 2002).

55 David Ricardo, "Letter to T.R. Malthus," October 9, 1820, in *The Works and Correspondence of David Ricardo*, ed. Piero Sraffa, vol. 8 (Cambridge: Cambridge University Press, 1951), 278–79.

56 Joseph Schumpeter, *Capitalism, Socialism and Democracy*, 3rd ed. (New York: Harper & Row, 1950), chap. 3.

57 Donald Devine, "The Nationalist Delusion," *American Conservative*, April 9, 2019.

58 Walter A. McDougall, *Promised Land, Crusader State* (Boston: Houghton Mifflin, 2007).

59 Angelo Codevilla, *Advice to War Presidents* (New York: Basic Books, 2009), 33–36.

60 Francis Fukuyama, "The End of History," *National Interest*, Summer 1989.

61 Donald Devine, "Can American's Foreign Policy Be Saved?" *American Conservative*, February 2, 2015; Barry R. Posen, *Restraint: A New Foundation for U.S. Grand Strategy* (Ithaca, N.Y.: Cornell University Press, 2014); Bret Stephens, *American Retreat: The New Isolationism and the Coming Global Disorder* (New York: Penguin, 2014); Angelo Codevilla, *To Make and Keep Peace: Among Ourselves and with All Nations* (Stanford, Calif.: Hoover Institution Press, 2014).

62 Pew Research Center, "Public Uncertain, Divided over America's Place in the World," May 5, 2016; Pew Research Center, "Foreign Policy: The Partisan Policy Divide Grows Even Wider," October 5, 2017; William A. Galston, "In defense of a reasonable patriotism," Brookings, July 23, 2018.

63 Sebastian Payne, "Obama: Islamists' rise was misjudged," *Washington Post*, September 29, 2014.

64 Tim Craig, Abby Phillip, and Joel Achenbach, "Journey of Tashfeen Malik from pharmacy student to San Bernardino terrorist," *Sidney Morning Herald*, December 7, 2015.

65 Ruth Gledhill, "Obama: Islam is a religion of 'peace, charity and justice,'" *Christianity Today*, February 8, 2016.

66 Jeff Seldin, "US Broadens War on Terror, Has Harsh Words for Iran," VOA News, October 4, 2018.

67 Lee Smith, "Iran's the Problem," *Weekly Standard*, February 24, 2014.

68 Angelo Codevilla, "While the Storm Clouds Gather," *Claremont Review of Books*, October 30, 2014; and personal discussions with the author.

69 Josh Meyer, "Bin Laden takes on capitalism," *Los Angeles Times*, September 8, 2007.

70 Paul Sonne and Karen DeYoung, "Trump wants to get the U.S. out of Syria's war, so he asked the Saudi King for $4 billion," *Washington Post*, March 16, 2018.

71 See Donald Devine, "Naming Our Middle East Enemy," *American Conservative*, December 12, 2015.

72 Nikolay Pakhomov, "What does Russia really want in Syria?" *National Interest*, November 18, 2015.

73 Alia Chughtai, "Syria's War: Allies and Opponents, *Al Jazeera*, April 16, 2018.

74   Angelo Codevilla, *No Victory, No Peace* (Lanham, Md.: Rowman & Littlefield, 2005), esp. 3–7.

75   Jeff Stein, "Can You Tell a Sunni from a Shiite?" *New York Times*, October 17, 2006.

76   Ibid.

77   Walter Russell Mead, "Putin Knows History Has Not Ended," *Wall Street Journal*, February 21, 2014.

78   Ibid.

79   Donald Devine, "Unfriendly turf for democracy," *Washington Times*, November 20, 2002.

80   Donald Devine, "Iraqization Process Well Underway," *Human Events*, November 24, 2003.

81   Robert Gates, former U.S. secretary of defense, wrote in his memoirs that even the world's dominant superpower makes the most important decisions in a crisis when there is great confusion and little time to deliberate. Disagreements tend to be resolved pragmatically on the fly by a half dozen executive officials, many of whom have little or no military or diplomatic background or knowledge. Robert Gates, *Duty: Memoirs of a Secretary at War* (New York: Vintage, 2014), 413.

82   Zeke J. Miller and Michael Sherer, "President Obama Attacks Republicans for Paris Response," *Time*, November 18, 2015.

83   Kristina Peterson, "House Speaker Ryan Calls for 'Pause' in Admitting Syrian Refugees," *Wall Street Journal*, November 17, 2015.

84   "Second Class Christians," Notable & Quotable, *Wall Street Journal*, February 28, 2014.

85   Jerry Markon, "Senior Obama officials have warned of challenges in screening," *Washington Post*, November 17, 2015.

86   Jeremy Diamond, "Obama calls for gun reforms in wake of San Bernardino shooting," CNN, December 2, 2015.

87   Colby Itkowitz and Hanna Knowles, "Trump takes aim at Schumer as lawmakers are split over order against Iran," *Washington Post*, January 4, 2020.

88   Peggy Noonan, "Whose Side Are We On?" *Wall Street Journal*, February, 21, 2014.

89   Bret Stephens, "Ukraine vs. Homo Sovieticus," *Wall Street Journal*, February 24, 2014.

90   William Booth and Will Englund, "Russia condemns Ukraine's new leadership," *Washington Post*, February 22, 2014.

91   Will Englund, "Ukraine delays naming new government," *Washington Post*, February 26, 2014.

92   *Authorization for the Use of Military Force*, Public Law 107–40, September 18, 2001, 115 Stat. 224.

93   Lawrence Kudlow, "We Will Never Destroy ISIS Without a Full-Blown Declaration of War," *RealClearPolitics*, January 23, 2016.

94   Marjorie Cohn, "Obama to Congress: Rubber-Stamp My Perpetual War," *HuffPost*, April 18, 2015, updated April 20, 2015.

95   Michael B. Mukasey and David B. Rivkin Jr., "Another Obama Collision With the Constitution," *Wall Street Journal*, February 20, 2015.

96   Curtis A. Bradley and Jack L. Goldsmith, "Congressional Authorization and the War on Terrorism," *Harvard Law Review*, 118:7 (May 2005), 2047–133. For more on the constitutionality of this longstanding practice, see David. J. Barron

and Martin S. Lederman, "The Commander in Chief at the Lowest Ebb—a Constitutional History," *Harvard Law Review*, 121:4 (February 2008), 941–1111.

97 Austin Wright, "Congress leaves Trump with unlimited war powers," *Politico*, December 29, 2016.

98 Many advocated NATO membership for Ukraine, but what would be the response by the United States if Canada decided to join a Russian military alliance?

99 This is a reality often ignored by modern nationalists such as Yoram Hazony; see Donald Devine, "The Virtue of Nationalism," *Imaginative Conservative*, October 29, 2018.

100 Schumpeter, *Capitalism, Socialism and Democracy*, 138.

101 Matthew White, Selected Death Tolls for Wars, Massacres, and Atrocities before the 20th Century, Necrometrics, http://necrometrics.com/pre1700a.htm; R. J. Rummel, *Death by Government* (New Brunswick: Transaction, 1994), 4, 70ff; Donald J. Devine, *America's Way Back: Reclaiming Freedom, Tradition, and Constitution* (Wilmington, Del.: ISI Books, 2013), 57.

102 Norman Davies, *Vanished Kingdoms: The History of Half-Forgotten Europe* (New York: Viking, 2012).

103 Gašper Završnik, "Brexit and Trump encouraged Eastern Europe populism," *Politico*, April 4, 2017, updated August 21, 2017.

104 Joseph C. Sternberg, "Europe's Nationalism Turns Out to Be Local," *Wall Street Journal*, May 5, 2019.

105 David P. Goldman, "Nationalism Is Dead. Long Live Nationalism," *Tablet*, August 14, 2018, https://www.tabletmag.com/jewish-news-and-politics/268478/nationalism-is-dead-long-live-nationalism.

106 Gallup International, "WIN/Gallup International's Global Survey Shows Three in Five Willing to Fight for Their Country," May 7, 2015.

107 F. H. Buckley, *American Secession: The Looming Threat of a National Breakup* (New York: Encounter Books, 2020), 12.

108 Scott Malone, "Angry with Washington, 1 in 4 Americans open to secession," Reuters, September 19, 2014.

109 John Zogby, "New Poll on Americans' Support for Secession," *Zogby Report*, October, 25, 2017; Jeremy Zogby, "Secessionist Sentiment Remains a Plurality Among Likely Voters," John Zogby Strategies, August 10, 2018, https://johnzogbystrategies.com/secessionist-sentiment-remains-a-plurality-among-likely-voters/.

110 Donald J. Trump, Inaugural Address, January 20, 2017, The White House, https://www.whitehouse.gov/briefings-statements/the-inaugural-address/.

111 Stephen Hawkins et al., *The Hidden Tribes of America: A Study of America's Polarized Landscape*, More in Common, New York, 2018, https://hiddentribes.us/pdf/hidden_tribes_report.pdf.

112 Daniel Yudkin, Stephen Hawkins, and Tim Dixon, *The Perception Gap: How False Impressions Are Pulling Americans Apart*, More in Common, New York, June 2019, 26–35, https://perceptiongap.us/media/zaslaroc/perception-gap-report-1-0-3.pdf./

113 Richard Wike and Katie Simmons, "Global Support for Principle of Free Expression, but Opposition to Some Forms of Speech," Pew Research Center, November 18, 2015.

114 Emily Ekins, "Poll: 62% of Americans Say They Have Political Views They're Afraid to Share," Cato Institute, July 2020.

115 Pew Research Center, "Public Trust in Government: 1958–2017," December 14, 2017.

116 Megan Brenan, "College Students See Less Secure First Amendment Rights," Gallup, March 13, 2018.

117 Aaron Zitner and Danny Dougherty, "Our Nation, Diverse and Divided: The Wall Street Journal poll looks back at 30 years of change," *Wall Street Journal*, December 27, 2019.

CHAPTER 3—RATIONALIZING NATION-STATE CAPITALISM

1 Karl Marx and Friedrich Engels, *The Communist Manifesto*, trans. Samuel Moore (Chicago: C. H. Kerr & Co., 1912).

2 Jonathan Steinberg, *Bismarck: A Life* (New York: Oxford University Press, 2011), chap. 9.

3 John Stuart Mill, *Principles of Political Economy with some of their Applications to Social Philosophy*, ed. William James Ashley (London: Longmans, Green & Co., 1920), Bk. II, chap. 12, sec. 2, https://oll.libertyfund.org/titles/101#Mill_0199_1101.

4 Donald J. Devine, *Does Freedom Work? Liberty and Justice in America* (Ottawa, Ill.: Caroline House, 1978).

5 Thomas West, "Poverty and Welfare in the American Founding," Heritage Foundation, May 19, 2015.

6 Karen Tumulty, "'Great Society' agenda led to great—and lasting—philosophical divide," *Washington Post*, January 8, 2014.

7 William Galston, "Where Left and Right Agree on Inequality," *Wall Street Journal*, January 15, 2014.

8 Dana Milbank, "GOP's war on 'War on Poverty,'" *Washington Post*, January 9, 2014.

9 Bruce Meyer and James Sullivan, "The Material Well-Being of the Poor and the Middle Class since 1980," American Enterprise Institute, October 25, 2011.

10 Brad Plumer, "Study finds poverty is a moving target," *Washington Post*, January 12, 2014.

11 John F. Early, "Reassessing the Facts about Inequality, Poverty, and Redistribution," Cato Institute, Policy Analysis no. 839, April 24, 2018.

12 Ibid. See also Donald Devine, "Puncturing the Inequality Fact Balloon," *Newsmax*, July 12, 2108.

13 Robert Samuelson, "How we won and lost the war on poverty," *Washington Post*, January 13, 2014.

14 Eric A. Hanushek et al., "The Achievement Gap Fails to Close: Half century of testing shows persistent divide between haves and have-nots," *Education Next*, Summer 2019.

15 Robert Rector and Rachel Sheffield, "The War on Poverty: After 50 Years," Heritage Foundation, 2014; Julie Bosman, "Obama Sharply Assails Absent Black Fathers," *New York Times*, June 6, 2008.

16 Rector and Sheffield, "The War on Poverty: After 50 Years."

17 Ari Fleischer, "How to Fight Income Inequality," *Wall Street Journal*, January 13, 2014.

18 Robert Rector, "How the War on Poverty Was Lost," *Wall Street Journal*, January 7, 2014.

19 Phil Gramm and John F. Early, "Government Can't Rescue the Poor," *Wall Street Journal*, October 11, 2018; Congressional Budget Office, "Illustrative Examples of Effective Marginal Tax Rates Faced by Married and Single Taxpayers," Supplemental Material for *Effective Marginal Tax Rates for Low- and Moderate-Income Workers*, Washington, D.C., November 2012.

20 Branko Milanovic, *The Haves and the Have-Nots: A Brief and Idiosyncratic History of Global Inequality* (New York: Basic Books, 2011).

21 Rector and Sheffield, "The War on Poverty: After 50 Years."

22 Something similar has actually been proposed. Simon L. Lewis and Mark A. Maslin, *The Human Planet: How We Created the Anthropocene* (New Haven: Yale University Press, 2018).

23 Galston, "Where Left and Right Agree on Inequality."

24 Harold Meyerson, "More free trade means more inequality," *Washington Post*, January 15, 2014.

25 World Bank, GINI index (World Bank estimate), http://data.worldbank.org/indicator/SI.POV.GINI.

26 Steven Radelet, "Progress in the global war on poverty," *Christian Science Monitor*, February 7, 2016.

27 David Dollar and Aart Kraay, "Growth Is Good for the Poor," Development Research Group, World Bank, March 2000.

28 Jacqueline A. Berrien, "Statement on 50th Anniversary of the Civil Rights Act of 1964," U.S. Equal Employment Opportunity Commission, July 2, 2014.

29 Ibid.

30 Cristal Hayes, "Trump asks Supreme Court to fast-track ruling on transgender military ban," *USA Today*, November 23, 2018.

31 Donald J. Devine, *The Political Culture of the United States: The Influence of Member Values on Regime Maintenance* (Boston: Little, Brown, 1972), 332–46.

32 Joseph Bottum, *An Anxious Age: The Post-Protestant Ethic and the Spirit of America* (Colorado Springs: Image, 2014), chap. 3.

33 Juliana Menasce Horowitz, Anna Brown, and Kiana Cox, "Race in America 2019," Pew Research Center, April 9, 2019.

34 Gallup, "Race Relations, In Depth Topics," http://www.gallup.com/poll/1687/race-relations.aspx.

35 Tom Worstall, "Gender Pay Gap Is the Result of Being Married," *Forbes*, October 1, 2015.

36 Audrey Carlsen et al., "#MeToo Brought Down 201 Powerful Men. Nearly Half of Their Replacements Are Women," *New York Times*, October 29, 2018. There is also a reaction from men fearing false charges: Kathleen Parker, "Me Too's unintended consequences," *Washington Post*, December 5, 2018; Tina McGregor, "#MeToo Reaction: More male managers avoid mentoring women or meeting alone with them," *Washington Post*, May 17, 2019; Lean In, "Working Relationships in the *MeToo Era: Key Findings," Online Poll by SurveyMonkey, 2019, https://leanin.org/sexual-harassment-backlash-survey-results.

37 Rector and Sheffield, "The War on Poverty: After 50 Years."

38 David Sherfinski, "O'Malley apologizes for saying all lives matter," *Washington Times*, July 20, 2015.

39 Bottum, *An Anxious Age*, chap. 5–6; Timothy P. Carney, *Alienated America: Why Some Places Thrive While Others Collapse* (New York: HarperCollins, 2019), chap. 7.

40 Karl Vick and Ashley Surdin, "Most of California's Black Voters Backed Gay Marriage Ban," *Washington Post*, November 7, 2008.

41 Freedom du Lac and Abby Ohlheiser, "Rachel Dolezal, ex-NAACP leader," *Washington Post*, June 6, 2015.

42 *Hollingsworth et al. v. Perry et al.*, https://www.supremecourt.gov/opinions/12pdf/12-144_8oko.pdf.

43 Barack Obama, Remarks by the President on the Supreme Court Decision on Marriage Equality, Office of the Press Secretary, The White House, June 26, 2015, https://obamawhitehouse.archives.gov/the-press-office/2015/06/26/remarks-president-supreme-court-decision-marriage-equality.

44 *Obergefell v. Hodges*, https://www.supremecourt.gov/opinions/14pdf/14-556_3204.pdf.

45 Chai Feldblum, *Moral Argument, Religion, and Same-Sex Marriage* (Lanham, Md.: Lexington Books, 2009), Conclusion.

46 Ryan T. Anderson, *Truth Overruled: The Future of Marriage and Religious Freedom* (Washington, D.C.: Regnery, 2015), chap. 4.

47 For all of these cases see ibid., chap. 1–2.

48 *Masterpiece Cakeshop v. Colorado Civil Rights Commission*, https://www.supremecourt.gov/opinions/17pdf/16-111_j4el.pdf.

49 Anderson, *Truth Overruled*, chap. 4.

50 John Safranek, "There is no right to same sex marriage," *Federalist*, May 7, 2015.

51 Anderson, *Truth Overruled*, 20, 109. He adds: "Whenever a baby is born, there is always a mother nearby.... Marriage increases the odds that the father of the child will be committed to the child's mother" and that both will be committed to the child. He maintains there is no such thing as "parenting" but only mothering and fathering, as men and women bring different things to the parenting enterprise (p. 25).

52 On natural-law rights, see Donald Devine, "Is Natural Law Sufficient to Defend the Founding?" *Imaginative Conservative*, July 26, 2010.

53 Anderson, *Truth Overruled*, 126.

54 James G. Dwyer, "Same-Sex Cynicism and the Self-Defeating Pursuit of Social Acceptance Through Litigation," *Southern Methodist University Law Review*, 68:1 (Winter 2015), 3–71.

55 David E. Bernstein, "Context Matters: A Better Libertarian Approach to Antidiscrimination Law," *Cato Unbound*, June 16, 2010.

56 Daniel Oliver, "Indiana Burning," *Candid American*, April 11, 2015.

57 Deborah Hellman and Sophia Moreau, eds., *Philosophical Foundations of Discrimination Law* (Oxford: Oxford University Press, 2013).

58 George Rutherglen, "Concrete or Abstract Conceptions of Discrimination Law," in *Philosophical Foundations of Discrimination Law*.

59 Michael Selmi, "Indirect Discrimination and the Anti-discrimination Mandate," in *Philosophical Foundations of Discrimination Law*.

60 Academy of European Law (ERA), "EU anti-discrimination law," Module 2, Exceptions to the principle of non-discrimination, http://www.era-comm.eu/anti-discri/e_learning/module2_2.html.

61 Paul Horwitz, "The Hobby Lobby Moment," *Harvard Law Review*, 128:1 (September 2014), 154–89.

62 Richard Cohen, "Richard Nixon's lasting damage to the GOP," *Washington Post*, August 4, 2014.

63 Jonathan Capehart, "Rep. Mo Brooks talks 'war on whites' as the GOP loses the battle for votes," *Washington Post*, August 4, 2014.

64 Dana Milbank, "A welcome end to American whiteness," *Washington Post*, August 6, 2014.

65 Darlena Cunha, "Ferguson: In Defense of Rioting," *Time*, November 25, 2014.

66 Dan Frosch, Zusha Elinson, and Erin Ailworth, "The Minneapolis Police Chief Promised Change. He Got a Disaster," *Wall Street Journal*, May 31, 2020.

67 Heather Mac Donald, "The Myths of Black Lives Matter," *Wall Street Journal*, February 11, 2016.

68 Juan Williams, "Ferguson and America's Racial Fears," *Wall Street Journal*, August 20, 2014.

69 Michael Gerson, "Paul's political trickery," *Washington Post*, August 19, 2014.

70 Jason L. Riley, *Please Stop Helping Us: How Liberals Make It Harder for Blacks to Succeed* (New York: Encounter Books, 2014).

71 Michael S. Rosenwald and Michael A. Fletcher, "Why couldn't $130 million transform one of Baltimore's poorest places?" *Washington Post*, May 2, 2015.

72 Emily Swanson and Bradley McCombs, "Support for gay marriage comes with caveats," AP-GfK Poll, February 5, 2015.

73 Pledge in Solidarity to Defend Marriage, https://www.wordfoundations.com/pledge-in-solidarity-to-defend-marriage/.

74 James Poulos, "Should Conservatives Practice Civil Disobedience?" *Atlantic*, June 6, 2015.

75 Sandhya Somashekhar, "Texas bill would limit marriage licenses," *Washington Post*, May 22, 2015.

76 Alex Dobuzinskis and Philip Pullella, "Pope's meeting with Kentucky clerk divides public after U.S. visit," Reuters, September 30, 2015.

77 Sophie McBain, "How conservative states are racing to end women's abortion rights," *New Statesman America*, December 7, 2018.

78 Mike Cason, "Alabama lawmakers pass bill to end marriage licenses," *Advance Local News* (AL.com), May 23, 2019.

79 Scott Clement and Jim Tankersley, "Divide over direction of country has widened vastly since 2008, poll finds," *Washington Post*, September 17, 2016.

80 Rich Morin, "Behind Trump's win in rural white America: Women joined men in backing him," Pew Research Center, November 17, 2016.

81 Gregory A. Smith and Jessica Martínez, "How the faithful voted: A preliminary 2016 analysis," Pew Research Center, November 9, 2017.

82 Fareed Zakaria, "The two sins dividing America," *Washington Post*, November 11, 2016.

83 Svetlana Alexievich, *Secondhand Time: The Last of the Soviets*, trans. Bela Shayevich (New York: Random House, 2016).

84 For a fuller discussion, see Donald Devine, "Secondhand Gulag," *Law and Liberty*, February 6, 2017.

85 David Filipov, "For Russians Stalin the 'most outstanding' figure in the world, followed by Putin," *Washington Post*, June 26, 2017.

86 Terry Miller, Anthony B. Kim, and James M. Roberts, *2019 Index of Economic Freedom* (Washington, D.C.: Heritage Foundation, 2019).

87 Pope Francis, *Laudato Si'*, Encyclical Letter on Care for Our Common Home, May 24, 2015.

88 Jim Yardley and Laurie Goodstein, "Pope Francis Calls for Swift Action on

Climate Change," *New York Times*, June 18, 2015; Laurie Goodstein and Justin Gillis, "On Planet in Distress, a Papal Call to Action," *New York Times*, June 18, 2015; E. J. Dionne Jr., "The pope, the saint and the climate," *Washington Post*, June 17, 2015; Francis X. Rocca, "Pope Blames Markets for Environment's Ills," *Wall Street Journal*, June 18, 2015.

89 The title page of *Laudato Si'* on the Vatican copy reads: "On Care for Our Common Planet," but the word "planet" does not appear in many other versions.

90 Thomas Robert Malthus, *An Essay on the Principle of Population* (London: J. Johnson, 1798; Library of Economics and Liberty, 2010), chap. 11, http://www.econlib.org/library/Malthus/malPop.html.

91 Rector and Sheffield, "The War on Poverty: After 50 Years."

92 The pope's later documents on the subject are somewhat more specific, but still rather broad. See Francis X. Rocca, "Vatican in Paper Supports Stricter Market Regulation," *Wall Street Journal*, May 18, 2018.

CHAPTER 4—THE EXPERT BUREAUCRACY SOLUTION

1 Max Weber, *Theory of Social and Economic Organization*, trans. A. M. Henderson and Talcott Parsons (New York: Free Press, 1947), 337. The ideal can be traced further back in time: Carl K. Y. Shaw, "Hegel's Theory of Modern Bureaucracy," *American Political Science Review*, 86:2 (June 1992), 381–89.

2 Ralph H. Gabriel, *The Course of American Democratic Thought* (New York: Randall Press, 1956), chap. 19.

3 The classic statement is from the Intergovernmental Personnel Act of 1970 (84 Stat. 1909), updating the original Pendleton Act of 1883. See also Congressional Research Service, Library of Congress, *History of Civil Service Merit Systems of the United States and Selected Foreign Countries, together with Executive Reorganization Studies and Personnel Recommendations*, compiled for the Subcommittee on Manpower and Civil Service, House of Representatives, 94th Congress (Washington, D.C.: Government Printing Office, December 31, 1976).

4 Donald J. Devine, *Reagan's Terrible Swift Sword* (Ottawa, Ill.: Jameson Books, 1991).

5 President Jimmy Carter, "Civil Service Reform Act of 1978: Statement on Signing S. 2640 Into Law," October 13, 1978, The American Presidency Project, https://www.presidency.ucsb.edu/documents/civil-service-reform-act-1978-statement-signing-s-2640-into-law.

6 Much of the following is from Donald Devine, Testimony Before the Subcommittee on Federal Workforce, U.S. Postal Service and the Census, Committee on Oversight and Government Reform, July 15, 2014; Devine, *Reagan's Terrible Swift Sword*.

7 Donald J. Devine, "Making Government Work: How Congress Can Really Reinvent Government," Heritage Foundation Backgrounder, August 24, 1995.

8 Donovan Slack and Bill Theobald, "Veterans Affairs pays $142 million in bonuses amid scandals," *USA Today*, November 11, 2015.

9 Lisa Rein, "Old civil service test becomes new tool," *Washington Post*, April 3, 2015.

10 Government Accountability Office, "Federal Workforce: Distribution of Performance Ratings Across the Federal Government, 2013," Washington, D.C., May 9, 2016, https://www.gao.gov/assets/680/676998.pdf.

11 Eric Yoder, "Study on Federal attrition," *Washington Post*, May 6, 2015.
12 Lisa Rein, "Discipline at EPA lax," *Washington Post*, June 8, 2015.
13 Donald J. Devine, "Reforming the Federal Bureaucracy," Heritage Foundation Backgrounder, December 10, 2018; Devine, *Reagan's Terrible Swift Sword*, chap. 8.
14 Lisa Rein, "Agencies cracking down on workers' misconduct," *Washington Post*, August 22, 2017.
15 Devine, *Reagan's Terrible Swift Sword*, chap. 9.
16 Joe Davidson, "Skepticism over 99% successful," *Washington Post*, June 17, 2016.
17 Lisa Rein, "Nearly all receive rave reviews," *Washington Post*, June 14, 2016. See also Donald Devine, "Making Government Unions More Responsible," *Newsmax*, June 19, 2018; Donald Devine, "Bloated Federal Leviathan Stagnation," *Newsmax*, October 10, 2018.
18 Eric Yoder, "Republicans push through change to personnel policies," *Washington Post*, July 8, 2016.
19 Executive Order on Modernizing and Reforming the Assessment and Hiring of Federal Job Candidates, The White House, June 26, 2020.
20 Pew Research Center, "Public Trust in Government: 1958–2019," April 11, 2019.
21 Pew Research Center, "Beyond Trust: How Americans View Their Government," November 23, 2015.
22 Pew Research Center, "The Public, the Political System and American Democracy," April 26, 2018.
23 Paul Light, "The real crisis in government," *Washington Post*, January 10, 2010.
24 Paul Demko, "Nearly $80 billion misspent on Medicare, Medicaid in 2014," *Modern Healthcare*, March 6, 2015.
25 Merrill Matthews, "We've Crossed the Tipping Point; Most Americans Now Receive Government Benefits," *Forbes*, July 2, 2014.
26 David Muhlhausen and Patrick Tyrrell, eds., *The 2013 Index of Dependence on Government*, 11th ed., Center for Data Analysis, Heritage Foundation, November 21, 2013, https://www.heritage.org/welfare/report/the-2013-index-dependence-government.
27 Ibid.; Department of Homeland Security, *National Incident Management System* (Washington, D.C.: Government Printing Office, March 1, 2004), Introduction.
28 James Q. Wilson, *Bureaucracy: What Government Agencies Do and Why They Do It* (New York: Basic Books, 1989), chap. 20.
29 U.S. Department of Health and Human Services and U.S. Department of Agriculture, *2015–2020 Dietary Guidelines for Americans* (8th ed.), December 2015, http://health.gov/dietaryguidelines/2015/guidelines/.
30 American College of Cardiology, "New ACC/AHA Cholesterol Guideline Allows for More Personalized Care; New Treatment Options," November 10, 2018, https://www.acc.org/latest-in-cardiology/articles/2018/11/07/15/19/sat-1130am-guideline-on-the-management-of-blood-cholesterol.
31 Peter Whoriskey, "Good news on Dietary Guidelines," *Washington Post*, January 8, 2016.
32 Ibid.
33 Centers for Disease Control and Prevention, "Sodium and the Dietary Guidelines," https://www.cdc.gov/salt/pdfs/Sodium_Dietary_Guidelines.pdf.
34 Robert Lowes, "Obama defends no-quarantine policy," *Medscape*, October 28, 2014, http://www.medscape.com/viewarticle/834028.

35 Scott Gottlieb and Tevi Troy, "Ebola Isn't Messaging Problem," *Wall Street Journal*, October 20, 2014; Centers for Disease Control, "Facts about Ebola in the U.S.," http://www.cdc.gov/vhf/ebola/pdf/infographic.pdf.

36 Margaret Chan, "WHO Director General addresses executive board," Director-General's Office, World Health Organization, January 25, 2016, http://www.who.int/dg/speeches/2016/executive-board-138/en/.

37 Denise Grady and Donald G. McNeil Jr., "Ebola Sample Is Mishandled at C.D.C. Lab in Latest Error," *New York Times*, December 24, 2014.

38 Dan Lamothe, "Army lab inadvertently shipped live anthrax," *Washington Post*, May 28, 2015.

39 Lena H. Sun and Brady Dennis, "Smallpox vials, decades old, found in storage room at NIH campus in Bethesda," *Washington Post*, July 8, 2014.

40 Lanny Bernstein, "Strife at agency threatens oversight of science studies," *Washington Post*, September 21, 2016.

41 Ibid.

42 Thomas M. Burton, "Mystery Fungus Sparks Crisis at NIH," *Wall Street Journal*, January 19, 2017.

43 Bernstein, "Strife at agency threatens oversight of science studies."

44 Christina Mlynski, "HUD watchdog sounds alarm," *HousingWire*, September 10, 2013, http://www.housingwire.com/articles/26753-hud-watchdog-sounds-alarm-on-fraud-activity.

45 Devlin Barrett and Damien Paletta, "Officials mask severity of hack," *Washington Post*, June 25, 2015; Donald Devine, "Big Data and Big Security Problems," *Newsmax*, May 10, 2019.

46 Charles S. Clark, "Lois Lerner and Other IRS Employees Will Not Be Prosecuted," *Government Executive*, October 23, 2015; Peter J. Kadzik, Assistant Attorney General, Letter to the Honorable Bob Goodlatte and the Honorable John Conyers, Jr., October 23, 2015, Office of Legislative Affairs, U.S. Department of Justice, https://www.govexec.com/media/gbc/docs/pdfs_edit/102315cc1.pdf.

47 Rachael Bade, "Thousands of lost Lerner IRS emails found by IG," *Politico*, November 23, 2014.

48 Robert W. Wood, "61% of IRS Employees Caught Willfully Violating Tax Law Aren't Fired, Many Get Promoted," *Forbes*, May 7, 2016.

49 Kevin McCoy, "Cyber hack got access to over 700,000 IRS accounts," *USA Today*, February 26, 2016.

50 Paul Krugman, "Vouchers for Veterans," *New York Times*, November 13, 2011.

51 Joe Davidson, "Cascade of failures," *Washington Post*, December 14, 2015.

52 Donovan Slack, "Trump signs VA law to provide more private health care choices," *USA Today*, June 6, 2018; Ben Kesling, "New VA Rates Expand Private-Care Access," *Wall Street Journal*, January 31, 2019.

53 "US government keeps 'boot on neck' of BP over spill," Reuters, May 24, 2010.

54 "An Animas River Accounting," Review & Outlook, *Wall Street Journal*, June 30, 2014.

55 Ruchir Sharma, "The Dollar—and the Fed—Still Rule," *Wall Street Journal*, July 28, 2016.

56 Federal Reserve, Transcript of Chairman Yellen's Press Conference, June 15, 2016, https://www.federalreserve.gov/mediacenter/files/FOMCpresconf20160615.pdf.

57 Brian S. Wesbury, Bob Stein, and Strider Elass, "Fed Funds Rate an Anachronism," *Monday Morning Outlook*, First Trust, March 24, 2016.

58 Brian S. Wesbury, Robert Stein, and Strider Elass, "It's Still the Fed, and It's Not Magic," *Monday Morning Outlook*, First Trust, September 26, 2016; Phil Gramm and Thomas R. Saving, "Has the Federal Reserve Lost Its Mojo?" *Wall Street Journal*, June 18, 2019.

59 Sharma, "The Dollar—and the Fed—Still Rule."

60 "Federal Reserve hints it could end asset purchases in 2014," BBC News, June 20, 2013; Paul Davidson, "Yellen gives no timetable to shrink assets," *USA Today*, February 15, 2017.

61 Ruchir Sharma, "Trump Tees Up a Necessary Debate on the Fed," *Wall Street Journal*, September 29, 2016.

62 Bank of International Settlement, 84th Annual Report, 2013/14, June 29, 2014, http://www.bis.org/publ/arpdf/ar2014e.htm.

63 Robert J. Samuelson, "The business cycle RIP," *Washington Post*, June 30, 2014.

64 Robert J. Samuelson, "Bankrupt Economics," *Washington Post*, July 29, 2008. The 1850 classic is Claude Frédéric Bastiat, *That Which Is Seen, and That Which Is Not Seen* (Cleveland: World Library Classics, 2010).

65 Ben S. Bernanke, Timothy F. Geithner, and Henry M. Paulson, *Firefighting: The Financial Crisis and Its Lessons* (New York: Penguin, 2019). See also Robert J. Samuelson, "What really caused the financial crisis," *Washington Post*, April 16, 2019.

66 Casey B. Mulligan, "A Recovery Stymied by Redistribution," *Wall Street Journal*, June 30, 2014.

67 Nick Timiraos, "Fed Tools for Bad Times Grow Dull," *Wall Street Journal*, December 9, 2019.

68 Òscar Jordà and Alan M. Taylor, "Riders on the Storm," prepared for the Federal Reserve Bank of Kansas City Economic Policy Symposium, Jackson Hole, Wyoming, August 22–24, 2019, 42; Robert J. Samuelson, "The Fed story we're missing," *Washington Post*, September 23, 2019.

69 Greg Ip, "What the Government Can and Can't Do About the Economic Fallout From the Coronavirus," *Wall Street Journal*, March 17, 2020; Donald Devine, "Who Is Opposed to What Free Market Capitalism?" *American Spectator*, September 14, 2020.

70 Heather Long, "The $2 trillion relief bill is massive, but it won't prevent a recession," *Washington Post*, March 16, 2020.

71 Lexington Law, "Important Welfare Statistics for 2020," January 3, 2020, https://www.lexingtonlaw.com/blog/finance/welfare-statistics.html.

72 Shelley K. Irving and Tracy A. Loveless, "Dynamics of Economic Well-Being: Participation in Government Programs, 2009–2012: Who Gets Assistance?" U.S. Census Bureau, May 28, 2015.

73 Trading Economics, "United States Labor Force Participation Rate, 2020," https://tradingeconomics.com/united-states/labor-force-participation-rate.

74 Adam Millsap, "How Higher Minimum Wages Impact Employment," *Forbes*, September 28, 2018.

75 Stastica, "Poverty rate in the United States from 1990 to 2018," https://www.statista.com/statistics/200463/us-poverty-rate-since-1990/.

76 Bureau of Industry and Security, Welfare and Employment, U.S. Department of Commerce; KFF, State Health Facts, "Distribution of Nonelderly Adults with Medicaid by Gender, 2018," https://www.kff.org/medicaid/state-indicator/distribution-by-gender-4/.

77 Congressional Budget Office, "Illustrative Examples of Effective Marginal Tax Rates Faced by Married and Single Taxpayers," Supplemental Material for *Effective Marginal Tax Rates for Low- and Moderate-Income Workers*, Washington, D.C., November 2012, https://www.cbo.gov/system/files/2018-09/43722-Supplemental_Material-MarginalTaxRates.pdf.

78 Shannon Mok, "How Taxes and Transfers Affect the Work Incentives of People With Low and Moderate Income," Congressional Budget Office, March 17, 2017, https://www.cbo.gov/publication/52472.

79 Council of Economic Advisers, "Government Employment and Training Programs: Assessing the Evidence on Their Performance," Executive Office of the President, Washington, D.C., June 2019.

80 U.S. Bureau of Labor Statistics, "Employment of Women on Non-Farm Payrolls," March 1, 2019.

81 Rick Reis, "Where the Guys Are: Males in Higher Education," *Tomorrow's Professor*, Stanford University, June 20, 2010, https://tomprof.stanford.edu/posting/1028.

82 U.S. Bureau of Labor Statistics, "Employment Status of the Civilian Population," March 6, 2019.

83 Nicholas Eberstadt, "Education and Men without Work," *National Affairs*, Spring 2020.

84 Ibid.

85 A. W. Geiger and Gretchen Livingston, "8 facts about love and marriage in America," *Fact Tank*, Pew Research Center, February 13, 2019.

86 Richard Fry, "New census data show more Americans are tying the knot, but mostly it's the college-educated," *Fact Tank*, Pew Research Center, February 6, 2014.

87 Centers for Disease Control and Prevention, Pandemic Influenza, https://www.cdc.gov/flu/pandemic-resources/index.htm.

88 Donald Devine, "Trump's Coronavirus Leadership," *American Spectator*, March 28, 2020.

89 Centers for Disease Control and Prevention, Coronavirus Disease 2019 (COVID-19), https://www.cdc.gov/coronavirus/2019-nCoV/index.html.

90 Carolyn Y. Johnson, "A 'negative' coronavirus test result doesn't always mean you aren't infected," *Washington Post*, March 23, 2020.

91 David Von Drehle, "Whatever happened to the CDC?" *Washington Post*, April 5, 2020.

92 Tommaso Ebhardt, Chiara Remondini, and Marco Bertacche, "99% of Those Who Died From Virus Had Other Illness, Italy Says," Bloomberg, March 18, 2020.

93 John P. A. Ioannidis, "A fiasco in the making? As the coronavirus pandemic takes hold, we are making decisions without reliable data," *STAT*, March 17, 2020.

94 Ibid.

95 Pew Research Center, "U.S. Public Sees Multiple Threats From the Coronavirus—and Concerns Are Growing," March 18, 2020.

96 Devine, "Trump's Coronavirus Leadership."

97 Doina Chiacu and Dan Whitcomb, "Trump backs off plan to reopen businesses by mid-April amid coronavirus warnings," Reuters, March 29, 2020.

98 William Wan et al., "Some Trump advisors doubt White House's estimates of virus deaths," *Washington Post*, April 3, 2020.

99 Ibid.

100 Douglas Dowley, "Cuomo: NY coronavirus projections 'all wrong,' too early to tell if reopening is working," Syracuse.com, May 25, 2020.

101 Elaine He, "The Results of Europe's Lockdown Experiment Are In," Bloomberg Opinion, May 20, 2020.

102 Rabail Chaudhry et al., "A country level analysis measuring the impact of government actions, country preparedness and socioeconomic factors on COVID-19 mortality and related health outcomes," *EclinicalMedicine*, July 21, 2020, https://www.thelancet.com/journals/eclinm/article/PIIS2589-5370(20)30208-X/fulltext.

103 Spencer S. Hsu, "FBI discloses errors in DNA analysis," *Washington Post*, May 30, 2015.

104 Harry T. Edwards and Jennifer L. Mnookin, "A wake-up call on the junk science infesting our courtrooms," *Washington Post*, September 20, 2016.

105 Ludwig von Mises, *Bureaucracy* (New Haven: Yale University Press, 1944), chap. 2.

106 James Q. Wilson, *Bureaucracy: What Government Agencies Do and Why They Do It* (New York: Basic Books, 1989).

107 Robert S. Mueller III, Director, Federal Bureau of Investigation, Statement Before the House Permanent Select Committee on Intelligence, Washington, D.C., October 6, 2011, https://archives.fbi.gov/archives/news/testimony/the-state-of-intelligence-reform-10-years-after-911; Donald Devine, "Return FBI to Law Enforcement Roots," *Newsmax*, April 12, 2018.

108 FBI National Press Office, "In Domestic Intelligence Gathering, the FBI Is Definitely on the Case," Federal Bureau of Investigation, March 21, 2007, https://archives.fbi.gov/archives/news/pressrel/press-releases/in-domestic-intelligence-gathering-the-fbi-is-definitely-on-the-case.

109 Alan M. Dershowitz, "Targeting Trump's lawyer should worry us all," *The Hill*, April 10, 2018.

110 Thomas J. Baker, "What Went Wrong at the FBI," *Wall Street Journal*, March 19, 2018.

111 Owen Scarborough, "Carter Page Exonerated by Mueller Report," *Washington Times*, April 18, 2019.

112 Office of the Inspector General, *Review of Four FISA Applications and Other Aspects of the FBI's Crossfire Hurricane Investigation*, U.S. Department of Justice, December 2019 (revised), https://www.justice.gov/storage/120919-examination.pdf.

113 Donald Devine, "No Bureaucratic Drainage in the Washington Swamp," *American Spectator*, January 1, 2020.

114 Philip Shenon, "Secret Court Says F.B.I. Aides Misled Judges in 75 Cases," *New York Times*, August 23, 2002.

115 Devlin Barrett, "Surveillance court demands answers from FBI for errors, omissions in Trump campaign investigation," *Washington Post*, December 17, 2019.

116 Josh Gerstein and Kyle Cheney, "Justice Department audit finds widespread flaws in FBI surveillance applications," *Politico*, March 31, 2020.

117 Devlin Barrett and Ellen Nakashima, "Audit of FBI surveillance finds chronic problems," *Washington Post*, March 31, 2020.

118 Gerstein and Cheney, "Justice Department audit finds widespread flaws in FBI surveillance applications."

119 Jan Crawford, "William Barr Interview," *CBS This Morning*, May 31, 2019.

120 Ibid.

121 Ibid.; Donald Devine, "Barr's Courageous Challenge to FBI Praetorian Guard," *Newsmax*, June 7, 2019.

122 Devine, *Reagan's Terrible Swift Sword*, chap. 11, 13.

123 Devlin Barrett and Dan Frosch, "FBI Head Defends Terror Probes," *Wall Street Journal*, September 28, 2016.

124 Mark Berman and Matt Zapotosky, "The FBI said it failed to act on a tip about the suspected Florida school shooter's potential for violence," *Washington Post*, February 16, 2018.

125 Ellen Nakashima, "N.Y. bombing suspect was twice flagged for questioning by customs officials," *Washington Post*, September 23, 2016.

126 Gregory Miller, "As the U.S. spied on the world, CIA and NSA bickered," *Washington Post*, March 9, 2020.

127 Ashley Halsey III, "Covert testing at airports shows TSA security problems continue, report says," *Washington Post*, November 10, 2017.

128 Frederick Kunkle, "FAA, airlines glitch," *Washington Post*, August 16, 2015.

129 Shane Harris, "Secret Service prostitution scandal," *Washington Post*, March 25, 2013.

130 Carol D. Leonnig and Jerry Markon, "Secret Service official wanted to embarrass congressman," *Washington Post*, October 1, 2015.

131 CNN Editorial Research, "White House Security Breaches Fast Facts," CNN Politics, March 25, 2020.

132 Carol D. Leonnig, "White House fence-jumper," *Washington Post*, September 29, 2014.

133 Leonnig and Markon, "Secret Service official wanted to embarrass congressman."

134 Devlin Barrett and David A. Fahrenthold, "Arrest at Mar-a-Lago stirs security questions," *Washington Post*, April 3, 2019.

135 Leonnig and Markon, "Secret Service official wanted to embarrass congressman."

136 Theodore Schleifer, "41 Secret Service employees disciplined after Chaffetz leak," CNN, May 26, 2016.

137 Clyde Wayne Crews Jr., *Ten Thousand Commandments 2016: An Annual Snapshot of the Federal Regulatory State* (Washington, D.C.: Competitive Enterprise Institute, 2016); Clyde Wayne Crews Jr., "One Nation, Ungovernable? Confronting the Modern Regulatory State," in *What America's Decline in Economic Freedom Means for Entrepreneurship and Prosperity*, ed. Donald J. Boudreaux (Fraser Institute and Mercatus Center at George Mason University, 2015), 117–81.

138 Clyde Wayne Crews Jr., "Trump Exceeds One-In, Two-Out Goals on Cutting Regulations, But It May Be Getting Tougher," *Forbes*, October 23, 2018.

139 Klaus Schwab and Xavier Sala-i-Martín, *The Global Competitiveness Report 2014–2015* (Geneva: World Economic Forum, 2014), United States, 379, http://reports.weforum.org/global-competitiveness-report-2014-2015/; Steven Globerman and George Georgopoulos, "Regulation and the International Competitiveness of the U.S. Economy," Mercatus Center, George Mason University, September 18, 2012, based on data from the Organization for Economic Cooperation and Development (OECD).

140 James Gattuso, "This legislation will curb bureaucracy power," Heritage

Foundation, April 27, 2016; Neil Siefring, "The REINS Act will keep regulations and their costs in check," *The Hill*, August 4, 2015.

141 GovTrack, "13 laws passed this session have utilized this previously-obscure 1996 provision," May 28, 2017; Clyde Wayne Crews Jr., "What Regulations Has the Trump Administration Eliminated This Year?" Competitive Enterprise Institute, October 17, 2018, https://cei.org/blog/what-regulations-has-trump-administration-eliminated-so-far.

142 Katie Weatherford, "Three Reasons the REINS Act Must Be Stopped (Again)," Center for Effective Government, January 23, 2015.

143 Aaron Greg, "First full audit of Pentagon notes compliance issues," *Washington Post*, November 26, 2018.

144 Clyde Wayne Crews Jr., *Mapping Washington's Lawlessness 2016: A Preliminary Inventory of "Regulatory Dark Matter"* (Washington, D.C.: Competitive Enterprise Institute, December 2015).

145 *Perez v. Mortgage Bankers Association*, http://www.scotusblog.com/case-files/cases/perez-v-mortgage-bankers-association/.

146 *Department of Transportation et al. v. Association of American Railroads*, https://www.supremecourt.gov/opinions/14pdf/13-1080_f29g.pdf.

147 U.S. Department of Justice, "Fact Sheet: President's Corporate Fraud Task Force Marks Five Years of Ensuring Corporate Integrity," July 17, 2007, https://www.justice.gov/archive/opa/pr/2007/July/07_odag_507.html.

148 John Reed Stark, "Mary Jo White Wants to Read Your Gmail," *Wall Street Journal*, June 17, 2016.

149 *Time*, February 13, 2012.

150 Christopher M. Matthews, "Ruling Puts Dent in Insider Probes," *Wall Street Journal*, December 11, 2014.

151 "An Outside the Law Prosecutor," Review & Outlook, *Wall Street Journal*, December 11, 2014.

152 Colby Hamilton, "Bharara praises Supreme Court decision on insider trading," *Politico*, December 6, 2016.

153 Christopher M. Matthews, "Insider Cases' Legal Basis Questioned," *Wall Street Journal*, April 23, 2014.

154 SI Staff, "Prosecutor urges insider trading conviction of pro gambler," *Sports Illustrated*, April 5, 2017.

155 David Rosenfeld, "Phil Mickelson as the SEC's Bogey," *Wall Street Journal*, June 16, 2016.

156 See Donald J. Devine, *America's Way Back: Reclaiming Freedom, Tradition, and Constitution* (Wilmington, Del.: ISI Books, 2013), 112–13.

157 Donald Devine, "Why Americans Fear Trial by Jury," *American Conservative*, December 31, 2019.

158 Matthews, "Ruling Puts Dent in Insider Probes."

159 Dennis Jacobe, "In U.S., 54% Have Stock Market Investments, Lowest Since 1999," Gallup: Economy, August 20, 2011.

160 James Madison, *Federalist* No. 62, *The Federalist Papers* (New York: New American Library, 1961).

161 Margaret Hartmann, "CNN Hires Preet Bharara, the Former Federal Prosecutor Fired by Trump," *Daily Intelligencer*, September 21, 2017.

162 Carrie Johnson and Brooke A. Masters, "Fraud Cases Focus on Top Executives," *Washington Post*, January 19, 2005.

163 Matthews, "Ruling Puts Dent in Insider Probes."

164 "An Outside the Law Prosecutor," Review & Outlook.

165 Woodrow Wilson, *Congressional Government* (Sidney: Westwood Press, 2017).

166 Lincoln Caplan, "What's really wrong with the Bush Justice Department," *Slate*, March 4, 2007.

167 Devine, *America's Way Back*, 165–73.

168 The American Presidency Project, Executive Orders, http://www.presidency. ucsb.edu/data/orders.php.

169 Ruth Marcus, "A pen, a phone and a precedent," *Washington Post*, February 5, 2014.

170 Brian Callanan, "A legal poison pill for Obamacare," *Wall Street Journal*, January 14, 2016.

171 Office of the Press Secretary, "U.S. Leadership and the Historic Paris Agreement to Combat Climate Change," Press Release, The White House, December 12, 2015.

172 Brent Kendall and Alicia Mundy, "EPA Carbon Rules to Spark Lawsuits," *Wall Street Journal*, May 30, 2014.

173 "Carbon Income Inequality," Review & Outlook, *Wall Street Journal*, June 3, 2014.

174 Ibid.

175 Devin Henry, "Trump Signs Bill Undoing Obama Coal Mining Rule," *The Hill*, February 16, 2017.

176 Joss Garman, "Macron's mistake: Taxing the poor to tackle climate change," *Politico*, December 7, 2018.

177 Bruce Katz, "The Metropolitan Revolution," Brookings, May 21, 2013.

178 Paul C. Light, "A Cascade of Failures: Why Government Fails and How to Stop It," Center for Effective Public Management, Brookings, July 2014.

179 Paul C. Light, "Your Tax Dollars Haplessly at Work," *Wall Street Journal*, September 17, 2014.

180 F. A. Hayek, "The Theory of Complex Phenomena," in *Studies in Philosophy, Politics and Economics* (Chicago: University of Chicago Press, 1967); Hayek, "Individualism: True and False," in *Studies on the Abuse and Decline of Reason*, vol. 13 of *The Collected Works of F. A, Hayek*, ed. Bruce Caldwell (Chicago: University of Chicago Press, 2010).

181 Donald Devine, "Understanding Genes, Decadence, and the Decline of Empires," *Imaginative Conservative*, June 1, 2020.

182 Friedrich A. Hayek, *Law, Legislation and Liberty*, vol. 3, *The Political Order of a Free People* (Chicago: University of Chicago Press, 1979), 175–76.

CHAPTER 5—SCIENTIFIC RATIONALIZATION

1 F. A. Hayek, "Kinds of Rationalism," in *Studies in Philosophy, Politics and Economics* (Chicago: University of Chicago Press, 1967), 85–95, esp. 95 for the quotation and 94 crediting Popper for the latter term. See also Hayek, "The Theory of Complex Phenomena," in *Studies in Philosophy, Politics and Economics* (Chicago: University of Chicago Press, 1967), 40, where he gives partial credit to Warren Weaver.

2 Michael Gerson, "The world we don't know," *Washington Post*, May 13, 2014.

3 Robert J. Samuelson, "We're on mission impossible to solve global warming," *Washington Post*, October 14, 2018.

4 Alfred North Whitehead, *The Concept of Nature* (1920; Mineola, N.Y.: Dover,

2004); Bertrand Russell, *Problems of Philosophy* (London: Oxford University Press, 1912), 23; Alfred North Whitehead, *Science and the Modern World* (New York: Macmillan, 1925), 4.

5 Sabine Hossenfelder, *Lost in Math: How Beauty Leads Physics Astray* (New York: Basic Books, 2018), esp. chap. 9.

6 Gerson, "The world we don't know."

7 Patrick Brennan, "Abuse from Climate Scientists Forces One of Their Own to Resign from Skeptic Group after a Week: 'Reminds Me of McCarthy,'" *National Review*, May 14, 2014.

8 Michael Polanyi, *Personal Knowledge: Towards a Post-Critical Philosophy* (Chicago: University of Chicago Press, 1958), chap. 10.

9 Thomas S. Kuhn, *The Structure of Scientific Revolutions* (Chicago: University of Chicago Press, 1962), esp. 4–9.

10 Jeremy White, "Fraud and Errors in Scientific Studies Skyrocket," *International Business Times*, August 10, 2011.

11 Lincoln Barnett, *The Universe and Dr. Einstein* (New York: William Morrow, 1957), chap. 3. See also Hayek, "The Theory of Complex Phenomena," 39–40.

12 The following comes from Hossenfelder, *Lost in Math*. See also Donald Devine, "Science Lost in Math," *Imaginative Conservative*, January 29, 2019.

13 Brian Greene, *The Elegant Universe: Superstrings, Hidden Dimensions, and the Quest for the Ultimate Theory* (New York: W. W. Norton, 2003), 5.

14 Brian Schmidt, "How Science Will Save the World," World Economic Forum, January 17, 2016.

15 M. G. Aartsen et al., "Searches for Sterile Neutrino with the IceCube Detector," *Physical Review Letters*, 117:7 (August 8, 2016).

16 Sarah Kaplan, "The strangest supernova: a star that keeps exploding—and surviving," *Washington Post*, November 10, 2017.

17 Anna Karnkowska et al., "A Eukaryote Without a Mitochondrial Organelle," *Current Biology*, 26:10 (May 23, 2016); John Gray, *Seven Types of Atheism* (New York: Farrar, Straus & Giroux, 2018).

18 Elizabeth Pennisi, "ENCODE Project Writes Eulogy for Junk DNA," *Science*, 3337 (September 7, 2012), 1159–61.

19 Amy Ellis Nutt, "New map ups number of brain areas," *Washington Post*, April 21, 2016.

20 National Human Genome Research Institute, "What Is the Human Genome Project?" National Institutes of Health, Bethesda, Md., https://www.genome.gov/ human-genome-project/What.

21 American Society of Gene and Cell Therapy, "Scientific Leaders Call for Global Moratorium on Germline Gene Editing," Waukesha, Wis., April 24, 2019.

22 Richard Harris, "Why Making a 'Designer Baby' Would Be Easier Said Than Done," *Shots*, NPR, May 2, 2019, https:// www.npr.org/sections/health-shots/2019/05/02/719665841/ why-making-a-designer-baby-would-be-easier-said-than-done.

23 Ibid.

24 Karl Popper, "Natural Selection and the Emergence of Mind," *Information Philosopher*, November 8, 1977, text and recording: http://www. informationphilosopher.com/solutions/philosophers/popper/natural _selection_and_the_emergence_of_mind.html.

25 Stephen Meyer, *Darwin's Doubt: The Explosive Origin of Animal Life and the Case for Intelligent Design* (New York: HarperCollins, 2013). For an interesting attempt to resolve the problem, see Simona Ginsburg and Eva Jablonka, *The Evolution of the Sensitive Soul: Learning and the Origins of Consciousness* (Cambridge, Mass.: MIT Press, 2019), esp. chap. 9.

26 Brendan Maher, "ENCODE: the human encyclopedia," *Nature*, September 5, 2012.

27 Alfred de Grazia, ed., *The Velikovsky Affair: The Warfare of Science and Scientism* (New York: University Books, 1966).

28 National Aeronautics and Space Administration, "Scientific Consensus: Earth's Climate is Warming," June 19, 2019, https://climate.nasa.gov/scientific-consensus/.

29 Holman W. Jenkins Jr., "As Al Gore Told Donald Trump," *Wall Street Journal*, December 10, 2016.

30 Holman W. Jenkins Jr., "Change Would Be Healthy at U.S. Climate Agencies," *Wall Street Journal*, February 4, 2017; Donald J. Devine, *America's Way Back: Reclaiming Freedom, Tradition, and Constitution* (Wilmington, Del.: ISI Books, 2013), 77–81.

31 Judith A. Curry, Statement to the Committee on Science, Space and Technology of the United States House of Representatives, Hearing on Climate Science: Assumptions, Policy Implications and the Scientific Method, March 29, 2017, available at: https://docs.house.gov/Committee/Calendar/ByEvent.aspx?EventID=105796; Chelsea Harvey, "These scientists want to create 'red teams' to challenge climate research, Congress is listening," *Washington Post*, March 29, 2017.

32 International Energy Agency, "Decoupling of global emissions and economic growth confirmed," March 16, 2016, https://www.iea.org/news/decoupling-of-global-emissions-and-economic-growth-confirmed.

33 David G. Victor and Charles F. Kennel, "Ditch the 2°C warming goal," *Nature*, October 1, 2014.

34 Gautam Naik, "Climate Experts Question Temperature Benchmark," *Wall Street Journal*, November 30, 2015.

35 See Nongovernmental International Panel on Climate Change (NIPCC), http://climatechangereconsidered.org/; Craig Idso and S. Fred Singer, eds., *Climate Change Reconsidered: 2009 Report of the Nongovernmental Panel on Climate Change* (Chicago: Heartland Institute, 2009); Craig D. Idso, Robert M. Carter, and S. Fred Singer, eds., *Climate Change Reconsidered II: Physical Science* (Chicago: Heartland Institute, 2013); Craig D. Idso, Sherwood B. Idso, Robert M. Carter, and S. Fred Singer, eds., *Climate Change Reconsidered II: Biological Impacts* (Chicago: Heartland Institute, 2014); Craig D. Idso, Robert M. Carter, and S. Fred Singer, eds., *Why Scientists Disagree about Global Warming: The NIPCC Report on Consensus*, Nongovernmental International Panel on Climate Change (Arlington Heights, Ill.: Heartland Institute, 2015); Roger Bezdek, Craig D. Idso, David R. Legates, and S. Fred Singer, eds., *Climate Change Reconsidered II: Fossil Fuels*, Nongovernmental International Panel on Climate Change (Arlington Heights, Ill.: Heartland Institute, 2019). See also Joseph Bass, "The Questionable Science Behind the Global Warming Scare," Heartland Institute, October 1, 1998.

36   On the importance of risk tolerance in Wildavsky's typology, see, Michael Thompson, Richard J. Ellis, and Aaron Wildavsky, *Culture Theory* (Boulder, Col.: Westview Press, 1990), chap. 1, esp. 27.

37   Ronald Bailey, *The End of Doom: Environmental Renewal in the Twenty-first Century* (New York: St. Martin's Press, 2015), xvii–xviii.

38   Dennis T. Avery, "Energy Secretary Admits We Don't Understand Climate Change," Climate Realists, March 30, 2010, http://climaterealists.com/index. php?id=5452.

39   Office of Energy Efficiency and Renewable Energy, "Reduce Climate Change," U.S. Department of Energy, https://www.fueleconomy.gov/feg/climate.shtml, accessed July 29, 2020.

40   Environmental Protection Agency, *The Plain English Guide to the Clean Air Act,* Office of Air Quality and Planning Standards, Research Triangle Park, N.C., April 2007, https://www.epa.gov/sites/production/files/2015-08/documents/peg. pdf.

41   "The VW Emissions Bug," Review & Outlook, *Wall Street Journal,* September 24, 2015.

42   Holman W. Jenkins Jr., "How to Settle the VW Scandal," *Wall Street Journal,* October 6, 2015.

43   Max Ehrenfreund, "Little evidence that emissions tests lead to better air quality," *Washington Post,* November 24, 2015.

44   D. F. Huelke, J. L. Moore, T. W. Compton, et al., "Upper extremity injuries related to airbag deployments," *Journal of Trauma,* 38:4 (1995), 482–88; E. L. Freedman, M. R. Safran, R. A. Meals, "Automotive airbag-related upper extremity injuries: a report of three cases," *Journal of Trauma,* 38:4 (1995), 577–81; C. Serra, O. Delattre, R. L. Despeignes, et al., "Upper limb traumatic lesions related to airbag deployment: a case report and review of literature," *Journal of Trauma,* 65:3 (2008), 704–7; M. Chong, G. Broome, D. Mahadeva, et al., "Upper extremity injuries in restrained front-seat occupants after motor vehicle crashes," *Journal of Trauma,* 70:4 (2011), 838–44.

45   Charles Murray, *Human Diversity: The Biology of Gender, Race, and Class* (New York: Hachette, 2020), esp. 213–26; Donald Devine, "Understanding Genes, Decadence, and the Decline of Empires," *Imaginative Conservative,* June 1, 2020.

46   Tom Burns, *Our Necessary Shadow: The Nature and Meaning of Psychiatry* (New York: Pegasus, 2014), from which the following is a summary. See also Donald Devine, "Psychiatry, Mental Illness, and the State," *Federalist,* September 25, 2014.

47   Ludy T. Benjamin and David B. Baker, *From Séance to Science: A History of the Profession of Psychology in America* (California: Wadsworth/Thomson Learning, 2004), 21–24.

48   Akinobu Takabayashi, "Surviving the Lunacy Act of 1890: English Psychiatrists and Professional Development during the Early Twentieth Century," *Medical History,* 61:2 (April 2017), 246–69, https://www.ncbi.nlm.nih.gov/pmc/articles/ PMC5426304/.

49   David Remnick, "25 Years of Nightmares: Victims of CIA-Funded Mind Experiments Seek Damages from the Agency," *Washington Post,* June 28, 1985.

50   Sarah Baughey-Gill, "When Gay Was Not Okay with the APA: A Historical Overview of Homosexuality and Its Status as Mental Disorder," *Occam's Razor,* Western Washington University, vol. 1 (2011), https://cedar.wwu.edu/orwwu/vol1/ iss1/2/.

51 Jon Xavier, "Google's Kurzweil says the machines will think for themselves by 2040, and oh—we'll be immortal," *Silicon Valley Business Journal*, February 10, 2014.

52 Ibid.

53 See Yuval Noah Harari, *Sapiens: A Brief History of Humankind* (New York: HarperCollins, 2015), chap. 20, esp. 413–16.

54 John Dewey, *Democracy and Education* (New York: Free Press, 1916), esp. chap. 7–9.

55 Lyndsey Layton, "How Bill Gates pulled off the swift Common Core revolution," *Washington Post*, June 7, 2014.

56 Ibid.

57 Emmett McGroarty and Jane Robbins, "Controlling Education from the Top: Why Common Core Is Bad for America," Pioneer Institute and American Principles Project, White Paper no. 87 (May 2012), https://appfdc.org/wp-content/uploads/2017/05/Controlling-Education-Common-Core-2013.pdf.

58 Layton, "How Bill Gates pulled off the swift Common Core revolution."

59 Emma Brown and Lyndsey Layton, "House approves rewrite of 'No Child Left Behind,'" *Washington Post*, July 8, 2015.

60 "Secretary Duncan wants DC kids to keep vouchers," *USA Today*, March 4, 2009; David N. Bass, "A private school on a public scale," *American Spectator*, February 17, 2015; Robert L. Luddy, *Entrepreneurial Life: The Path from Startup to Market Leader* (Raleigh, N.C.: CaptiveAire, 2018), chap. 10; Mark Walsh, "RAND study balances the debate on school choice," *Education Week*, December 12, 2001.

61 Organization for Economic Cooperation and Development (OECD), "New approach needed to deliver on technology potential in schools," September 15, 2015, https://www.oecd.org/education/new-approach-needed-to-deliver-on-technologys-potential-in-schools.htm.

62 David A. Mindell, *Our Robots, Ourselves: Robotics and the Myths of Autonomy* (New York: Viking, 2015).

63 Arthur Herman, "The Quantum Computing Threat to American Security," *Wall Street Journal*, November 10, 2019.

64 Rachel Botsman, *Who Can You Trust? How Technology Brought Us Together and Why It Could Drive Us Apart* (New York: Hachette, 2017), 150–74.

65 Barnett, *The Universe and Dr. Einstein*; Hayek, "The Theory of Complex Phenomena," 44; K. R. Popper, "On Sources of Knowledge and Ignorance," *Proceedings of the British Academy*, 46 (1960), 69; Warren Weaver, "A Scientist Ponders Faith," *Saturday Review*, January 3, 1959.

66 Daniele Fanelli, "How Many Scientists Fabricate and Falsify Research? A Systematic Review and Meta-Analysis of Survey Data," *PLoS ONE*, 4:5 (May 29, 2009).

67 Adam Marcus and Ivan Oranski, "More scientific papers should be retracted," *Washington Post*, December 27, 2018.

68 Caroline Y. Johnson, "Harvard investigation finds fraudulent heart data," *Washington Post*, October 18, 2018.

69 John P. A. Ioannidis, "Why Most Published Research Findings Are False," *PLoS Med*, 2:8 (August 30, 2005).

70 Joel Achenbach, "No, science's reproducibility problem is not limited to psychology," *Washington Post*, August 28, 2015.

71 Joel Achenbach and Laurie McGinley, "Researchers struggle to replicate 5 influential cancer experiments from top labs," *Washington Post*, January 18, 2017.

72 Joel Achenbach, "Many scientific studies can't be replicated," *Washington Post*, August 27, 2015.

73 Achenbach and McGinley, "Researchers struggle to replicate 5 influential cancer experiments from top labs."

74 Social Science Replication Project, *Nature Human Behaviour*, August 27, 2018, https://www.nature.com/collections/nfkchhxllx.

75 Karl Popper, *The Poverty of Historicism*, 2nd ed. (London: Routledge, 1957), 130–43; and see Bruce Caldwell, Introduction to *Studies on the Abuse and Decline of Reason*, vol. 13 of *The Collected Works of F. A. Hayek*, ed. Caldwell (Chicago: University of Chicago Press, 2010), 36–37.

76 Alan Greenspan, *The Map and the Territory: Risk, Human Nature, and the Future of Forecasting* (New York: Penguin, 2013).

77 Andy Kessler, "'The Fed Is Flying Blind on Inflation," *Wall Street Journal*, May 14, 2019.

78 George Gilder, *Knowledge and Power: The Information Theory of Capitalism and How It Is Revolutionizing Our World* (Washington, D.C.: Regnery, 2013).

79 Thomas Nagel, *Mind and Cosmos: Why the Materialist Neo-Darwinian Conception of Nature Is Almost Certainly False* (New York: Oxford University Press, 2012).

80 Quoted by Andrew Ferguson, "The Heretic," *Weekly Standard*, March 25, 2013.

81 Alvin Plantinga, *Where the Conflict Really Lies: Science, Religion, and Naturalism* (New York: Oxford University Press, 2011), 258.

82 Albert Camus, "Homage to an Exile," Speech delivered 7 December 1955 at a banquet in honor of President Eduardo Santos, editor of *El Tiempo*, driven out of Colombia by the dictatorship, in *Resistance, Rebellion, and Death: Essays*, trans. Justin O'Brien (New York: Vintage International, 1995), 101.

83 John Gray, *Seven Types of Atheism* (New York: Farrar, Straus & Giroux, 2018), esp. chap. 3.

84 Hayek, "Kinds of Rationalism," 95.

CHAPTER 6—MORALIZING CAPITALISM

1 Pope Francis, *Laudato Si'*, Encyclical Letter on Care for Our Common Home, May 24, 2015.

2 Joseph Ratzinger (Pope Benedict XVI), *"In the Beginning ...": A Catholic Understanding of the Story of Creation and the Fall*, 1986, trans. Boniface Ramsey (Grand Rapids, Mich.: William B. Eerdmans, 1995), 33.

3 Gerard M. Verschuuren, *The Myth of an Anti-Science Church: Galileo, Darwin, Teilhard, Hawking, Dawkins* (Kettering, Ohio: Angelico Press, 2018), 39.

4 Ratzinger, *"In the Beginning,"* 93–94; See also Donald Devine, "Jonah Goldberg's Intricate Story of Romantic Suicide," *Imaginative Conservative*, June 7, 2018; Jonah Goldberg, *Suicide of the West: How the Rebirth of Tribalism, Populism, Nationalism, and Identity Politics Is Destroying America* (New York: Crown Forum, 2018), chap. 11.

5 Joseph Cardinal Ratzinger (Pope Benedict XVI), *Truth and Tolerance: Christian Belief and World Religions* (San Francisco: Ignatius Press, 2004), 253–55.

6 Pope Paul VI, *Populorum Progressio*, Encyclical Letter on the Development of Peoples, March 26, 1967.

7   Donald J. Devine, *Does Freedom Work? Liberty and Justice in America* (Ottawa, Ill.: Caroline House Books, 1978), 137–56.

8   Pope John Paul II, *Centesimus Annus*, Encyclical Letter on the Hundredth Anniversary of *Rerum Novarum*, May 1, 1991, §§ 15, 48.

9   Jorge Mario Bergoglio and Abraham Skorka, *On Heaven and Earth: Pope Francis on Faith, Family, and the Church in the 21st Century* (New York: Transaction, 2013), 160.

10 When Bergoglio was elected to the papacy, Rev. Albert Mohler, president of the Baptist Theological Seminary in Louisville, commented in an interview that the new pope was not a philosopher like John Paul nor a theologian like Benedict, and predicted that he would be "more pastoral" and emotive than his predecessors. Interview, "Dr. Albert Mohler Reflects on Pope Francis," *Hugh Hewitt*, March 14, 2013, http://www.hughhewitt.com/dr-albert-mohler-reflects-on-pope-francis/.

11 Charles Koch, "This is the one issue where Bernie Sanders is right," *Washington Post*, February 19, 2016.

12 Pope Francis, *Laudato Si'*, § 189.

13 See Sheldon Richman, "Adam Smith vs. Crony Capitalism," *Reason*, March 9, 2012; Donald Devine, *America's Way Back: Reclaiming Freedom, Tradition, and Constitution* (Wilmington, Del.: ISI Books, 2013), 16–19.

14 Anna Wilde Matthews, "Insurers Flag Deepening Losses on Health Law," *Wall Street Journal*, February 11, 2016.

15 Devine, *America's Way Back*, 38–39.

16 Ibid., 17–19, 28–31.

17 George J. Stigler, "The Theory of Economic Regulation," *Bell Journal of Economics and Management Science*, 2:1 (Spring 1971), 3–21.

18 Amy Goldstein and Scott Wilson, "Obama launches attack on health insurance companies," *Washington Post*, March 9, 2010.

19 Economic Club of Washington, "Health Care and Medical Innovation," C-SPAN, April 3, 2018, https://www.c-span.org/video/?443426-1/health-care-medical-innovation. Interestingly, the participants also agreed that the best prospect for better health and lower costs was by reducing obesity, which is statistically associated with various illnesses and is the most serious preventable problem. But politicians and bureaucrats do not like telling people that the way to better health is to control what they eat. Donald Devine, "Ugly Idea of Death Panels Revived," *Newsmax*, May 9, 2018.

20 Clyde Wayne Crews Jr., *Mapping Washington's Lawlessness 2016: A Preliminary Inventory of "Regulatory Dark Matter"* (Washington, D.C.: Competitive Enterprise Institute, December 2015).

21 Joseph Schumpeter, *Capitalism, Socialism and Democracy*, 3rd ed. (New York: Harper & Row, 1950), 137.

22 Murray N. Rothbard, *Classical Economics*, vol. 2 of *An Austrian Perspective on the History of Economic Thought* (Auburn, Ala.: Ludwig von Mises Institute, 2006), 103–13.

23 Adam Smith, *The Theory of Moral Sentiments* (New York: Penguin, 2010), Part II, sec. 2; Donald Devine, "Adam Smith and the Problem of Justice in Capitalist Society," *Journal of Legal Studies*, 6:2 (June 1977).

24 Devine, "Adam Smith and the Problem of Justice in Capitalist Society"; Devine, *America's Way Back*, 94–95; Devine, *Does Freedom Work?* 12–17.

25 Arthur C. Brooks, *Who Really Cares? The Surprising Truth about Compassionate Conservatism* (New York: Basic Books, 2006).

26 F. A. Hayek, *The Constitution of Liberty*, ed. Ronald Hamowy, vol. 17 of *The Collected Works of F. A. Hayek* (Chicago: University of Chicago Press, 1960), 231.

27 Schumpeter, *Capitalism, Socialism and Democracy*, chap. 12–14.

28 See especially Harold J. Berman, *Law and Revolution: The Formation of the Western Legal Tradition* (Cambridge, Mass.: Harvard University Press, 1983).

29 Deirdre McCloskey, *The Bourgeois Virtues: Ethics for an Age of Commerce* (Chicago: University of Chicago Press, 2006), 8.

30 Robert Sirico, "The Pope's Green Theology," *Wall Street Journal*, June 18, 2015.

31 Laurence Chandy and Geoffrey Gertz, "Missing poverty's new reality," *Washington Post*, June 25, 2011.

32 Pope John Paul II, *Centesimus Annus*, § 48.

33 Joe Davidson, "EPA employees hearing agenda," *Washington Post*, May 19, 2016.

34 Yuval Levin, *The Great Debate: Edmund Burke, Thomas Paine, and the Birth of Left and Right* (New York: Basic Books, 2014).

35 Ibid., 128.

36 This is mainly from F. A. Hayek, "Kinds of Rationalism," in *Studies in Philosophy, Politics and Economics* (Chicago: University of Chicago Press, 1967), but appears throughout his work.

37 Levin, *The Great Debate*, 7.

38 For this and the following, see Mark Lilla, *The Shipwrecked Mind: On Political Reaction* (New York: New York Review Books, 2016).

39 Donald J. Trump, Inaugural Address, January 20, 2017, The White House.

40 Daniel McCarthy, "Why Libertarians Are Wrong," *Spectator*, February 21, 2019; Daniel McCarthy, "A New Conservative Agenda," *First Things*, March 2019. The U.S. service sector had been larger than the industrial sector since 1840, and by the twenty-first century it accounted for 80 percent of employment and 70 percent of value-added. Louis D. Johnson, "History Lessons: Understanding the decline in manufacturing," *MinnPost*, February 22, 2012. Even China's service sector had been larger than its manufacturing since 2015. Mark Magnier, "As Growth Slows, China Highlights Transition from Manufacturing to Service," *Wall Street Journal*, January 19, 2016.

41 Donald Devine, "The Nationalist Delusion," *American Conservative*, April 9, 2019. The U.S. already led all nations in service-sector exports. IndexMundi, Manufacturing, value added (current US$)—Country Ranking, https://www.indexmundi.com/facts/indicators/NV.IND.MANF.CD/rankings.

42 Devine, "The Nationalist Delusion."

43 Frank S. Meyer, "Western Civilization: The Problem of Political Freedom," in *In Defense of Freedom and Related Essays* (Indianapolis: Liberty Fund, 1996), esp. 210.

44 Donald Devine, "The Enduring Tension That Is Modern Conservatism," *Law and Liberty*, May 20, 2015.

45 Neal B. Freeman, "Conservatism's Future after Trump," *National Review*, July 6, 2016; Donald Devine, "How Should Conservatives Respond to President Trump's Nationalism?" *Imaginative Conservative*, January 30, 2017.

46 Irving Kristol, *Two Cheers for Capitalism* (New York: Basic Books, 1978).

47 Russell Kirk, *The Conservative Mind: From Burke to Eliot* (Chicago: Regnery, 1953). For his fusionist orientation, see especially Russell Kirk, "The Problem of Tradition," in *A Program for Conservatives* (Chicago: Regnery, 1954).

48 Devine, "The Enduring Tension That Is Modern Conservatism."

49 See especially Meyer, "Western Civilization: The Problem of Political Freedom."

50 William F. Buckley Jr., *Up From Liberalism* (New York: McDowell, Obolensky, 1959), 123–24, 193, 197.

51 Meyer, "Western Civilization: The Problem of Political Freedom," esp. 220–24; E. J. Dionne Jr., *Why Americans Hate Politics* (New York: Touchstone, 1992), 161.

52 The Philadelphia Society, "The Roots of Conservatism and Its Future," conference in Philadelphia, Pennsylvania, March 27–29, 2015.

53 Ronald Reagan, Address to the Conservative Political Action Conference, March 20, 1981, Young America's Foundation, https://www.yaf.org/news/35th-anniversary-of-president-reagans-our-time-is-now-address/.

54 Ibid.

55 St. Augustine, *The City of God* (New York: Penguin Classics, 1984), Bk. XIX, chap. 27.

56 Meyer, "Western Civilization: The Problem of Political Freedom," 209–24.

57 Meyer, "In Defense of Freedom," in *In Defense of Freedom and Related Essays* (Indianapolis: Liberty Fund, 1996), 80.

58 Ibid., 85.

59 William F. Buckley Jr., "The Courage of Friedrich Hayek," in *Let Us Talk of Many Things* (Rosedale, Calif.: Primus Publishing, 2000), 223–34; Frank S. Meyer, "Champion of Freedom," *National Review*, May 7, 1960.

60 F. A. Hayek, "Individualism: True and False," in *Studies on the Abuse and Decline of Reason*, ed. Bruce Caldwell, vol. 13 of *The Collected Works of F. A. Hayek* (Chicago: University of Chicago Press, 2010), 149.

61 Hayek, *The Constitution of Liberty*, 61.

62 Friedrich A. Hayek, *Law, Legislation and Liberty*, vol. 3, *The Political Order of a Free People* (Chicago: University of Chicago Press, 1979), 176. See also Eric Voegelin, *The New Science of Politics*, in *Modernity Without Restraint*, ed. Manfred Henningsen, vol. 5 of *The Collected Works of Eric Voegelin* (Columbia: University of Missouri Press, 2000), 90–108.

63 Hayek, *The Constitution of Liberty*, 162–68.

64 Daniel Hannan, "Magna Carta: Eight Centuries of Liberty," *Wall Street Journal*, May 29, 2015; Daniel Hannan, *Inventing Freedom: How the English-Speaking Peoples Made the Modern World* (New York: HarperCollins, 2013), 112–16.

65 Daniel McCarthy emphasizes this point in "An Autopsy: Why Liberalism Failed," *National Interest*, July–August 2018.

66 Elements of Magna Carta liberties also appeared in the founding charters of the colonies of Connecticut, Georgia, Maine, and Massachusetts. Tully Vaughan, "Magna Carta and the Colonies," The Baronial Order of Magna Carta, https://www.magnacharta.com/bomc/magna-charta-and-the-colonies-ii/.

67 Cf. Ofir Haivry and Yoram Hazony, "What Is Conservatism?" *American Affairs Journal*, Summer 2017; Donald Devine, "Anglifying American Conservatism," *Law and Liberty*, November 16, 2017.

68 Friedrich A. Hayek, *The Road to Serfdom* (New York: Routledge, 1944). Voegelin's American influence did not start until the 1950s. Manfred Henningsen, Editor's Introduction to Voegelin, *Modernity Without Restraint*, 1–2.

69 Arthur M. Schlesinger Jr., *The Disuniting of America: Reflections on a Multicultural Society*, rev. ed. (New York: W. W. Norton, 1998), 11–17.

70 Murray N. Rothbard, "Frank S. Meyer: The Fusionist as Libertarian Manqué," *Modern Age*, Fall 1981, 352–63.

71 Clark Ruper, "The Death of Fusionism," *Cato Unbound*, May 10, 2013.

72 Michael J. Gerson, *Heroic Conservatism: Why Republicans Need to Embrace America's Ideals (And Why They Deserve to Fail If They Don't)* (New York: HarperCollins, 2007), 60–61.

73 Ibid., 16–18, 22–23, 275; Michael Gerson, "Where the new right has failed," *Washington Post*, October 25, 2016.

74 Michael J. Gerson, "Open-Arms Conservatism," *Washington Post*, October 31, 2007.

75 Gerson, *Heroic Conservatism*, 61–62.

76 Henry Olsen and Peter Wehner, "If Ronald Reagan Were Alive Today, He Would Be 103 Years Old," *Commentary*, November 1, 2014. Gerson made a similar point in *Heroic Conservatism*, 158–59.

77 Sung Deuk Hahm et al., "The Influence of the Gramm-Rudman-Hollings Act on Federal Budgetary Outcomes, 1986–1989," *Journal of Policy Analysis and Management*, 11:2 (Spring 1992), 207–34.

78 Devine, *America's Way Back*, 188–89.

79 Donald Devine, "Post-Election Propaganda Begins," *Federalist*, November 5, 2014; Peter Wehner, "Clarifying the Reagan Record (and Correcting Don Devine)," *Commentary*, November 14, 2014. See also Henry Olsen, "Reagan's Real Legacy: A Response to Don Devine," *National Review*, November 10, 2014.

80 Office of Management and Budget, 2015 FY Budget (Washington, D.C.: Government Printing Office, 2016), Historical Table 5.6.

81 Office of Management and Budget, 1999 FY Budget: Budget Authority (Washington, D.C.: Government Printing Office, 2000), Table 8.9, 141.

82 Earlier estimates also showed net savings for Outlays under Reagan. Office of Management and Budget, 2011 FY Budget: Outlays (Washington, D.C.: Government Printing Office, 2012), www.whitehouse.gov/sites/default/files/omb/budget/fy2011/assets/hist06z1.xls.

83 Ronald Reagan, Inaugural Address, January 20, 1981, The Public Papers of Ronald Reagan, Reagan Presidential Library.

84 Reagan, Address to the Conservative Political Action Conference.

85 Ronald Reagan, Remarks at the Dedication of the James Madison Memorial Building of the Library of Congress, November 20, 1981, The Public Papers of Ronald Reagan, Reagan Presidential Library.

86 Jim Hardy, "As Congress pushes for Medicaid change, could a version of Reagan's grand bargain simplify the program?" Deloitte Center for Health Solutions, July 12, 2017.

87 Ronald Reagan, Address to Members of the Royal Institute of International Affairs in London, June 3, 1988, The Public Papers of Ronald Reagan, Reagan Presidential Library.

88 Edmund Burke, Letter to Richard Shackleton, in Joseph L. Pappin III, *The Metaphysics of Edmund Burke* (New York: Fordham University Press, 1993), 104.

89 Leo Strauss, *The Rebirth of Classical Political Rationalism*, ed. Thomas Prangle (Chicago: University of Chicago Press, 1989), 270.

90 Emma Brown, "Obama administration spent billions to fix failing schools, and it didn't work," *Washington Post*, January 9, 2017; see also Eric A. Hanushek and Paul E. Peterson, "Is the U.S. Catching Up?" *Education Next*, Fall 2012.

91 Patricia A. Davis, "Medicare Financial Status," Congressional Research Service, August 9, 2017, 12.

92 M. Stephen Weatherford, "The President, the Fed and the Financial Crisis," *Presidential Studies Quarterly*, 43:2 (June 2013), 320.

93 Robert J. Samuelson, "Greenspan's grim forecast for growth," *Washington Post*, November 21, 2016, where the years in question are discussed.

94 The speech was condensed into a column by David Keene, "At CPAC, rebuilding that shining city upon a hill," *Washington Times*, March 10, 2014.

95 Michael Thompson, Richard J. Ellis, and Aaron Wildavsky, *Culture Theory* (Boulder, Col.: Westview Press, 1990).

96 "The Sharon Statement," Sharon, Connecticut, September 11, 1960, at Young America's Foundation, https://www.yaf.org/news/the-sharon-statement/.

97 Robert L. Bartley, *The Seven Fat Years: And How to Do It Again* (New York: Free Press, 1992).

98 Devine, *America's Way Back*, 186–91, 207–8.

99 Max Weber, *The Theory of Social and Economic Administration*, trans. A. M. Henderson and Talcott Parsons (New York: Free Press, 1947), III, v, 11–12A.

100 Charles C. W. Cooke, *The Conservatarian Manifesto: Libertarians, Conservatives, and the Fight for the Right's Future* (New York: Crown Forum, 2015), 16.

101 Kenneth Minogue, "The Conditions of Freedom and the Condition of Freedom," in *On Liberty and Its Enemies*, ed. Timothy Fuller (New York: Encounter Books, 2017), 99–126.

102 Minogue, "Can Scholarship Survive the Scholars?" in *On Liberty and Its Enemies*, 135–37.

103 Minogue, "The Conditions of Freedom and the Condition of Freedom."

104 Minogue, "Morals and the Servile Mind," in *On Liberty and Its Enemies*, 237–47.

105 Ibid.

106 Minogue, "The Self-Interested Society," in *On Liberty and Its Enemies*, 300.

107 Minogue, "Two Concepts of the Moral Life," in *On Liberty and Its Enemies*, 190.

108 Minogue, "The Irresponsibility of Rights," in *On Liberty and Its Enemies*, 248–52. Minogue actually uses the term "monarchy" for the first institution, but includes the United States as "belonging to the class."

109 Minogue, "The Self-Interested Society," in *On Liberty and Its Enemies*, 302–4; Donald Devine, "Minogue on States, Institutions, and the Enemies of Liberty," *Law and Liberty*, June 11, 2018.

110 Pope Francis, *Laudato Si'*, § 229.

CHAPTER 7—THE PLURALIST ADMINISTRATION SOLUTION

1 Stephen M. Griffin, *Broken Trust: Dysfunctional Government and Constitutional Reform* (Lawrence, Kans.: University Press of Kansas, 2015), chap. 2.

2 For a more detailed discussion and citations, see Donald J. Devine, *America's Way Back: Reclaiming Freedom, Tradition, and Constitution* (Wilmington, Del.: ISI Books, 2013), 17–20, 28.

3 Griffin, *Broken Trust*, 48.

4 Griffin, *Broken Trust*, 18, and chap. 1 generally.

5 Ibid., chap. 4–5.

6 Ibid., 142–46.

7 Ibid., 149.

8 Ibid., 154.

9 Ibid., 151–55.
10 John Dewey, *Liberalism and Social Action* (1935; Amherst, N.Y.: Prometheus Books, 2000), 7; Devine, *America's Way Back*, 223–25.
11 Donald Devine, "The Problem of Question Form in Describing Public Opinion," *Polity*, 12:3 (Spring 1980).
12 Woodrow Wilson, "The Study of Administration," *Political Science Quarterly*, 2:2 (June 1887), 197–222.
13 Paul Light, "The real crisis in government," *Washington Post*, January 10, 2010.
14 Griffin, *Broken Trust*, 18–22.
15 Ibid., 21–22.
16 National Commission on the Public Service, *Report and Recommendations of the National Commission on the Public Sector* (Washington, D.C.: Government Printing Office, 1989); National Commission on the Public Service, *Urgent Business for America: Revitalizing the Federal Government for the 21st Century* (Washington, D.C.: Government Printing Office, January 2003).
17 Robert Samuelson, "The Deadlock of Democracy," *Washington Post*, November 1, 2004; Donald Devine, "Illuminating the Intricacies of Swamp Management," *American Spectator*, July 4, 2018.
18 James Madison, *The Federalist Papers*, No. 10.
19 Kendra Bischoff and Sean F. Reardon, "Residential Segregation by Income, 1970–2009," US2010 Project, Russell Foundation and American Communities Project of Brown University, October 16, 2013, http://cepa.stanford.edu/content/residential-segregation-income-1970-2009.
20 Dissimilarity is the proportion of individuals of either group that would have to change to integrate a neighborhood. Isolation measures the tendency for members of one group to live in neighborhoods where their share of the population is above the citywide average. Edward Glaeser and Jacob Vigdor, "The End of the Segregated Century: Racial Separation in America's Neighborhoods, 1890–2010," Center for State and Local Leadership, Manhattan Institute, January 2012, https://media4.manhattan-institute.org/pdf/cr_66.pdf.
21 Richard Rothstein, "The Racial Achievement Gap, Segregated Schools, and Segregated Neighborhoods—A Constitutional Insult," Economic Policy Institute, November 12, 2014.
22 Government Accountability Office, "K–12 Education: Better Use of Information Could Help Agencies Identify Disparities and Address Racial Discrimination," Washington, D.C., April 2016.
23 Will Stancik, "School Segregation Is Not a Myth," *Atlantic*, March 14, 2018.
24 Dorceta Taylor, *The State of Diversity in Environmental Organizations*, prepared for Green 2.0, July 2014, http://orgs.law.harvard.edu/els/files/2014/02/FullReport_Green2.0_FINALReducedSize.pdf.
25 Ryan D. Enos, "Causal effect of intergroup contact on exclusionary attitudes," *Proceedings of the National Academy of Sciences*, 111:10 (March 11, 2014), 3699–704.
26 Robert Sapolsky, "Spanish Speakers Set Off Bias in Suburbs," *Wall Street Journal*, August 7, 2014.
27 Robert D. Putnam, "*E Pluribus Unum*: Diversity and Community in the Twenty-first Century," The 2006 Johan Skytte Prize Lecture, *Scandinavian Political Studies*, 30:2 (2007).
28 James Q. Wilson, "Bowling with Others," *Commentary*, October 10, 2001.

29 Pew Research Center, "Political Polarization in the American Public," June 12, 2014. See also Donald J. Devine, *The Attentive Public: Polyarchical Democracy* (Chicago: Rand McNally, 1970).

30 Pew Research Center, "Political Polarization in the American Public."

31 Pew Research Center, "Political Typology Reveals Deep Fissures on the Right and Left," October 14, 2017; Donald Devine, "Division and Repair of American Politics," *American Spectator*, December 12, 2017.

32 James Madison, *The Federalist Papers*, No. 10.

33 Pew Research Center, "Political Polarization in the American Public."

34 Jeffrey M. Jones, "U.S. Muslims Most Approving of Obama, Mormons Least," Gallup Presidential Job Approval Center, July 11, 2014.

35 Ibid.

36 Frank Newport, "Highly Religious, White Protestants Firm in Support for Trump," Gallup Polling Matters, April 9, 2019; Devine, "Division and Repair of American Politics."

37 Anti-Defamation League, "Poll: Americans Believe Religion Is 'Under Attack'—Majority Says Religion Is 'Losing Influence' in American Life," November 21, 2005, http://archive.adl.org/presrele/relchstsep_90/4830_90. html#.V-ox9EoVCM8.

38 Jay Alan Sekulow, "Religious Liberty and Expression Under Attack: Restoring America's First Freedoms," Heritage Foundation, Legal Memorandum no. 88, October 1, 2012.

39 Chai Feldblum, *Moral Argument, Religion and Same-Sex Marriage* (Lanham, Md.: Lexington Books, 2009), Conclusion.

40 Richard Dawkins, *The Root of All Evil: The God Delusion*, https://www.youtube. com/watch?v=8nAos1M-_Ts; Richard Dawkins, *The God Delusion* (New York: Houghton Mifflin, 2006), chap. 8.

41 Svetozar Pejovich, *Law, Informal Rules and Economic Performance: The Case for Common Law* (Northampton, Mass.: Edward Elgar Publishing, 2008), chap. 3.

42 Joseph Schumpeter, *Capitalism, Socialism and Democracy*, 3rd ed. (New York: Harper & Row, 1950), 157, 160.

43 Allan C. Carlson, "A City on a Hill—With Transgender Toilets?" *Chronicles*, March 2017.

44 Rod Dreher, "Hobby Lobby and 'The Bad Old Days,'" *American Conservative*, July 22, 2014.

45 Pew Research Center, "A Barometer of Modern Morals," March 28, 2006.

46 Jay Michaelson, "Were Christians Right About Gay Marriage All Along?" *Daily Beast*, May 27, 2014.

47 Justin Raimondo, "The Libertarian Case Against Gay Marriage," *American Conservative*, April 1, 2011. Raimondo later came to support formal marriage for gays, but was concerned that the *Obergefell* decision would intrude on religious believers' rights. Justin Raimondo, "Recanting the Libertarian Case Against Gay Marriage," *American Conservative*, July 2, 2015.

48 Richard Reeves, "How to Save Marriage in America," *Atlantic*, February 13, 2014.

49 Masood Farivar, "Black Lives Matter Gains Concessions in DNC Platform," VOA News, August 19, 2020.

50 James Davison Hunter, *To Change the World: The Irony, Tragedy, and Possibility of Christianity in the Late Modern World* (New York: Oxford University Press, 2010).

51 Rod Dreher, "The Future of Christians in Post-Christian America," *American Conservative*, July 4, 2014.
52 Ibid.
53 *United States v. Windsor*, https://www.supremecourt.gov/opinions/12pdf/12-307_6j37.pdf.
54 Dreher, "The Future of Christians in Post-Christian America."
55 Ryan T. Anderson, *Truth Overruled: The Future of Marriage and Religious Freedom* (Washington, D.C.: Regnery, 2015), chap. 3–4.
56 Richard, Cohen, "Trump has taught me to fear my fellow Americans," *Washington Post*, May 30, 2016.
57 William Galston, "Sex and Citizens," *Wall Street Journal*, May 31, 2016.
58 Nicholas Kristof, "A Confession of Liberal Intolerance," *New York Times*, May 17, 2016.
59 Pew Research Center, "Beyond Distrust: How Americans View Their Government," November 23, 2015. See also Pew Research Center, "Public Trust in Government Remains Near Historic Lows as Partisan Attitudes Shift," May 3, 2017.
60 Pew Research Center, "Beyond Distrust."
61 Megan Brenan, "Americans' Trust in Government to Handle Problems at New Low," Gallup Politics, January 31, 2019.
62 Ludwig von Mises, *Bureaucracy* (New Haven: Yale University Press, 1944), Part II.
63 Robert Skidelsky, *The Road from Serfdom: The Economic and Political Consequences of the End of Communism* (New York: Penguin, 1995).
64 Niall Ferguson, *Civilization: The West and the Rest* (New York: Penguin, 2012).
65 Ibid., 292.
66 Aaron B. Wildavsky, *The Politics of the Budgetary Process*, 4th ed. (Boston, Little, Brown, 1984).
67 Committee for a Responsible Federal Budget, "Mandatory Spending Continues to Drive Spending Growth," January 25, 2017, https://www.crfb.org/blogs/mandatory-spending-continues-drive-spending-growth.
68 John Mauldin, "How I Learned to Love the Debt," *Thoughts from the Frontline*, April 26, 2019, https://www.mauldineconomics.com/frontlinethoughts/how-i-learned-to-love-the-debt.
69 Ferguson, *Civilization: The West and the Rest*, 292–310.
70 Donald Devine, "Uncontrollable Debt's Painful Lessons," *Newsmax*, September 10, 2018.
71 Gary North, "Guns or Granny: The Looming Political Battle of the West," *GaryNorth.com*, May 20, 2017.
72 Coral Davenport, "McConnell Urges States to Defy U.S. Plan to Cut Greenhouse Gas," *New York Times*, March 4, 2015.
73 Ibid.
74 Hugh Hewitt, "The lawless obstruction of Beltway elites," *Washington Post*, March 9, 2017.
75 James Madison, "Notes of the Constitutional Convention, July 17, 1787," in *The Records of the Federal Convention*, ed. Max Farrand (New Haven: Yale University Press, 1911), vol. II, 25–36.
76 James Madison, *The Federalist Papers*, No. 51.
77 Virginia Resolutions, 1798, from *The American Republic: Primary Sources*,

ed. Bruce Frohnen (Indianapolis: Liberty Fund, 2002), available at http://oll. libertyfund.org/pages/1798-virginia-resolutions.

78 Kentucky Resolutions, 1798, from *The American Republic: Primary Sources*, available at http://oll.libertyfund.org/pages/1798-kentucky-resolutions.

79 Andrew Jackson, Bank Veto Message, July 10, 1832, Presidential Speeches, Miller Center of Public Affairs, University of Virginia, http://millercenter.org/scripps/ archive/speeches/detail/3636.

80 Abraham Lincoln, Speech on the Dred Scott Decision, June 26, 1857, Teaching American History, https://teachingamericanhistory.org/library/document/ speech-on-the-dred-scott-decision/.

81 Robert A. Dahl, "Decision-Making in a Democracy: The Supreme Court as a National Policy-Maker," *Journal of Public Law*, 6:2 (Fall 1957), 279–93.

82 For documentation, see Devine, *America's Way Back*, 163–66.

83 For example, see Ruth Marcus, "A pen, a phone and a precedent," *Washington Post*, February 5, 2014; Mike DeBonis and Damian Paletta, "Senate GOP changes tax bill to add Obamacare mandate appeal," *Washington Post*, November 14, 2017.

84 William A. Galston, "The Supreme Court's Stunning Decision on the Affordable Care Act," Brookings, June 28, 2012.

85 Amber Phillips, "Republicans want to change the 14th Amendment. But that often requires war, crisis or death," *Washington Post*, August 19, 2015. Justice Neil Gorsuch upbraided a state solicitor general for even questioning national preemption in setting legal fines: "Here we are in 2018 still litigating incorporation of the Bill of Rights. Really, come on general." Mark Sherman, "High court likely to say states can't levy excessive fines," *The Star*, November 28, 2018.

86 Alexander Hamilton, *The Federalist Papers*, No. 85.

87 Jared Diamond, *Guns, Germs and Steel: The Fates of Human Societies* (New York: W. W. Norton, 1997), chap. 14.

88 Peter H. Wilson, *Heart of Europe: A History of the Holy Roman Empire* (Cambridge Mass.: Harvard University Press, 2016).

89 Donald Devine, "The Decentralized yet Durable Empire," *Law and Liberty*, February 6, 2017.

90 Donald Devine, "The Virtue of Nationalism," *Imaginative Conservative*, October 29, 2018.

91 Johannes Althusius, *Politica* (1614), abridged trans. by Frederick S. Carney (Indianapolis: Liberty Fund, 1995). In his Foreword, Daniel J. Elazar traces the concept back to the Hebrew scriptures. See also Elazar, "Althusius and Federalism as Grand Design," 1990, Jerusalem Center for Public Affairs, http:// www.jcpa.org/dje/articles2/althus-fed.htm; Elazar, "Viewing Federalism as Grand Design," in *Federalism as Grand Design: Political Philosophers and the Federal Principle*, ed. Daniel J. Elazar (Lanham, Md.: University Press of America, 1987), 1–11.

92 Alain de Benoist, "The First Federalist: Johannes Althusius," *Telos*, Winter 2000.

93 Henry Heller, *The Birth of Capitalism: A 21st Century Perspective* (London: Pluto Press, 2011), 61.

94 John Emerich Edward Dalberg-Acton, 1st Baron Acton, *The History of Freedom and Other Essays* (London: Macmillan, 1909), chap. 2.

95 Ibid., 36.

96 John C. Eastman, "Federal Court Precedent: A Defense of Judge Roy Moore and the Alabama Supreme Court," *Public Discourse*, March 16, 2015.

97 Jean-Jacques Rousseau, *The Social Contract*, in *The Social Contract and Discourses*, trans. G. D. H. Cole (New York: E. P. Dutton, 1950), Bk. III, chap. 1.

98 William Han, "Is Democracy Illusory? Of Kenneth Arrow and the Marquis de Condorcet," *Exile's Bazaar*, February, 27, 2017, http://www.exilesbazaar.com/home/is-democracy-illusory-of-kenneth-arrow-and-the-marquis-de-condorcet. See also Kenneth J. Arrow, *Social Choice and Individual Values* (New York: Wiley, 1951).

99 Wilfred McClay and Ted V. McAllister, eds., *Why Place Matters: Geography, Identity, and Civic Life in Modern America* (New York: Encounter Books, 2014), Preface.

100 Pope Francis, *Laudato Si'*, Encyclical Letter on Care for Our Common Home, May 24, 2015, §§ 156–57; Catechism of the Catholic Church, Part 3, Section 1, Chapter 2, Article 1.

101 Emily Badger, "Why Trump's Use of the Words 'Urban Renewal' Is Scary for Cities," *New York Times*, December 7, 2016.

102 Federal Bureau of Investigation, Latest Crime Statistics, September 26, 2016; National Center for Victims of Crime, Urban and Rural Crime, 2015.

103 Kevin Rector, "Baltimore, Chicago, D.C. drove half national increase in homicides in 2015," *Baltimore Sun*, April 20, 2016; Andrew Grawert and James Cullen, "Crime in 2015: A Final Analysis," Brennan Center for Justice, April 20, 2016.

104 Heather Mac Donald, "The Black Body Count Rises as Chicago Police Step Back," *Wall Street Journal*, September 12, 2016.

105 Matt Furber, John Eligon, and Audra D. S. Burch, "Minneapolis Police, Long Accused of Racism, Face Wrath of Wounded City," *New York Times*, May 27, 2020.

106 Bureau of Justice Assistance, *Understanding Community Policing: A Framework for Action*, U.S. Department of Justice, August 1994.

107 Robert McGurn, "The Secrets of New York Policing Success," *Wall Street Journal*, January 27, 2017; Dana Rubinstein and Jeffrey A. Mays, "Nearly $1 Billion Is Shifted From Police in Budget That Pleases No One," *New York Times*, June 30, 2020.

108 Peter Hermann and Clarence Williams, "Justice system in D.C. is 'broken,' Lanier says," *Washington Post*, September 2, 2016.

109 Ibid.

110 Jane Jacobs, *The Life and Death of Great American Cities* (New York: Vintage, 1961).

111 Communications with police division headquarters would be critical. The professionals might look down on the constables and be concerned about volunteers or low-paid individuals "replacing" them, but the best headquarters police today depend on local informers. We now see challenges between volunteer and paid fire departments, but many of the volunteer forces work well and should be learned from. Mutual respect is crucial. Being unarmed, or perhaps with only short-distance stun guns, constables must rely on the professionals, who in turn need eyes and ears in the community. A good constable would know the locals who could help the police in an emergency.

Perhaps hundreds of people could be in his or her network and provide the basis for an unarmed posse to calm riots.

112 Christopher Chantrill, Current Government Spending in the US, http://www. usgovernmentspending.com/total_spending_1900USrn (accessed April 24, 2017); Tax Policy Center, *Briefing Book: A Citizens' Guide to the Fascinating (though Often Complex) Elements of the Federal Tax System* (Washington, D.C.: Tax Policy Center, Urban Institute and Brookings Institution, 2016), 12–13, https://www.taxpolicycenter.org/sites/default/files/briefing-book/tpc-briefing-book_0.pdf.

113 Donald Devine, "American Culture and Public Administration," *Policy Studies Journal*, 11:2 (December 1982), 255–60.

114 Vincent Ostrom, *The Intellectual Crisis in American Public Administration* (Tuscaloosa: University of Alabama Press, 1973).

115 Wilson, "The Study of Administration," 18.

116 In 1942, there were 35,139 municipalities and townships in the United States; in 2007, there were 36,021; in 2012, there were 35,884. U.S. Bureau of the Census, *Historical Statistics of the United States: Colonial Times to 1970*, Bicentennial Edition, Part 2 (Washington, D.C., 1975), chap. Y, 1086; U.S. Census Bureau, "Census Bureau Reports There Are 89,004 Local Governments in the United States," Newsroom, August 30, 2012, https://www.census.gov/newsroom/releases/archives/governments/cb12-161.html. The latter figure includes special-purpose districts of all types that are independent of general-purpose local governments.

117 Vincent Ostrom, *The Meaning of American Federalism: Constituting a Self-Governing Society* (San Francisco: Institute for Contemporary Studies Press, 1991).

118 Charles M. Tiebout, "A Pure Theory of Local Expenditures," *Journal of Political Economy*, 64:5 (October 1956); Donald J. Devine, *Does Freedom Work? Liberty and Justice in America* (Ottawa, Ill.: Caroline House Books, 1978), 56–59 and chap. 5.

119 Stephen Goldsmith, *Putting Faith in Neighborhoods: Making Cities Work through Grassroots Citizenship* (Noblesville, Ind.: Hudson Institute, 2002), esp. chap. 4.

120 U.S. Advisory Commission on Intergovernmental Relations, *Residential Community Associations: Private Governments in the Intergovernmental System*, Washington, D.C., 1989.

121 John C. Bollens and Henry J. Schmandt, *The Metropolis* (New York: Harper & Row, 1965), 154–82.

122 Staffan Canbäck, Phillip Samouel, and David Price, "Do Diseconomies of Scale Impact Firm Size and Performance? A Theoretical and Empirical Overview," *Journal of Managerial Economics*, 4:1 (February 2006), 27–70.

123 Karen Shanton, "The Problem of African American Underrepresentation in City Councils," *Demos*, October 30, 2014; Matt Pierce, "Ferguson Officials Now Mostly Black Like City," *Los Angeles Times*, February 26, 2016.

124 George W. Liebmann, *Neighborhood Futures: Citizen Rights and Local Control* (New Brunswick, N.J.: Transaction, 2004).

125 Devine, *America's Way Back*, 205.

126 Alexis de Tocqueville, *Democracy in America*, trans. Henry Reeve, vol. 2 (London: Saunders & Otley, 1840), v.

127 Ibid.; Devine, *Does Freedom Work?* chap. 5.

128 Pew Research Center, "About half of Americans who know at least some of their neighbors talk to them weekly," May 15, 2018.

129 U.S. Bureau of Labor Statistics, "Volunteering in the United States—2015," February 25, 2016.

130 Robert Putnam and David E. Campbell, *American Grace* (New York: Simon & Schuster, 2010), 19, 463–79. See also Timothy P. Carney, *Alienated America: Why Some Places Thrive While Others Collapse* (New York: HarperCollins, 2019), chap. 7–8.

131 There are scores more examples in my earlier books: *Does Freedom Work?*; *Reagan's Terrible Swift Sword*; and *America's Way Back.*

132 F. H. Buckley, *American Secession: The Looming Threat of a National Breakup* (New York: Encounter Books, 2020), Appendix.

133 This was well illustrated with the red/blue national map of voting statistics by county. Brilliant Maps, "2016 US Presidential Election Map By County & Vote Share," November 29, 2016, https://brilliantmaps.com/2016-county-election-map/.

134 Yuval Levin, *The Fractured Republic: Renewing America's Social Contract in the Age of Individualism* (New York: Basic Books, 2016).

135 John Micklethwait and Adrian Wooldridge, *The Fourth Revolution: The Global Race to Reinvent the State* (New York: Penguin, 2014), 216–19.

136 Gallup, "Trust in Government, In Depth Topics," 2018, http://www.gallup.com/poll/5392/trust-government.aspx; Pew Research Center, "Trust, Facts and Democracy," April 26, 2018; Devine, "Division and Repair of American Politics."

137 This might require some rethinking of the Supreme Court's narrow view of the Fourteenth Amendment incorporation doctrine; see Robert Natelson, "With Civil Forfeiture Decision, Justice Thomas Again Shows He's the Supreme Court's Sole Originalist," *Daily Caller*, February 25, 2019. See also Michael S. Greve, *The Upside-Down Constitution* (Cambridge, Mass.: Harvard University Press, 2002), chap. 8; Donald Devine, "Minogue on States, Institutions and the Enemies of Liberty," *Law and Liberty*, June 11, 2018.

138 Tocqueville observed that a localized system of governance had the result of making citizens mindful of the national interest: "In the United States the interests of the country are everywhere kept in view; they are an object of solicitude to the people of the whole Union, and every citizen is as warmly attached to them as if they were his own." *Democracy in America*, vol. I, chap. 5.

CHAPTER 8—THE CIVILIZATIONAL CHOICE

1 Glenn R. Parker, Review of *The Political Culture of the United States* by Donald J. Devine, *American Political Science Review*, 69:1 (March 1975), 267–68.

2 Donald J. Devine, *The Political Culture of the United States: The Influence of Member Values on Regime Maintenance* (Boston: Little, Brown, 1972).

3 Charles Murray, "Trump's America," *Wall Street Journal*, February 12, 2016. See also Samuel P. Huntington, *Who Are We? The Challenges to National Identity* (New York: Simon & Schuster, 2005).

4 Murray, "Trump's America." See also Daniel Markovits, *The Meritocracy Trap: How America's Foundational Myth Feeds Inequality, Dismantles the Middle Class, and Devours the Elite* (New York: Penguin, 2019); Timothy P. Carney, *Alienated America: Why Some Places Thrive While Others Collapse* (New York: HarperCollins, 2019), chap. 11.

5 Murray, "Trump's America."

6 Ibid.

7 Ronald Brownstein, "A Referendum on America's Identity," *Atlantic*, July 13, 2016;

Ronald Brownstein, "Culture Is Replacing Class As the Key Political Divide," *Atlantic*, June 30, 2016; Lynn Vavereck "Younger Americans Are Less Patriotic, at Least in Some Ways," *New York Times*, July 4, 2014.

8 Walter Lippmann, *Essays in the Public Philosophy* (New Brunswick, N.J.: Transaction, 1955).

9 Ibid., 87; see also Eric Voegelin, *The New Science of Politics*, in *Modernity Without Restraint*, ed. Manfred Henningsen (Columbia: University of Missouri Press, 2000), chap. 6. Lippmann acknowledges Voegelin's influence.

10 Lippmann, *Essays in the Public Philosophy*, 72.

11 Ibid., 61.

12 John Nicholas Gray, *The Soul of the Marionette: A Short Inquiry into Human Freedom* (New York: Farrar, Straus & Giroux, 2015), 165.

13 Ibid., 160.

14 Steven Pinker, *The Better Angels of Our Nature: Why Violence Has Declined* (New York: Viking Penguin, 2011).

15 John Gray, "Steven Pinker is wrong about violence and war," *Guardian*, March 13, 2015; Donald Devine, "Finding Meaning in a Purposeless Universe," *American Spectator*, May 11, 2018.

16 Gray, *The Soul of the Marionette*, 165; see also Eric Voegelin, *Science, Politics and Gnosticism*, in *Modernity Without Restraint*; John Gray, *Seven Types of Atheism* (New York: Farrar, Straus & Giroux, 2018), chap. 3.

17 Gray, *The Soul of the Marionette*, 28.

18 Ibid., 10–11.

19 Ibid., 151.

20 Ibid., 19.

21 Ibid., 9–10.

22 Ibid., 15.

23 Ibid., 156–58.

24 Ibid., 33.

25 Ibid., 7.

26 Ibid., 161–62

27 Ibid., 7.

28 Niccolò Machiavelli, *The Prince*, chap. 18.

29 Michael J. Abramowitz, "Democracy in Crisis: Freedom in the World, 2018," Freedom House, https://freedomhouse.org/report/freedom-world/freedom-world-2018.

30 Yuval Levin, *The Fractured Republic: Renewing America's Social Contract in the Age of Individualism* (New York: Basic Books, 2016), chap. 1.

31 Jaweed Kaleem, "Under Obama, the Justice Department aggressively pursued police reforms. Will it continue under Trump?" *Los Angeles Times*, February 5, 2017.

32 U.S. Department of Justice, "Attorney General Jeff Sessions Delivers Remarks at National Association of Attorneys General Annual Winter Meeting," Washington, D.C., February 28, 2017, https://www.justice.gov/opa/speech/attorney-general-jeff-sessions-delivers-remarks-national-association-attorneys-general.

33 Ibid.; see also Erica Werner and Alan Fram, "Sessions suggests police need less Federal scrutiny," *Washington Post*, February 28, 2017.

34 Wesley Lowery, "Federal civil rights group urges U.S. to increase scrutiny of local police," *Washington Post*, November 16, 2018.

35 U.S. Department of Justice, "Attorney General William P. Barr Announces Launch of Operation Legend," July 8, 2010, https://www.justice.gov/opa/pr/attorney-general-william-p-barr-announces-launch-operation-legend.

36 Sari Horwitz and Matt Zapotosky, "Sessions backed states' rights as a senator. As Trump's attorney general, it's complicated," *Washington Post*, March 9, 2018.

37 Charles Lane, "Liberals learn to love states' rights," *Washington Post*, March 16, 2017.

38 Posse Comitatus came under increased challenge in the wake of the 9/11 attacks. Timothy M. MacArthur and Leigh Winstead, "Skirting the constraints of Posse Comitatus," *The Hill*, March 4, 2017. Congress amended the Insurrection Act in 2006, following Hurricane Katrina, to permit national armed forces to restore order and enforce law in emergencies, but the amendment was repealed in 2008.

39 Harvey Silverglate, *Three Felonies a Day: How the Feds Target the Innocent* (New York: Encounter Books, 2009).

40 Charles Murray, *By the People: Rebuilding Liberty Without Permission* (New York: Crown Forum, 2015).

41 John Emerich Edward Dalberg-Acton, 1st Baron Acton, Creighton Correspondence, Letter I, April 5, 1887, in *Lectures on Modern History*, ed. John Neville Figgis and Reginald Vere Laurence (London: Macmillan, 1906).

42 Bob Kerrey, "How did Department of Justice get the Trump-Russia investigation so wrong?" *Omaha World-Herald*, March 29, 2019; Jan Crawford, "William Barr Interview," *CBS This Morning*, May 31, 2019; Donald Devine, "Barr's Courageous Challenge to FBI Praetorian Guard," *Newsmax*, June 7, 2019.

43 Donald Devine, "Manafort Indictment Late," *Newsmax*, November 8, 2017; Donald Devine, "Roger Stone Raid Shows FBI out of Control," *Newsmax*, February 1. 2018.

44 See pp. 136–44 above.

45 James Hohmann, "Anxiety over pace of change evident in populist movements," *Washington Post*, January 15, 2019.

46 John Dewey, *Democracy and Education* (New York: Free Press, 1916), esp. 88.

47 Roberto Stefan Foa and Yascha Mounk, "The Danger of Deconsolidation: The Democratic Disconnect," *Journal of Democracy*, 27:3 (July 2016), 5–17.

48 Peter Beinart, "A Violent Attack on Free Speech at Middlebury," *Atlantic*, March 6, 2017.

49 James Hohmann, "Poll commissioned by Bush, Biden shows Americans losing faith in democracy," *Washington Post*, June 27, 2018; George W. Bush Presidential Center, "The Democracy Project: Reversing a Crisis of Confidence," June 26, 2018.

50 Pew Research Center, "Five Facts About Prayer," May 4, 2016; Donald Devine, "Losing Their Religion, Really?" *American Conservative*, November 27, 2019.

51 John Micklethwait and Adrian Wooldridge, *God Is Back: How the Global Revival of Faith Is Changing the World* (New York: Penguin, 2009).

52 Gallup International, "Losing Our Religion? Two Thirds of People Still Claim to Be Religious," June 8, 2015; Pippa Norris and Ronald Inglehart, "God, Guns, and Gays: Religion and Politics in the U.S. and Western Europe," John F. Kennedy School of Government, Harvard University, September 6, 2004, Fig. 8, http://www.tinyurl.com/y5z7dlva.

53 Pew Research Center, "The Changing Global Religious Landscape," April 5, 2017; Pew Research Center, "The Future of World Religions, 2010–2050," April 2, 2015. See also "Michael Lipka and David McClendon, "Why people with no religion are projected to decline as a share of the world's population," Pew Research Center, April 2017.

54 In a note, Pew acknowledged that China was a "wild card," since fear of the authorities might suppress self-reporting of religious belief. Earlier Pew studies found from one-third to a majority in China saying religion was somewhat important in their lives. Russell Heimlich, "Religious in China," Pew Research Center, July 29, 2008. A later Freedom House study estimated there might be 185 to 250 million Buddhists and Taoists in China, who are actually supported by authorities, plus 70 million Christians and 20 million Muslims, curbed in one way or another, and Pentecostal Christians, Tibetan Buddhists, Uighur Muslims, and Falun Gong, all severely persecuted. Freedom House, "The Battle for China's Spirit: Religious Revival, Repression, and Resistance under Xi Jinping," Special Report, February 2017. See also "China Bans On-line Bible Sales as It Tightens Religious Controls," *New York Times*, April 5, 2018.

55 Pew Research Center, "The Future of World Religions, 2010–2050."

56 Pew Research Center, "Attitudes of Christians in Western Europe," May 29, 2018.

57 Richard Dawkins, *The Root of All Evil: The God Delusion*, https://www.youtube.com/watch?v=8nAosiM-_Ts; Richard Dawkins, *The God Delusion* (New York: Houghton Mifflin, 2006), 51–59, 323–26, 341–48, 354–79.

58 Donald Devine, *America's Way Back: Reclaiming Freedom, Tradition, and Constitution* (Wilmington, Del.: ISI Books, 2013), 127–30. See also Alicia Adsera, "Religion and Changes in Family-Size Norms in Developed Countries," *Review of Religion Research*, 47:3 (March 2006), 271–86.

59 Robert D. Putnam and David E. Campbell, *American Grace: How Religion Divides and Unites Us* (New York: Simon & Schuster, 2010), chap. 13; Rodney Stark, *For the Glory of God: How Monotheism Led to Reformations, Science, Witch-Hunts, and the End of Slavery* (Princeton: Princeton University Press, 2002), 375ff; Patrick Fagan, "Why Religion Matters Even More: The Impact of Religious Practices on Social Stability," Heritage Foundation, December 16, 2006; Patrick Fagan, "95 Social Science Reasons for Religious Worship and Practice," Marri Research (Marriage and Religion Research Institute), October 16, 2012.

60 Angelina E. Theodorou, "Americans are in the middle of the pack globally when it comes to importance of religion," Pew Research Center, December 2015.

61 Pew Research Center, Religion and Public Life, "Importance of Religion in One's Life," http://www.pewforum.org/religious-landscape-study/importance-of-religion-in-ones-life/.

62 Hamza Sharan, "Super Bowl ratings slide to lowest since 2009," *Washington Post*, February 5, 2019; Joe Otterson, "TV Ratings: Super Bowl Slips to 103 Million," *Variety*, February 5, 2018; Will Leitch, "No One Is Going to Sports in Person, and No One Seems to Care," *Intelligencer*, June 11, 2018; Carol Pipes, "No Place Like Church for the Holidays," *Lifeway Research*, December 14, 2015; Jeffrey M. Jones, "Christmas Strongly Religious for Half in U.S. Who Celebrate It," Gallup, December 24, 2010.

63 David Martin, *Religion and Power: No Logos without Mythos* (New York: Routledge, 2016). Cf. Eric Voegelin, *The Political Religions, in Modernity Without*

Restraint, ed. Manfred Henningsen (Columbia: University of Missouri Press, 2000), 30–33, 70–71.

64 Rodney Stark cites Gallup data showing that 28 percent of Austrians said they believed in fortune tellers and 33 percent in lucky charms. One-fifth of Swedes believed in reincarnation. Half of Icelanders believed in elves and trolls. As G. K. Chesterton quipped, the result when people reject God is not that they will believe in nothing, but that they can believe in anything. Rodney Stark, *The Triumph of Faith: Why the World Is More Religious than Ever* (Wilmington, Del.: ISI Books, 2015).

65 Voegelin, *The New Science of Politics*, chap. 6.

66 Pew Research Center, "The Future of World Religions, 2010–2050."

67 John D. Martin, "Reports of Christianity's Death in Europe Greatly Exaggerated," *Federalist*, March 23, 2018.

68 Martin, *Religion and Power*, 33–34.

69 Frank S. Meyer, "Western Civilization: The Problem of Political Freedom," in *In Defense of Freedom and Related Essays* (Indianapolis: Liberty Fund, 1996), 14–15.

70 Alec Ryrie, *Protestants: The Faith That Made the Modern World* (New York: Viking Penguin, 2017), esp. chap. 16.

71 Martin, *Religion and Power*, 37.

72 Jean-Jacques Rousseau, *The Social Contract*, in *The Social Contract and Discourses*, trans. G. D. H. Cole (New York: E. P. Dutton, 1950), 329–33; Donald J. Devine, *In Defense of the West: American Values Under Siege* (Dallas: University Press of America, 2004), 145–47.

73 Matthew Kaminski, "Democracy May Have Had Its Day," Weekend Interview with Donald Kagan, *Wall Street Journal*, April 26, 2013.

74 Ronald Reagan, Address to Members of the Royal Institute of International Affairs in London, June 3, 1988, The Ronald Reagan Presidential Library; Devine, *America's Way Back*, 20.

75 Pew Research Center, "Attitudes of Christians in Western Europe."

76 Pew Research Center, "Being Christian in Western Europe," May 29, 2018.

77 Samuel P. Huntington, *The Clash of Civilizations and the Remaking of World Order* (New York: Touchtone, 1996).

78 Donald Devine, "Is God Dead ... or Is It Nietzsche?" *Imaginative Conservative*, August 10, 2018.

79 Nicholas Wade, *The Faith Instinct: How Religion Evolved and Why It Endures* (New York: Penguin, 2009).

80 Ibid., chap. 2. See also Herbert S. Terrace, *Why Chimpanzees Can't Learn Language and Only Humans Can* (New York: Columbia University Press, 2019); Ian Tattersall, *Becoming Human: Evolution and Human Uniqueness* (New York: Harcourt Brace, 1998), chap. 2

81 Wade, *The Faith Instinct*, chap. 3

82 J. Budziszewski, *What We Can't Not Know* (Dallas: Spence Publishing, 2003).

83 Ibid., 3.

84 Ibid., 234.

85 Donald J. Devine, "Adam Smith and the Problem of Justice in Capitalist Society," *Journal of Legal Studies*, 6:2 (June 1977), 399–410; Donald J. Devine, *Does Freedom Work? Liberty and Justice in America* (Ottawa, Ill.: Caroline House Books, 1978), chap. 2.

86 Andrew Ferguson, "The Heretic," *Weekly Standard*, March 25, 2013.
87 Jamie Weinstein, "Krauthammer: I don't believe in God but I fear him," Video, *Daily Caller*, December 22, 2013.
88 Nicholas G. Hahn III, "George Will: The RealClearReligion Interview," *RealClearReligion*, September 22, 2014.
89 Kenneth Minogue, "Individualism and Its Contemporary Fate," in *On Liberty and Its Enemies*, ed. Timothy Fuller (New York: Encounter Books, 2017), 141; Donald Devine, "Minogue on States, Institutions and the Enemies of Liberty," *Law and Liberty*, June 11, 2018.
90 Martin, *Religion and Power*, 147.
91 Larry Siedentop, *Inventing the Individual: The Origins of Western Liberalism* (Cambridge, Mass.: Harvard University Press, 2014), chap. 1.
92 René Girard, *Things Hidden Since the Foundations of the World*, trans. and with commentary by Stephen Bann and Michael Metteer (Stanford: Stanford University Press, 1978), 59; Voegelin, *The New Science of Politics*, 89; Stanley Parry, "Reason and the Restoration of Tradition," in *What Is Conservatism?* ed. Frank S. Meyer (New York: Holt, Rinehart & Winston, 1964).
93 See Meyer, "Western Civilization: The Problem of Political Freedom"; M. Stanton Evans, *The Theme Is Freedom* (Washington, D.C.: Regnery, 1994).
94 Siedentop, *Inventing the Individual*, 83, Harold J. Berman, *Law and Revolution: The Formation of the Western Legal Tradition* (Cambridge, Mass.: Harvard University Press, 1983).
95 Siedentop, *Inventing the Individual*, chap. 14–16.
96 Ibid., chap. 17. Cf. Rodney Stark, *How the West Won: The Neglected Story of the Triumph of Modernity* (Wilmington, Del.: ISI Books, 2014), chap. 13.
97 Kai Qin Chan, Eddie Mun Wai Tong, and Yan Lin Tan, "Taking a Leap of Faith: Reminders of God Lead to Greater Risk Taking," *Social Psychological and Personality Science*, 5:8 (October 9, 2014), 901–9; Daniella M. Kupor, Kristin Laurin, and Jonathan Levav, "Anticipating Divine Protection? Reminders of God Can Increase Nonmoral Risk Taking," *Psychological Science*, 26:2 (April 13, 2015), 374–84.
98 See Stark, *How the West Won*, chap. 5 and chap. 11–13.
99 Especially in England and the United States, particularly through the legal philosopher H. L. A. Hart. See Devine, *America's Way Back*, chap. 6, esp. 111–12.
100 Bryan Appleyard, "The God Wars," *New Statesman*, February 28, 2012.
101 Friedrich Nietzsche, *The Anti-Christ*, trans. H. L. Mencken (New York: Knopf, 1920), preface.
102 Ibid., 5–6 (sec. 6–7).
103 Ibid., 3 (preface).
104 Ibid., 58 (sec. 57).
105 Plato, *The Republic* (New York: Scribner & Son, 1871), Bk. X.
106 Arthur Schoenberg, *Theory of Harmony*, trans. Roy E. Carter (1911; Berkeley: University of California Press, 1983).
107 Robert Reilly, *Surprised by Beauty: A Listener's Guide to the Recovery of Modern Music* (San Francisco: Ignatius Press, 2016).
108 Ibid., 23.
109 Ibid., 332.
110 Ibid., 295.

111 Borstlap calls Beethoven's style a fusion of "Apollonian decorum and Dionysian violence." John Borstlap, *The Classical Revolution: Thoughts on New Music in the 21st Century* (New York: Scarecrow Press, 2013), 28.

112 Reilly, *Surprised by Beauty*, 338.

113 For example, Patrick J. Deneen, *Why Liberalism Failed* (New Haven: Yale University Press, 2018), 31–34; Donald Devine, "The Fusionist Fight over Everything," *Imaginative Conservative*, November 4, 2019.

114 Jared Diamond, *The World Until Yesterday* (New York: Penguin, 2012), chap. 9.

115 René Girard, *I See Satan Fall Like Lightning*, trans. James D. Williams (Maryknoll, N.Y.: Orbus Books, 2001), 178–81.

116 George Orwell, *1984* (New York: Harcourt Brace Jovanovich, 1949), Part I, sec. 1; Joshua Mitchell, "Why Conservatives Struggle with Identity Politics," *National Affairs*, 40 (Summer 2019).

117 René Girard correctly argued that Judaism by the time of Deutero-Isaiah had anticipated Christian views about scapegoats as innocent victims: Girard, *Things Hidden Since the Foundations of the World*, Bk. 2, chap. 1. See also Ben Shapiro, *The Right Side of History* (New York: Broadside Books, 2019), chap. 2.

118 Gray, *The Soul of the Marionette*, 165.

119 Robert Reich, *Tales of a New America* (New York: Times Books, 1987), chap. 1, sec. 2.

120 F. A. Hayek, *The Fatal Conceit: The Errors of Socialism*, ed. W. W. Bartley III, vol. 1 of *The Collected Works of F. A. Hayek* (Chicago: University of Chicago Press, 1988), 136–40.

121 F. A. Hayek, *The Constitution of Liberty*, ed. Ronald Hamowy, vol. 17 of *The Collected Works of F. A. Hayek* (Chicago: University of Chicago Press, 1960), chap. 12.

122 Minogue, "Individualism and Its Contemporary Fate."

123 John Locke, *The Reasonableness of Christianity*, in *The Works of John Locke*, vol. 7 (London: Tegg et al., 1823), 150.

124 Jonah Goldberg, *Suicide of the West: How the Rebirth of Tribalism, Populism, Nationalism, and Identity Politics Is Destroying America* (New York: Crown Forum, 2018), 3, 330.

125 Ibid., 331–34.

126 Pierre Manent, *Beyond Radical Secularism: How France and the Christian West Should Respond to the Islamic Challenge* (South Bend, Ind.: St. Augustine Press, 2016), 6, 57, 60–66, 112–15.

127 Richard Dawkins, "Gaps in the Mind," in *The Great Ape Project: Equality Beyond Humanity*, ed. Paola Cavalieri and Peter Singer (New York: St. Martin's Griffin, 1993), 81–87; Peter Singer, *Practical Ethics* (New York: Cambridge University Press, 1980), chap. 3.

128 Richard Lough, "Captive orangutan has human right to freedom, Argentine court rules," Reuters, December 21, 2014.

129 Karin Brullard, "Chimpanzees are not 'persons,' appeals court says," *Washington Post*, June 12, 2017.

130 Lauren Steussy et al., "Are SeaWorld's killer whales slaves?" NBC News, February 6, 2012. See also Garrett Epps, "Dangerous Metaphors: Are SeaWorld's killer whales slaves?" *Atlantic*, November 2, 2011.

131 Brian Clark Howard, "SeaWorld to end controversial orca shows and breeding," *National Geographic*, March 17, 2014.

132 Grennan Milliken, "Are viruses alive? New evidence says yes," *Popular Science*, September 25, 2015.

133 D. Keith Mano, *The Bridge: A Novel about the Last Man on Earth* (New York: Doubleday, 1971). See also Pascal Bruckner, *The Fanaticism of the Apocalypse: Save the Earth, Punish Human Beings*, trans. Steven Rendall (Malden, Mass.: Polity Press, 2013).

134 Jeremy Waldron, *God, Locke and Equality: Christian Foundations of John Locke's Political Thought* (Cambridge: Cambridge University Press, 2002), 232. Dennis Prager traces the source of Locke's concept of rights to the book of Genesis, in *Genesis: God, Creation, and Destruction*, The Rational Bible (Washington, D.C.: Regnery, 2019), chap. 1. See also Donald Devine, "Critiquing Robert Kagan's Enlightenment Liberalism," *Imaginative Conservative*, May 6, 2019.

135 J. Leslie Hall, "The Religious Opinions of Thomas Jefferson," *Sewanee Review*, 21:2 (April 1913), 173.

136 Joseph Schumpeter, *Capitalism, Socialism and Democracy*, 3rd ed. (New York: Harper & Row, 1950), 289–96.

INDEX

Abraham, 294
Abu Ghraib, 156
Achieve Inc., 187, 188
Acton, John Emerich Edward Dalberg
  Acton, 1st Baron, 260, 283
Acton Institute, 206
Adler, Alfred, 181
Adnani, Abu Mohammad al-, 142
Affordable Care Act (Obamacare):
  and corporate lobbyists, 202–3; and
  executive orders, 153; as inefficient,
  126; religious exemptions to, 241;
  scope of, 154; state challenges to, 261
Afghanistan, 68, 69, 76
Age of Reason (Paine), 208
Age of Transition, An (Dyer), 12
Akhromeyev, Sergey Fedorovich, 108
Akiva, Rabbi, 40–41
Alabama, 103; Birmingham
  demonstrations, 90
'Alam, Abdallah, 69
Albert of Saxony, 39
Albertus Magnus, 39
Alchian, Armen A., 37
Alexander, Lamar, 189
Alexander the Great, 276, 295
Alexievich, Svetlana, 105–9
Alien and Sedition Acts, 255
Allen, Tom, 234–35
Althusius, Johannes, 61–62, 63, 260
Ambrose, Saint, 59, 295
American Bar Association, 218
American College of Cardiology, 122
American Conservative Union, 215, 224
American Federation of Government
  Employees, 119

American Hospital Association, 122
American Psychiatric Association, 185
American Revolution, 40, 208
American Society of Gene and Cell
  Therapy, 172
Anderson, Ryan T., 95–96
anthrax, 124
Anti-Christ, The (Nietzsche), 297–98
Anti-Defamation League, 241
Aquinas, Thomas. *See* Thomas
  Aquinas, Saint
Argentina, 16–19, 201
Aristotle, 39, 57; on first cause, 292;
  on inequality, 50, 294; and Locke,
  32–33; on natural sociability, 58; and
  self-preservation, 50
Arkani-Hamed, Nima, 169
Arlene's Flowers, 94
Armstrong, Karen, 39–40, 45, 64
artificial intelligence, 186–87
associational freedom, 20–21, 97, 269,
  271, 274
Athelstan, 289
Athens (ancient), 50, 55–56, 211–13,
  294–95, 297
Attila, 59
Augsburg Settlement, 61–62
Augustine, Saint, 12, 43, 211, 212, 216
Awlaki, Anwar al-, 142

Bacchae, The (Euripides), 56–57
Bahrain, 14
Bailey, Ronald, 176–77
Baker, Thomas J., 138
Bald, Garry, 69
Baltimore (riots in), 101–2